Coal Science

VOLUME 1

CONTRIBUTORS

Robert M. Davidson
Warren R. Grimes
Richard C. Neavel
H. L. Retcofsky
Leon M. Stock

Coal Science

VOLUME 1

Edited by

MARTIN L. GORBATY
Corporate Research–Science Laboratories
Exxon Research and Engineering Company
Linden, New Jersey

JOHN W. LARSEN
Department of Chemistry
The University of Tennessee
Knoxville, Tennessee

IRVING WENDER
Chemical and Petroleum Engineering Department
The University of Pittsburgh
Pittsburgh, Pennsylvania

1982

ACADEMIC PRESS
A Subsidiary of Harcourt Brace Jovanovich, Publishers
New York   London
Paris   San Diego   San Francisco   São Paulo   Sydney   Tokyo   Toronto

CHEMISTRY

7117 - 6330

ACADEMIC PRESS, INC.
111 Fifth Avenue, New York, New York 10003

United Kingdom Edition published by
ACADEMIC PRESS, INC. (LONDON) LTD.
24/28 Oval Road, London NW1 7DX

ISBN 0–12–150701–7

ISSN 0730–5397
This publication is not a periodical and is not
subject to copying under CONTU guidelines.

PRINTED IN THE UNITED STATES OF AMERICA

82 83 84 85   9 8 7 6 5 4 3 2 1

Contents

Contributors vii

Preface ix

Coal Plasticity Mechanism: Inferences from Liquefaction Studies

RICHARD C. NEAVEL

    I. Introduction 1
   II. Coal Constitution 2
  III. Liquefaction 3
  IV. Relationship between Liquefaction and Plasticity 7
   V. Role of Vehicle 8
  VI. Oxidation and "Low-Rank" Oxygen 10
 VII. Hydrogenation and Dehydrogenation 12
VIII. Heating Rate 13
  IX. Chemical Additives................................. 15
   X. Summary.. 16
       References 17

The Physical Structure of Coal

WARREN R. GRIMES

    I. Introduction 21
   II. Brief Overview of Physical Structure................. 23
  III. The Porous Nature of Coal......................... 25
  IV. Effects of Heat on Coal Structure................... 32
       References 39

Magnetic Resonance Studies of Coal

H. L. RETCOFSKY

I.	Introduction	43
II.	The Aromaticity of Coal	44
III.	Free Radicals in Coal	50
IV.	Preliminary ESR and NMR Investigations of Chinese Coals: A Cautionary Note on Coal Structure Studies	69
V.	ESR Studies of Respiratory-Size Coal Particles	73
VI.	Summary	80
	References	80

Molecular Structure of Coal

ROBERT M. DAVIDSON

I.	Introduction	84
II.	Molecular Weight of Coal	87
III.	Carbon Aromaticity	98
IV.	Aromatic Ring Structures	108
V.	Aliphatic Structures	115
VI.	Low Molecular Weight Compounds	120
VII.	Free Radicals	126
VIII.	Functional Groups and Heteroatoms	132
IX.	Structural Changes with Rank	136
X.	Macromolecular Skeletal Structures	142
XI.	Conclusions	154
	References	155

The Reductive Alkylation Reaction

LEON M. STOCK

I.	Introduction	161
II.	The Chemistry of the Reaction	162
III.	Methods and Procedures	187
IV.	Applications of Reduction, Alkylation, and Reductive Alkylation	218
	References	279
Index		283

Contributors

Numbers in parentheses indicate the pages on which the authors' contributions begin.

Robert M. Davidson (83), IEA Coal Research, London SW1W 0EX, England

Warren R. Grimes (21), Chemical Technology Division, Oak Ridge National Laboratory, Oak Ridge, Tennessee 37830

Richard C. Neavel (1), Exxon Research and Engineering Company, Baytown, Texas 77520

H. L. Retcofsky (43), United States Department of Energy, Pittsburgh Energy Technology Center, Pittsburgh, Pennsylvania 15236

Leon M. Stock (161), Department of Chemistry, University of Chicago, Chicago, Illinois 60637

Preface

The ultimate need to replace liquid and gaseous fuels now derived from petroleum and natural gas resources with synthetic fuels from coal is generally recognized. Although technological bases already exist for utilizing coal directly in combustion and for converting it to liquid and gaseous fuels, we feel that considerable improvements in these technologies (such as greater overall thermal efficiency, hydrogen utilization, and selectivity to specific products) are needed. These improvements, along with new, more efficient routes to coal utilization, will only come about if we are able to gain more fundamental knowledge of coal in terms of its structure and reactivity.

Because coal is such a heterogeneous material, coal science in its broadest sense embraces many scientific disciplines—including chemistry (organic, inorganic, and physical), physics, and engineering (chemical and mechanical). Consequently, the literature of coal science is vast and complex, containing a great deal of information—but only rarely in such a form that this information is logically assembled, reviewed, and evaluated. Because of the current interest in coal, we felt that *critical* reviews written by experts and aimed at the professional chemist or engineer now working (or contemplating working) in coal science would be both timely and useful. These reviews would not only pull together what has been reported in the past into a coherent picture, they would also point out the original reports' significance as well as areas where more research is needed. Ultimately, coal science should furnish data and clues for solving or bypassing many of the problems now associated with coal conversion technologies.

This is the first in a series devoted to presenting and evaluating selected fundamental scientific areas involved with our understanding of coal structure, reactivity, and utilization. Included are reviews describing current state-of-the-art knowledge of coal's organic and physical structure, and contributions on plasticity mechanisms and reductive alkylation chemistry covering some important aspects of coal reactivity. Lately, new spectroscopic instrumentation has been developed which may help answer age-old but vital questions about coal structure and reactivity. An article on magnetic resonance studies of coal illustrates the power of this technique and shows where we stand today as well as indicating what more can be done. It is our hope that this and subsequent volumes will not only be a valuable source of information and a guide to the coal literature, but will also stimulate research and serve as a basis for further advances in science and technology.

The editors would like to thank the contributing authors for their time and diligence in helping to make this volume a reality. We may have acted as catalysts, but the major share of the credit belongs to them.

Coal Plasticity Mechanism: Inferences from Liquefaction Studies

RICHARD C. NEAVEL

Exxon Research and Engineering Company
Baytown, Texas

I.	Introduction	1
II.	Coal Constitution	2
III.	Liquefaction	3
IV.	Relationship between Liquefaction and Plasticity	7
V.	Role of Vehicle	8
VI.	Oxidation and "Low-Rank" Oxygen	10
VII.	Hydrogenation and Dehydrogenation	12
VIII.	Heating Rate	13
IX.	Chemical Additives	15
X.	Summary	16
	References	17

I. INTRODUCTION

When heated in an inert atmosphere to ~350–450°C, particles of certain coals soften and become deformable, characteristics that are attributable to plastic (actually thermoplastic) substances. Coals exhibiting this property are called *plastic* coals. Pyrolytic decomposition of the coal substance in the 350–450°C range results in the formation of volatile "gases" within the plastic particles, which lead to vacuole development and swelling. Because particles of such coals are essentially viscous liquids in their plastic state, they can coalesce or agglomerate to form an indivisible mass or "cake"; hence, plastic coals are often referred to as *caking* or *agglomerating* coals. As a result of progressive chemical reactions and loss of volatile compounds, the plastic state is a transient phenomenon, even under isothermal conditions, and the agglomerated mass ultimately solidifies into a semicoke, which can be sintered to a true coke. Only those plastic or caking coals which can be converted to a commercially usable furnace coke are properly referred to as *coking coals*. Thus, in brief summary: plastic capability is inherent in some coals (plastic coals), agglomeration can result from plastic development (caking coals), and coke is a product of some plastic coals (coking coals).

Whereas plasticity and agglomeration can be profoundly significant in most process uses of coal, they do not appear to exert any direct influence on liquefaction processing. However, because plastic development is equivalent to the formation of liquids, a study of the mechanism of coal liquefaction divulges information about the mechanism of plastic development. Conversely, it seems likely that understanding of plastic development may lead to insights into liquefaction mechanisms. An exposition of the relationships between plastic development and liquefaction is the purpose of this article.

II. COAL CONSTITUTION

Plasticity is exhibited only by the macerals exinite (a minor constituent of typical coals) and vitrinite; moreover, except for a few very unusual coals, only vitrinite in coals of bituminous rank has plastic capability. In this discussion, when plasticity of coal is referred to, it should be inferred that we mean vitrinite. Vitrinite is composed of packets (*micelles*) of more or less aligned molecular units (*lamellae*) of variable structures typified by condensed ring systems connected by bridging atoms. Attached to the rings are various functional groups ($=O$, $-COOH$, $-OH$, $-C_nH_m$). Some of the ring carbons may be saturated with hydrogen (hydroaromatic structures). The model attributed to P. H. Given, and shown in Fig. 1a, is the generally most accepted representation of a "typical" lamellar unit, although it is important to recognize that among vitrinites from coals of different ranks (and even within a given vitrinite) there are significant variations on this theme. Lamellar size, degree of condensation, aromaticity, heteroatom content, and functional group characteristics all vary. As shown in Fig. 1b, imperfect "packing" of the molecular units and the micelles leads to microporosity. The lamellae and micelles are bound together by hydrogen bonds, van der Waals forces, and occasional structural (covalent) bridges such as etheric oxygen and (poly)methylenes. About 20% or less of unbonded, relatively low molecular weight material (often referred to as "bitumen") is readily extracted through the pore system of plastic coals by solvents (Dryden, 1963).

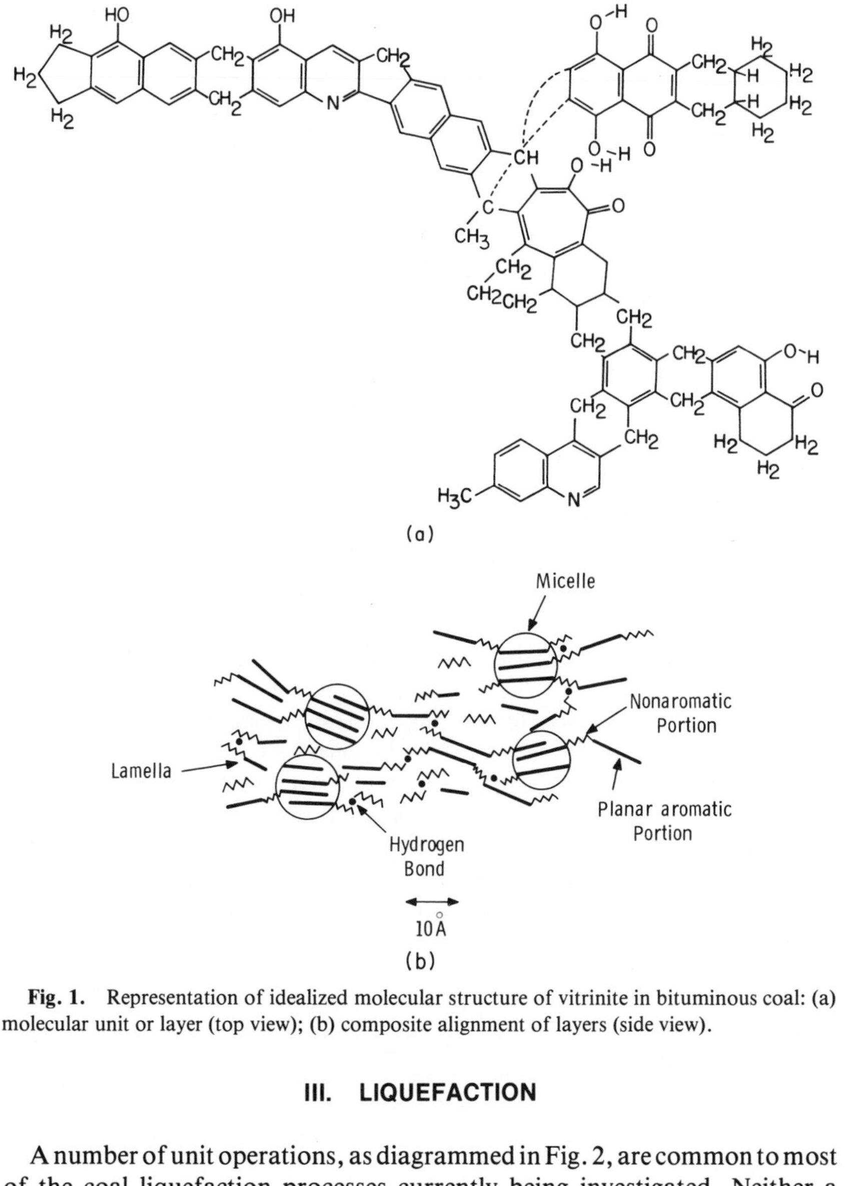

(a)

(b)

Fig. 1. Representation of idealized molecular structure of vitrinite in bituminous coal: (a) molecular unit or layer (top view); (b) composite alignment of layers (side view).

III. LIQUEFACTION

A number of unit operations, as diagrammed in Fig. 2, are common to most of the coal liquefaction processes currently being investigated. Neither a catalyst nor molecular hydrogen is necessary in the liquefaction reactor; hence, liquefaction (as distinguished from reactions of liquefied coal) is *not* a hydrogenation reaction in the classic sense of adding hydrogen to carbon via a

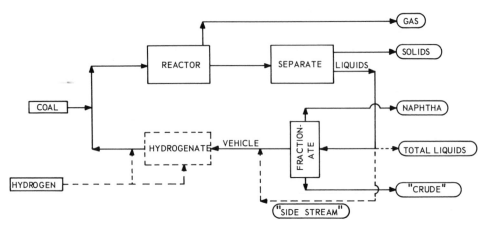

Fig. 2. Principal unit operations in coal liquefaction processes. Reactor: 400–500°C, 100–2000 psi, may contain catalyst. Separation: filter, centrifuge, distill, coke, flocculate. Hydrogenation unit and/or gas in reactor optional.

reaction involving catalytic dissociation of molecular hydrogen. A recycled mixture of partially hydrogenated, multiring compounds, of which tetralin (Fig. 3) is a model, is used as a slurrying vehicle that acts as both a solvating agent and a hydrogen-donating reactant. If a hydroaromatic hydrogen atom is abstracted from each of the saturated carbons (labeled 1–4 in Fig. 3), the tetralin molecule is converted to naphthalene, a stable aromatic species. Hydrogen transfer from the vehicle is critical to the liquefaction operation; a major theme of this article is that hydrogen transfer also appears to be an important reaction within and between coal molecules themselves when coal is pyrolyzed in the absence of a vehicle. Note that, according to the Given model, coal contains hydroaromatic hydrogen such as that on tetralin (see Fig. 1a). It has also been reported that vitrinite was found to be an excellent hydrogen donor in reactions involving pyrolytic decomposition of thermally cleavable pure compounds (Collins *et al.,* 1977).

We conducted liquefaction experiments in which samples of a vitrinite-rich fraction of a mildly plastic, Illinois high-volatile C bituminous coal were reacted with twice their weight of tetralin in small sealed bombs, which were heated rapidly in a sand bath to 400°C and maintained for various residence

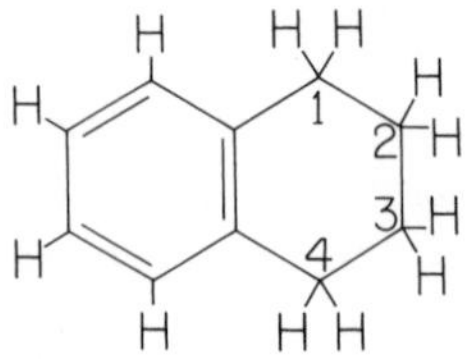

Fig. 3. Molecular structure of tetralin.

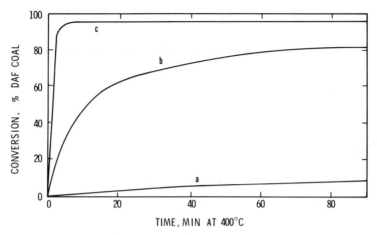

Fig. 4. Conversion of coal reacted at 400°C in tetralin: (a) noncondensable gas; (b) benzene-soluble material + gas; (c) pyridine-soluble material + gas. Daf = dry ash-free.

times. As shown graphically in Fig. 4, with increasing residence time some noncondensable gases are formed (curve a), and increasing amounts of coal can be solubilized (at room temperature after quench cooling) in benzene (curve b) or pyridine (curve c).

Microscopic observation of the coal particles after quenching indicates that they bloat very rapidly into the form of (imperfect) cenospheres; these then disintegrate (disperse in the vehicle) within less than 10 min. Initially, the vehicle appears to serve solely as a solvating agent; however, with increasing residence time (and, hence, increasing conversion of the coal to molecules potentially soluble in benzene), hydrogen is abstracted from the tetralin (which is thereby converted to naphthalene) in amounts that increase exponentially in proportion to coal conversion. The coal-derived products of this combination dissolution–hydrogen transfer process are substances with a wide range of molecular weights, from noncondensable gases to heavy tars (plus unconvertible mineral matter and fusinite particles). Additional information on these experiments is given by Neavel (1976).

Within the first few minutes of reacting coal in tetralin, about 95% of the vitrinite is rendered pyridine-soluble (curve c, Fig. 4). Even when coal is reacted in naphthalene, which has no hydrogen-donating properties, about 80–85% is rendered pyridine-soluble within a few minutes (curve a, Fig. 5). Thus, although minimal chemical reaction occurs within the first few minutes at 400°C (as inferred from low gas production, low conversion to benzene solubles, and low hydrogen consumption), the coal micelles are sufficiently "loosened up" (probably by weakening of van der Waals forces and hydrogen bonds) to be potentially mobile, as demonstrated by pyridine ex-

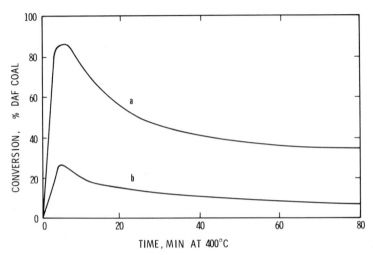

Fig. 5. Conversion of coal reacted at 400°C in naphthalene: (a) pyridine-soluble material; (b) benzene-soluble material. Daf = dry ash-free.

tractability. The cleavage of relatively weak ether systems may also contribute to this initial mobility. Among others, Lazarov and Angelova (1968) have discussed the role of ethers in coal decomposition. However, if the reaction is continued beyond about 5 min with no hydrogen donor present (i.e., in naphthalene, curve a, Fig. 5), the potentially mobile (i.e., pyridine-soluble) material reacts chemically to form a substance no longer extractable in pyridine and, by inference, much less mobile.

As Curran *et al.* (1967) demonstrated and as our experiments have confirmed, when a hydrogen-donating vehicle is present, free radicals (unpaired electrons) formed by the pyrolytic bond rupture of coal molecules abstract hydroaromatic hydrogen from the solvating vehicle, thereby forming a stable C:H bond of the reactive free radicals. The reaction is shown diagrammatically in Fig. 6. In this fashion, the thermal rupture of bridging bonds, followed by hydrogen stabilization of the resulting radicals, leads to progressive reduction of molecular size ("depolymerization") and hence conversion to "liquid-sized," benzene-soluble molecules. That the donor hydrogen is not transferred by a concerted reaction unassociated with thermally induced depolymerization is implied by the fact that neither the yield nor the rate of formation of benzene-soluble liquids appears to be enhanced by higher concentrations of donor or by better donors (i.e., dihydronaphthalene). In passing, it should be noted that Orchin (1944), in contrast to some references to his work, did not find significant direct transfer of hydrogen from tetralin to anthracene around 400°C, except when the reaction was Pd catalyzed.

When heated in a vehicle that is not a donor, not only does pyridine solubil-

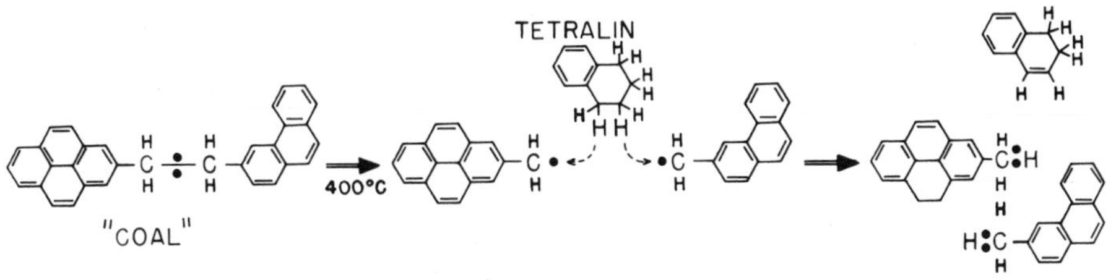

Fig. 6. Diagrammatic representation of mechanism of hydrogen transfer promoted by thermal bond rupture.

ity fall off after reaching a maximum, but the relatively small quantity of benzene-soluble liquids formed within the first few minutes also disappears (curve b, Fig. 5). It is assumed that the initially soluble materials increase in molecular size (thereby being rendered insoluble) as a result of a form of polymerization resulting from free-radical attack. Thus, if radicals formed by thermal depolymerization are not stabilized by hydrogen, they will react to form larger molecules. Microscopic observation of repolymerization-formed pyridine insolubles from liquefaction indicates that they are spherical and probably equivalent to the pseudocrystalline mesophase found in semicokes (Brooks and Taylor, 1958; Ladner and Stacey, 1963; Marsh, 1973; Shibaoka and Ueda, 1978).

IV. RELATIONSHIP BETWEEN LIQUEFACTION AND PLASTICITY

Wiser (1968) appears to have been the first investigator to call attention explicitly to the relationship between liquefaction and pyrolysis. Brown and Waters (1966), in a thoroughly documented report, proposed that low molecular weight, hydrogen-rich bitumens indigenous to coal probably serve as solvating and as hydrogen-donating agencies during coal pyrolysis. Following the lead of the above-cited authors, I propose that the development of plasticity in coals is essentially a transient, hydrogen-donor liquefaction process in which the solvating and hydrogen-donating "vehicle" is supplied by the coal itself, and it is the progressive reduction of transferable hydrogen inventory that leads to progressive decay of fluidity, even at isothermal conditions, as shown in Fig. 7.

Three conditions appear to be necessary and sufficient for the formation of low molecular weight substances by liquefaction of coals: (1) the presence in the coal of lamellae-bridging structures that can be thermally ruptured; (2) a supply of hydroaromatic hydrogen; and (3) an initial intrinsic potential for micellar and lamellar mobility (not related to the rupture of chemical bonds),

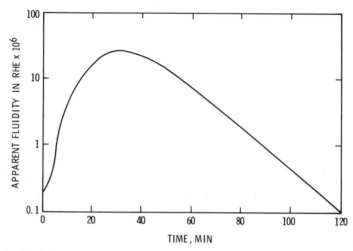

Fig. 7. Variation of apparent fluidity (in rheological units) of Silkstone coal with time at 407°C. (From Fitzgerald, 1956.)

which provides opportunity for free radicals formed by bond rupture to contact potentially transferable hydrogen. These conditions also appear to be necessary and sufficient for the development of plasticity in coal. Inherent coal properties, alterations to the coal, or variations in processing conditions, which especially affect Conditions (2) and (3), can exert significant influences (beneficial or deleterious) on plastic development.

V. ROLE OF VEHICLE

Paucity of vehicle seems to be deleterious to both liquefaction and plastic development. Evidence applicable to liquefaction has already been cited, and additional information published by Consolidation Coal Company (1968) further substantiates the importance of vehicle quantity and quality. Similarly, the extraction of only a minor proportion of the indigenous low molecular weight substances (bitumens) from plastic coal by solvents leaves a residue incapable of becoming plastic when heated (Brown and Waters, 1966; Pieron et al., 1959; Pieron and Rees, 1960). Many earlier investigators (for reviews, see Loison et al., 1963; Howard, 1963) suggested that these extractable bitumens contained the "coking principle" but their exact role was difficult to define. Teichmüller (1973) also imputes a significant role to "extractable bitumens" in the development of plasticity. Lazarov and Angelova (1968) attributed a significant role to pyrolytically formed chloroform extractables in plastic development; however, Berkowitz (1967) has called attention to the fact that during isothermal pyrolysis chloroform solubles disappear before maximum fluidity is achieved.

These bitumens, which have low melting points (150–180°C) according to Brown and Waters (1966), probably serve as the initial vehicle that solvates ("lubricates" per Berkowitz, 1960) the micelles as they become thermally loosened. Their higher than average hydroaromatic hydrogen content (see Table 2 of Brown and Waters, 1966) allows them to serve as mobile hydrogen donors to stabilize free radicals produced thermally before general plasticity commences. Even if these relatively small molecules were to cross-link (polymerize) with the free radicals formed by the decomposition of larger molecules, the net product might still be relatively smaller than the original coal molecule. Once general mobility of micellar units begins (i.e., once the coal begins to melt), then free radicals that continue to form would have improved chances to contact donor hydrogen on nonbitumen molecules. At that point, the original bitumen and vehicle would be of less consequence [in fact, Brown and Waters (1966) note that they decompose around 480°C]. Without the initial vehicle, however, there would be no transiently necessary solvating and hydrogen-donating substance, and therefore bond rupture at temperatures preceding general softening would simply lead to polymerization such that plasticity would never "get off the ground."

Apparently, in the liquefaction system, a vehicle like naphthalene serves as a solvating vehicle, but not as a hydrogen donor. It may serve as a hydrogen shuttle, however, facilitating the transfer of hydroaromatic hydrogen from the coal to pyrolytically formed free radicals. Naphthalene might also cross-link with free radicals, still producing a relatively low molecular weight product. As mentioned above, bitumen fractions that are not potential donors themselves could serve the same function.

Whereas the removal of potential vehicles serves to destroy or diminish plastic capability, the retardation of vehicle loss during pyrolysis, for example, by imposed pressures or larger particle size, enhances the development of plasticity. Skylar and Shustikov (1962) found that H_2 evolution was less whereas benzene solubility was greater if coals were heated to 400°C at higher pressures or if larger sizes were employed. Such coals mildly pyrolyzed under pressure also exhibited higher fluidity when subsequently tested at higher temperatures in a plastometer. Imposed pressures likewise, and for evidently similar reasons (i.e., maintaining the vehicle in contact with the coal), enhance liquefaction (Consolidation Coal Co., 1968). As might be expected, reduced pressure during pyrolysis, which could be expected to promote escape of the vehicle, reduces plastic capability, even to the extent of destroying it completely in some plastic coals (Loison *et al.,* 1963).

Numerous investigators, as exemplified by Berkowitz (1960, 1979), subscribe to the hypothesis that the magnitude of the porosity of any coal influences the rate of loss of volatile materials during pyrolysis, thereby dominantly influencing the development of plasticity. That this could be a factor cannot be denied; however, the more or less parallel changes of both physical properties (especially porosity) and chemical properties with rank make it difficult to conclude unequivocally which of these has the greater effect on plasticity. It is the hypothesis of this article that a substantial number of chemical phenomena can be cited as affecting the development of plasticity, and that these phenomena must be considered to be at least as important as porosity. It is my personal conviction that chemical properties play the dominant role in establishing whether or not a coal is plastic, as a result of, among other things, the observed lack of bloating exhibited by nonplastic, low-rank coal particles in the liquefaction system.

VI. OXIDATION AND "LOW-RANK" OXYGEN

Oxidation of coal preceding liquefaction can be deleterious to liquid yield, as the data plotted in Fig. 8 indicate. Wood *et al.* (1970) report that preoxidation reduces liquid yield when catalyst-impregnated coal is pyrolyzed in hydrogen. Oxidation also reduces pyrolytic tar yield while increasing the

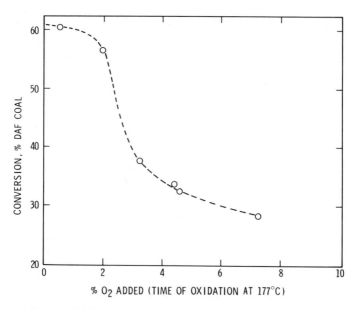

Fig. 8. Influence of oxidation on yield of benzene solubles obtainable by reacting Illinois coal in tetralin for 30 min at 400°C. Daf = dry ash-free.

molecular weight of pyrolysis tar (Wachowska and Pawlak, 1977). It is also well known that oxidation affects plasticity adversely, and deoxygenation has been reported to enhance plasticity (Rembashevski and Kisina, 1967). Although selected studies (Ignasiak *et al.*, 1972, 1974a,b; Wachowska *et al.*, 1974) have led to some insight, the role of oxygen is still not well understood. Most severely affected are the lower rank plastic coals (Yohe, 1958), in which oxidation appears to be essentially equivalent to reducing the coal rank. It is probable that low-rank (i.e., subbituminous) coals are prevented from becoming plastic because of (among other factors) their high oxygen contents, especially as oxygenated functional groups. Berkowitz *et al.* (1974) recently called attention to the apparent deleterious effect of —OH substituents on plastic development in western Canadian coals. Thus, a discussion of the effects of added oxygen on plasticity can reasonably be considered with a discussion of the effects of inherent oxygen.

Oxygen-related effects seem principally to involve cross-linking reactions. Cross-linking may be inherent in low-rank, high-oxygen coals, as Given *et al.* (1959) contended; however, the fact that significant proportions (~ 70–80%) of low-rank coals can be dispersed in appropriate solvents [e.g., ethanolamine, van Krevelen (1961)] implies that cross-linking does not bond micellar units throughout the coal. Instead, oxygen-related cross-linking seems to result from reactions promoted by mild energy input, such as the low temperatures to which the coal is transiently exposed while progressing from ambient temperature to "softening" temperature. Even mechanical energy, such as cavitation caused by pumping coal in a water slurry, can affect such cross-linking (Parkash *et al.*, 1971). Goodarzi and Murchison (1972) studied several coals optically as they were progressively heated and concluded that lower rank coals were more readily cross-linked during heating. Cross-linking could result from the formation of stable ethers, or the process could be more complicated, involving functional group condensations or the formation of unstable ethers that then proceed to cleave at still higher temperatures, leading to carbon-to-carbon polymerization. It has been claimed that the controlled cleavage of etheric bonds in oxidized coals restores lost plasticity, even though —OH and —COOH content remains high (Wachowska *et al.*, 1979).

We have observed that, although rate and ultimate conversion to gases plus benzene solubles during liquefaction is about equal for Illinois bituminous coal and Wyoming subbituminous coal, the rate and ultimate conversion to pyridine solubles is significantly lower for the high-oxygen, subbituminous coal [dmmf (dry mineral-matter-free basis) oxygen 18% for Wyoming coal, 12% for Illinois coal]. Thus, in high-oxygen coals strong mobility-restricting cross-links can be formed, even when donatable hydrogen is available.

We also have observed that when coal is heated in the inlet of a mass spectrometer, oxygenated substances (CO, CO_2, H_2O) are evolved in amounts that increase progressively as the coal temperature is raised from 150 to 350°C. Ignasiak *et al.* (1974b) reported that 53% of ^{18}O added through mild oxidation evolved as water when the oxycoal was heated to only 115°C during pyridine extraction. Wachowska *et al.* (1974) reported the formation of weak ether links during oxidation. When such materials crack in still nonmobile coal molecules, free radicals must be created. If transferable hydrogen is not proximally available, these free radicals must stabilize by "polymerization," forming mobility-retarding cross-links. Marsh and Walker (1979) discuss the reduced anisotropy of cokes from oxidized coals, contending that oxidation leads to cross-linking, which retards mobility that would otherwise result in the formation of anisotropic domains.

Another deleterious effect of low-temperature cleavage of oxygen bonds could be inefficient hydrogen consumption, for if donatable hydrogen were proximal to low-temperature-formed free radicals, it would probably transfer to and stabilize them. Hydrogen consumed in this fashion would not be available for the stabilization of depolymerization fragments at higher temperatures, and thus the hydrogen would be inefficiently employed with respect to aiding plastic development. The reduction in plasticity resulting from adding oxidized char (Wolfson *et al.,* 1964) or Fe_2O_3 (Loison *et al.,* 1963) to coal is most likely caused by consumption of hydrogen by these substances during pyrolysis.

To some extent, sulfur (or at least some molecular forms of sulfur) may behave like oxygen in coal. Ignasiak *et al.* (1978) have speculated that sulfur in the well-known Rasa lignite behaves similarly to weak etheric oxygen bridges, which cleave thermally, resulting in reduction in molecular size and the significant thermoplasticity of this coal.

VII. HYDROGENATION AND DEHYDROGENATION

Tar yields from pyrolysis of coal are directly proportional to hydro-aromatic hydrogen content, and hydrogenation increases tar yield, whereas dehydrogenation suppresses it (Lahiri and Mazumdar, 1967; Bodily *et al.,* 1974). Moderate hydrogen pressure (1000 psi) promotes agglomeration of nominally noncaking subbituminous coal at 500–600°C (Kawa *et al.,* 1959). Hydrogenation, catalytically promoted at 300–350°C, has been reported to enhance plasticity when the coal is heated to higher temperatures (Loison *et al.,* 1963). According to Shibaoka and Ueda (1978), hydrogen pressure during pyrolysis has also been found to retard or prevent mesophase development, which can be taken to be analogous to polymerization as discussed here. These observations indicate that hydrogen exerts a significant influence on plastic development.

Brooks and Sternhell (1958) reported that the reaction of low-rank (nonplastic) coals with alcoholic or aqueous alkali at 190°C for 12 hr resulted in significant addition of hydrogen and the production of a thermoplastic material. Hence, hydrogen addition seems capable of converting noncaking coals to caking coals.

Other experiments suggest that the addition of hydrogen to coal at very mild conditions, which are unlikely to result directly in depolymerization, that is, through lithium promotion (Given et al., 1959) or electrochemically (Sternberg et al., 1967), apparently reduces the coplanarity of lamellae and results in significant (75–85%) pyridine solubility and moderate (25–30%) benzene solubility.

It seems likely that all three of the conditions hypothesized as necessary and sufficient for plastic development would be favorably affected by the addition of hydrogen: (1) adding hydrogen would convert strong double bonds to weaker, potentially cleavable single bonds; (2) added hydrogen would be available for stabilizing free radicals formed by any thermal bond rupture; and (3) micellar mobility would be promoted (as implied by increased extractability of such hydrogenated coals). Some insight into the relative significance of hydrogen in terms of these three conditions can be inferred from our observation that exinite, which contains more hydroaromatic hydrogen than associated vitrinite (Given et al., 1960; Ladner and Stacey, 1963), liquefies at a *slower* rate than associated vitrinite, although ultimately all exinite is converted. We take this to mean that the rate of depolymerization of H_2C-CH_2 bonds is relatively slow (at 400–450°C) and therefore hydrogen appears not to enhance depolymerization *per se* so much as it deters or defers the repolymerization of free radicals formed by the rupture of inherently weaker bonds.

VIII. HEATING RATE

Our experiments, as well as those by Hill et al. (1966) and Curran et al. (1967), indicate that there is a finite rate function associated with the rupture of chemical bonds that leads to liquefaction. Hill's data also indicate that there is an asymptotic conversion limit related to any (isothermal) reaction temperature. The latter observation leads to the conclusion that, as Hill contended, there is a range of bond strengths indigenous to coal, and these different kinds of bonds decompose thermally at their own characteristic rate. We have already discussed the cracking of weak oxygen bonds (ethers) and note that liquefaction above about 450°C results in a high gas production (Nasritdinov, 1968), which implies the rupture of progressively "different" (viz. stronger) bonds.

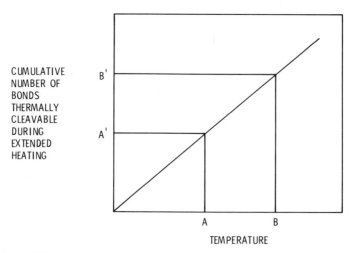

Fig. 9. Schematic representation of hypothesized bond strength distribution in coal.

When nominally plastic coals are heated very slowly to and beyond the nominal softening temperature, they fail to become plastic (Loison *et al.,* 1963). The reason for this behavior can be understood by reference to Fig. 9. Raising the temperature very slowly to B (assumed to be the nominal softening temperature) would provide ample opportunity for the rupture of all bonds weaker than, say, A'. That most free radicals so formed would not be inherently stable (in other words, that they have a short half-life) is implied by the fact that the "polymerization" of pyridine solubles to an insoluble form during "liquefaction" in naphthalene proceeds at about that same rate as that by which hydrogen-stabilized benzene solubles are formed in the tetralin vehicle. Thus, the radicals formed before general molecular mobility (melting) must stabilize by either hydrogen abstraction or repolymerization. Because there would be only a finitely limited probability of donor hydrogen being proximal to the radicals, their statistically dominant fate would be repolymerization into an extended system that could no longer be made mobile by heating to or above the nominal softening temperature.

On the other hand, rapid elevation of the temperature to or above B (the nominal softening temperature) would provide only limited time for bonds such as those represented by A' to rupture. Their subsequent rupture would not be deleterious once general softening occurred to allow free radicals formed by their rupture to contact indigenous donor hydrogen.

Experimental evidence reported by Rembashevski and Kisina (1967) bears upon the concept of cross-linking caused by heating below the melting point. They found that the amount of material extractable in anthracene oil at 350°C decreased as a tested coal was preheated in an inert atmosphere to progressively higher temperatures (below 350°C). In addition, these same investigators found that the production of benzene-soluble materials decreased with increasing preheat temperature. We infer that the potential for the formation of extractable (and presumably mobile) material was reduced through cross-linking, brought about at the lower temperatures.

Thus, if the rate of decomposition without attendant formation of "liquids" is high at relatively low temperatures (likely to be true where relatively weak covalent bonds, e.g., oxygen and thiol ethers, are abundant), then recombination would result before melting occurs. Should the decomposition be retarded or the melting temperatures be reduced, otherwise nonplastic coals could be converted to plastic coals. Certain forms of alkylation seem to reduce the melting point and could also serve to stabilize systems that might decompose in their natural state (Wachowska et al., 1979).

IX. CHEMICAL ADDITIVES

At present, the information available about the influence on plasticity of various inorganic reagents added to coal is somewhat conflicting. It is claimed that Na_2CO_3 decreases plastomer-measured fluidity without significantly affecting the temperatures characteristic of fluid development (Patrick and Shaw, 1972). It is claimed that NaOH decreases the free-swelling index (Crewe et al., 1975) and agglomerating propensities (Wachowska and Pawlak, 1977) of nominally agglomerating coals. Ting (1973) has reported an unusual caking lignite that loses its plastic capacity when Na is added. Given et al. (1975) report a higher liquefaction yield from a lignite impregnated with NaCl. Our own limited experience in this area, which we shall not detail here, also leads to conflicting findings. "Cross-linking" is alluded to as a factor in the reported effects of alkali additives, and it might be pertinent to recognize that alkalis do promote methylene condensation in phenol/aldehyde reactions (Carswell, 1947). Marsh and Walker (1979) suggest that alkali salts catalyze dehydrogenation, and this could be expected to reduce plasticity. It is more likely, however, that alkalis and alkaline earth elements simply form salts of oxygenated acidic functionalities, thereby increasing their melting points and reducing micellar or lamellar mobility.

Other instances of effects of alkali and alkaline earth elements on plasticity are as follows: (1) when lignites are autoclaved with an aqueous solution of CaO for 3 hr at 340°C, the reacted lignite loses significant oxygen and is converted to a caking material (Khemchandani et al., 1974); and (2) sodium apparently acts in some fashion to promote liquid formation from lignite reacted at ~350–400°C under substantial CO and H_2O pressure (Appell et *al.,* 1972). The former reaction could be similar to the hydrolytic decomposition of cured phenophasts in aqueous NaOH at 350°C reported by Carswell (1947); the latter could involve some intermediate reaction, which stabilizes oxygen, thus preventing condensation, followed by a similar hydrolytic decomposition. Another possibility might be that NaOH promotes cleavage of ethers, leading to molecular fragmentation. Makabe and Ouchi (1979) have demonstrated that the use of NaOH in ethanol leads to ether cleavage.

Numerous other additives and chemical reactions have been reported to influence plasticity. The addition of BF_3 (Chakrabartty and Berkowitz, 1972) and HCl (Ignasiak *et al.,* 1974a) destroys plasticity. Halogens are known to promote hydrocarbon cracking; if they were to do so during coal pyrolysis, the effect could be deleterious to plasticity if the free radicals formed were to exceed significantly the capacity of the coal to provide stabilizing hydrogen. The effect would be especially deleterious if the reaction occurred before temperatures were sufficiently high to provide some mobility, allowing contact between free radicals and indigenous potential-donors.

X. SUMMARY

Coal liquefaction experiments, in which coal is heated to ~400°C while slurried in various organic liquid vehicles, indicate that the formation of low molecular weight products from high molecular weight coal proceeds via the thermal rupture of relatively weak chemical bonds ("depolymerization"), followed in the optimum case by stabilization of the resulting free radicals by hydrogen atoms abstracted from hydroaromatic structures. In particular, the hydroaromatic hydrogens on partially saturated, multiring, organic molecules (e.g., tetralin) in the vehicle are readily abstracted in this process. If abstractable hydrogen is not readily available, some of the pyrolytically formed free radicals either recombine or "attack" adjacent molecules, forming stable, higher molecular weight materials ("repolymerization"). Thermally promoted plasticity of vitrinite in bituminous coals apparently involves these same chemical reactions; however, whereas in liquefaction the slurrying vehicle is the major source of transferable hydrogen, in inert-atmosphere pyrolysis the coal itself is the source of such hydrogen.

In the ideal case, plastic development proceeds as follows:

1.   General mobility of micellar units (physical melting) commences around 350–400°C as van der Waals forces and hydrogen bonds become weakened, the bitumen serving as a necessary solvating vehicle and hydrogen donor early on during this phase. Some minimal cleavage of covalent bonds (especially ethers) may be involved in this phase.

2.   Viscosity decreases progressively as mean molecular size is reduced at a rate reflecting the thermal rupture of covalent bonds that bridge stable molecular units, but only so long as the free radicals thereby formed are stabilized by indigenous donatable (hydroaromatic) hydrogen.

3.   As the limited donor-hydrogen inventory becomes consumed through the transfer of hydrogen and loss of volatile hydrogen-rich species, free radicals, which continue to be formed, increasingly "stabilize" by repolymerization. Viscosity, having reached a minimum, then increases progressively as the molecular weight of the residual materials increases.

4.   Repolymerization then becomes the dominant fate of free radicals, and the metaplast solidifies as a semicoke. In coals where premature crosslinking is not significant, molecular moieties align to form anisotropic mesophase structures.

Three properties of coal appear to be necessary and sufficient for plastic development: (1) lamellae-bridging structures that can be thermally broken; (2) an indigenous supply of hydroaromatic hydrogen; and (3) an intrinsic capability of micelles and lamellae to become mobile (to "melt") independently of quantitatively significant bond rupture. Enhancement or destruction of these properties leads to enhanced or reduced plastic capability.

Bitumen removal, high concentration of relatively weak bonds (especially oxygen related), slow heating rates, and possibly some additives (i.e., halogens) all appear to promote cross-linking before the temperature is reached at which general mobility (melting) can take place. Subsequent contact between indigenous donatable hydrogen and any free radicals formed at higher temperature is thereby prevented; this, in turn, leads to reduced plastic capability.

Imposed reduction of the donor hydrogen content can lead to reduced plastic capability by allowing the "premature" repolymerization of thermally produced free radicals. Direct dehydrogenation, bitumen removal, and oxidation can all result in such reduced availability of donor hydrogen, thereby reducing plastic capability.

ACKNOWLEDGMENT

Discussions with Dr. Ron Liotta of the Corporate Pioneering Research Laboratory (Linden, New Jersey) of Exxon Research and Engineering have helped me to crystallize my thinking on several aspects of the phenomena discussed herein. I hereby express my gratitude to him.

REFERENCES

Appell, H. R., Wender, I., and Miller, R. D. (1972). *Inf. Circ.—U.S., Bur. Mines* **8543**, 32–39.
Berkowitz, N. (1960). *Prepr. Blast Furnace, Coke Oven, Raw Mater. Conf., AIME* pp. 1–18.

18   Richard C. Neavel

Berkowitz, N. (1967). *Proc. Symp. Sci. Tech. Coal, 1967* pp. 149–155.
Berkowitz, N. (1979). *Proc. Robert A. Welch Found. Conf. Chem. Res.* **22.**
Berkowitz, N., Fryer, J. F., Ignasiak, B. S., and Szladow, A. J. (1974). *Fuel* **53,** 141.
Bodily, D. M., Lee, S. H. D., and Wiser, W. H. (1974). *Prepr. Pap.—Am. Chem. Soc., Div. Fuel Chem.,* **19**(1), 163–166.
Brooks, J. D., and Sternhell, S. (1958). *Fuel* **37,** 124–126.
Brooks, J. D., and Taylor, G. H. (1958). *Chem. Phys. Carbon* **4,** 243–286.
Brown, H. R., and Waters, P. L. (1966). *Fuel* **45,** 17–60.
Carswell, T. A. (1947). "Phenoplasts: Their Structure, Properties, and Chemical Technology." Wiley (Interscience), New York.
Chakrabarrty, S. K., and Berkowitz, N. (1972). *Fuel* **51,** 44–46.
Collins, C. J., Benjamin, B. M., Raaen, V. F., Maupin, P. H., and Roark, W. H. (1977). *Prepr. Pap.—Am. Chem. Soc., Div. Fuel Chem.,* **22**(5), 98–102.
Consolidation Coal Co. (1968). Project gasoline pre-pilot plant phase 1 research on CSF Process. *Res. Dev. Rep.—U.S., Off. Coal Res.* **39**(2).
Crewe, G. F., Gat, V., and Dhir, V. K. (1975). *Fuel* **54,** 20–23.
Curran, G. P., Struck, R. T., and Gorin, E. (1967). *Ind. Eng. Chem. Process Des. Dev.* **6,** 166–173.
Dryden, I. G. C. (1963). *In* "Chemistry of Coal Utilization" (H. H. Lowry, ed.), Suppl. Vol., pp. 232–295. Wiley, New York.
Fitzgerald, D. (1956). *Trans. Faraday Soc.* **52,** 362–369.
Given, P. H., Lupton, V., and Peover, M. E. (1959). *Proc.—Resid. Conf. Sci. Use Coal, 1958* pp. A38–A42.
Given, P. H., Peover, M. E., and Wyss, W. F. (1960). *Fuel* **39,** 323.
Given, P. H., Cronauer, D. C., Spackman, W., Lovell, H. L., Davis, A., and Biswas, B. (1975). *Fuel* **54,** 34–39.
Goodarzi, F., and Murchison, D. G. (1972). *Fuel* **51,** 322–328.
Hill, G. R., Hariri, H., Reed, R. I., and Anderson, L. L. (1966). *Adv. Chem. Ser.* **55,** 427–447.
Howard H. C. (1963). *In* "Chemistry of Coal Utilization" (H. H. Lowry, ed.), Suppl. Vol., pp. 340–394. Wiley, New York.
Ignasiak, B. S., Clugston, D. M., and Montgomery, D. S. (1972). *Fuel* **51,** 76–80.
Ignasiak, B. S., Szladow, A. J., and Montgomery, D. S. (1974a). *Fuel* **53,** 12–15.
Ignasiak, B. S., Szladow, A. J., and Berkowitz, N. (1974b). *Fuel* **53,** 229–231.
Ignasiak, B. S., Fryer, J. K., and Jadernik, P. (1978). *Fuel* **57,** 578–584.
Kawa, W. R., Hiteshue, W., Budd, W. A., Friedman, S., and Anderson, R. B. (1959). *Bull.—U.S., Bur. Mines* **579.**
Khemchandani, G. W., Ray, T. B., and Sarhar, S. (1974). *Fuel* **53,** 163–167.
Ladner, W. R., and Stacey, A. E. (1963). *Fuel* **42,** 75–83.
Lahiri, A., and Mazumdar, B. K. (1967). *Proc. Symp. Sci. Tech. Coal, 1967* pp. 175–184.
Lazarov, L., and Angelova, G. (1968). *Fuel* **47,** 342–352.
Loison, R., Peytavy, A., Boyer, A. F., and Grillot, R. (1963). *In* "Chemistry of Coal Utilization " (H. H. Lowry, ed.), Suppl. Vol., pp. 150–201. Wiley, New York.
Makabe, M., and Ouchi, K. (1979). *Fuel Process. Technol.* **2**(2), 131–141.
Marsh, H. (1973). *Fuel* **52,** 205–212.
Marsh, H., and Walker, P. L., Jr. (1979). *Fuel Process. Technol.* **2,** 61–75.
Nasritdinov, S. (1968). *Inst. Khim., Akad. Nauk Uzb. SSR.*
Neavel, R. C. (1976). *Fuel* **55,** 237–242.
Orchin, M. (1944). *J. Am. Chem. Soc.* **66,** 535–538.
Parkash, S., Szladow, A. J., and Berkowitz, N. (1971). *CIM Bull.,* Nov.
Patrick, J. W., and Shaw, F. H. (1972). *Fuel* **51,** 69–75.

Pieron, E. D., and Rees, O. W. (1960). *Circ.—Ill. State Geol. Surv.* [N.S.] **288,** 1–11.

Pieron, E. D., Rees, O. W., and Clarke, G. L. (1959). *Circ.—Ill. State Geol. Surv.* [N.S.] **269,** 1–36.

Rembashevski, A. G., and Kisina, A. M. (1967). *Khim. Tverd. Topl. (Leningrad)* **4,** 26–34; *Chem. Abstr.* **68,** 4710.

Sanada, Y., Furuta, T., Kimira, H., and Hanada, H. (1973). *Fuel* **52,** 143–148.

Shibaoko, M., and Ueda, S. (1978). *Fuel* **58,** 667–672.

Skylar, M. G., and Shustikov, V. I. (1962). *Khim. Tekhnol. Topl. Masel* **7,** 39–42.

Sternberg, H. W., Markly, R. E., Delle Donne, C. L., and Wender, I. (1967). *Rep. Invest.—U.S., Bur. Mines* **RI-7017.**

Teichmüller, M. (1973). *Adv. Org. Geochem.* 379–407.

Ting, F. T. C. (1973). *Proc., Bienn. Conf. Inst. Briquet. Agglom.* **13,** 251–254.

van Krevelen, D. W. (1961). "Coal." Elsevier, Amsterdam.

Wachowska, H., and Pawlak, W. (1977). *Fuel* **56,** 342–343.

Wachowska, H. M., Nandi, B. N., and Montgomery, D. S. (1974). *Fuel* **53,** 212–219.

Wachowska, H. M., Nandi, B. N., and Montgomery, D. S. (1979). *Fuel* **58,** 257–263.

Wiser, W. (1968). *Fuel* **47,** 475–486.

Wolfson, D. C., Birge, G. W., and Walter, J. G. (1964). *Rep. Invest.—U.S., Bur. Mines* **RI-6354.**

Wood, R. E., Anderson, L. L., and Hill, G. R. (1970). *Colo. Sch. Mines Q.* **65,** No. 4, 201–216.

Yohe, G. R. (1958). *Rep. Invest.—Ill., State Geol. Surv.* **207.**

The Physical Structure
of Coal

WARREN R. GRIMES
Chemical Technology Division
Oak Ridge National Laboratory
Oak Ridge, Tennessee

I.   Introduction 21
II. Brief Overview of Physical Structure 23
III. The Porous Nature of Coal 25
     A. Porosity of Coal 25
     B. The Internal Surface of Coal 26
     C. Pore Size Distribution in Coal 31
IV. Effects of Heat on Coal Structure 32
     A. Changes in Pore Structure on Heating 32
     B. Devolatilization 33
     C. Plastic Behavior 35
     References 39

I. INTRODUCTION

Coal is a generic term applied to widespread and widely differing natural rocks that consist primarily of organic material admixed with smaller quantities of mineral matter. The mineral matter may occur both as discrete mineral species and mineral phases of extremely various particle size and (in markedly smaller quantities) as species that are chemically complexed by the organic matter. The organic components, derived from complex plant materials by complicated chemical and biochemical reaction processes, are in the main composed of large, polymeric molecules whose chemical structures

21

ISBN 0-12-150701-7

are variable and nonrepetitive. The physical and chemical structures of these organic components are responsible for the properties of coal.

Physical structure is taken to represent the spatial arrangement of the large, complex molecules, along with a few much smaller ones, in coals and in solid coal products. All physical properties of coals and the complex changes in these properties on heating are, to a considerable degree, consequences of this physical structure. This is not meant to imply that the physical and the chemical structures of coal are completely independent or that chemical properties are without influence on physical properties and their complex temperature dependence. But many facets of the physical structure of coal may be studied with profit without a detailed knowledge of the chemical structure of the large coal molecules.

A great body of evidence shows that the chemical structures of the organic molecules vary greatly, and particularly with rank, among the coals; indeed, it appears that the chemical structures of coals—and even coals of substantially the same rank—show more differences than similarities. On the other hand, the physical structures of coals appear to show more similarities than differences. Although there are numerous anomalous examples, the physical properties of coal generally vary so systematically with coal rank that many can be reasonably predicted from the carbon content or the volatile material content of the coal.

It should be noted that the physical properties of coal, and those many aspects of coal behavior that depend upon the physical structure, seldom change monotonically with coal rank. Many physical properties show maximum or minimum values at carbon contents between 85 and 90%, that is, near the point of transition from bituminous to anthracitic coals.

Many properties and many aspects of coal behavior have been studied over the past several decades. Several of these, that is, infrared (IR) absorption spectra, electron spin resonance (ESR), and nuclear magnetic resonance (NMR), are valuable chiefly for the light they throw on the chemical structure of the organic coal components. Many properties, although they furnish some information about physical structure, are measured primarily because they serve more utilitarian ends. Hardness, friability, grindability, and impact resistance are useful in coal comminution practice, and compressive strength is important to underground mine design and mine safety assessment. Plastic behavior and agglomeration propensity are essential in the formation of metallurgical cokes and are detrimental in some other coal uses or coal conversion technologies. But several properties, X-ray diffraction, porosity and pore structure, swelling in solvents, refraction and reflection of visible and ultraviolet (UV) light, shear moduli, and electrical conductivity and dielectric constant, are useful primarily for the light that they shed on the physical structure of coal and of solid coal products.

II. BRIEF OVERVIEW OF PHYSICAL STRUCTURE

Coals are such heterogeneous materials that statements about their composition or structure mean little unless one specifies whether they apply to macroscopic, microscopic, or submicroscopic domains.

On the macroscopic scale, a dominant feature of bituminous coal is its banded appearance. These bands consist of irregularly alternating layers of differing composition as well as differing appearance. It has long been recognized that there are four such components, now referred to as lithotypes, in coals. They are [after Stopes (*1*), whose terminology has been adopted in the petrographic nomenclature]: vitrain, clarain, durain, and fusain. The banded appearance is, in general, much less apparent in less mature coals, but examination of the youngest coals under the microscope reveals entities that correspond to those in the bituminous materials.

At the microscopic level these lithotypes, such as vitrain, are found to be complex aggregates. The ultimate microscopic organic constituents of coal, as now internationally codified (*2-5*), are the macerals. Some 14 of these, which differ in appearance, optical properties, and chemical composition, are known (*4*). In this international codification, the macerals fall into three groups (known as vitrinite, exinite, and inertinite), and these maceral groups occur in a total of seven associations known as microlithotype groups.[1] The three maceral groups have considerably different chemical compositions (with H/C ratios highest in exinites and lowest in inertinites) in high-volatile bituminous coals. The chemical and optical differences among the maceral groups narrow as coal rank increases further; in anthracites (> 93% C) the maceral groups become essentially impossible to distinguish.

At the molecular level, it seems certain that the macerals consist, in large part, of organic macromolecules (*5*), although much smaller volatile or easily extractable molecules are present to the extent of a few percentage points of the total carbon atoms. Much chemical information about the macromolecules has been obtained over the years, but the detailed chemical structures of these macromolecules are unknown and—except in some general statistical sense—may even be unknowable. The identities of the smaller molecules constitute a much simpler problem, and considerable information exists about them (*5-14*).

The maceral groups (and, in particular, vitrinite, which is by far the most studied) appear to be composed of poorly aligned lamellae consisting of the coal macromolecules. From studies of coals by X-ray crystallographic techniques, it is inferred (*5,16,17*) that the diameters of the lamellae increase from

---

[1] A different system of the nomenclature, specified in ASTM (D2796-72), is used in North America. In that system, six macerals (of which vitrinite and exinite are two) are recognized.

about 7 to near 7.5 Å as the carbon content of the coal rises from 80 to 91.5% and then increases rapidly in the anthracitic range. It appears, however, that X-ray study of coals is relatively unrewarding, and it is true that the X-ray diffraction data have been successively reinterpreted to yield smaller and smaller sizes of the structural units. How much real information has been (or can be) obtained from interpretation of X-ray diffraction patterns of coal remains uncertain.

Binding forces among the macromolecules have been ascribed to molecular entanglement, to weak noncovalent associative forces (such as hydrogen bonds or acid–base interactions) or to highly cross-linked structures (6,18–20). It seems likely that all these mechanisms play a part. Evidence for the entangled molecules includes the fact that the residue of coal after exhaustive Soxhlet extraction with pyridine continues to be extracted further with pyridine aided by ultrasonic vibration. Acidic and basic moieties are present in coal-derived liquids. They must be present in coal, and their interaction could provide cross-linkages of some strength. However, coal is an elastic solid. When it is subjected to a constant tensile stress, it stretches for a time and then remains at that length; when the stress is removed, the coal returns to its original length (21,22). It is difficult to reconcile this behavior with any other than that of a three-dimensional network with some degree of cross-linking between the macromolecules. Imperfect packing of the lamellae and of the micelles is responsible for the long known (5) microporosity of coal. The equally well-known macroporosity (5) may be inherent in this submicroscopic structure of coal or it may result from microcracks and fissures in the coal macro- or microstructure.

This physical structure of coal—as qualitatively sketched above—may also, in combination with some recently learned chemical facts (23,24), lead to qualitative mechanisms for the "melting,"[2] agglomeration, and plastic behavior exhibited by some coals on pyrolysis (5,25,26), or during hydroliquefaction (23,27–29).

Coal seems, at our present state of knowledge, to consist of a network of clusters—somewhat like beads on a string—held together by mechanisms that include occasional chemical cross-links. The clusters appear to be relatively small (7–8 Å in diameter and containing 16–18 carbon atoms) and consist of aromatic and hydroaromatic rings and condensed rings. The cross-linkages apparently consist at least in part of ether linkages, methylene groups, and some longer (but apparently less than four-carbon) alkanes.

---

[2] "Melting" of coal is, to a considerable extent, a misnomer. The fact that the pore structure of coal persists through the fluid stage certainly seems to rule out simple, complete melting.

III. THE POROUS NATURE OF COAL

That coals are inherently porous materials has been apparent for many years (5). The delicately structured pore systems give coals all the properties of viscoelastic colloids (30). The void volume—and the more difficultly determinable details of the pore structure—influence the behavior of coal more directly than does virtually any other property.

It should be noted at this stage that the entire question of porosity and pore structure poses fundamental problems. Modern studies strongly suggest, as noted above and as indicated in various places below, that coal is a viscoelastic solid; it is neither rigid nor fluid, but it is deformable to the extent permitted by distortion of the cross-linkages from their equilibrium configurations. As a consequence, the porosity of coal can be appreciably altered by adsorption of some of the fluids that may be used to measure the porosity. The results obtained are therefore likely to be strong functions of the probe used for the measurement. Moreover, it now seems certain that the interconnecting pores in coal are very far from cylindrical. They more closely resemble very thin slits of considerable width and they contain even more narrow portions incapable of accommodating more than a monolayer of adsorbate without distortion. For these reasons, equilibrium sorption–desorption data with appropriate vapors, whether interpreted by the Brunauer–Emmett–Teller or the Dubinin–Polanyi model, can provide valuable information on relative surface areas of coals and solid coal products, but they may be incapable of establishing absolute values of the internal surface area.

A. Porosity of Coal

Total void volume of coal (the porosity) can be estimated from the capacity bed moisture content of coal—that is, from the quantity of water the coal holds at nearly 100% relative humidity in the undisturbed seam. Mined coal loses water upon exposure to ambient air, but capacity bed moisture can be virtually restored by equilibration at 30°C in a dessicator at 97% relative humidity maintained by saturated aqueous K_2SO_4 solution (31). Porosity so estimated ranges from a high of more than 30% in lignites through a low of about 1% in coals with 87–90% C and rises again to about 10% in the more mature anthracites (32).

The porosity of coals and of solid coal products is generally determined from absolute density measurements made using helium and mercury as displacement fluids. Helium (with its very small atomic diameter and its virtual lack of sorption on coal) is supposed to penetrate the entire pore structure (5,33,34), although more recent X-ray studies (35) suggest that anthracites have some blind pores inpenetrable to He. Mercury at normal pressures does not penetrate pores smaller than 10 μm. Total available pore volume (in pores less than 10 μm diameter) accordingly is given by the difference between the specific volume in mercury and in helium. Values obtained in this way follow the same trend as (and are similar to) those obtained from measurement of capacity bed moisture.

Apparent densities, measured by the displacement of convenient liquids (water, methanol), depend on how completely the pore space is permeated and the extent of liquid–surface interaction. Apparent densities measured with H_2O as the displacement fluid are generally slightly higher for lower rank coals and slightly lower for higher rank coals than those measured with He. Those measured with methanol are generally higher than those obtained with He over all coal ranks (33); these consistently higher densities obtained in methanol have been attributed (33) to compression of the liquid in the small coal pores. Methanol (see the subsequent discussion) is strongly adsorbed by coal, and the density of the first adsorbed layer of methanol is some 20% greater than that of liquid methanol (36).

B.   The Internal Surface of Coal

The internal surface area of coal is determined from (1) measurements of the heat generated when coal is treated with liquids, (2) determination of the quantities of adsorbed gases or vapors as functions of temperature and adsorbate pressure, and more recently, (3) by careful examination of scattering by X rays incident at very small angles.

1.   Heats of Wetting

The earliest attempts to determine the internal surface area of coals used measurements of heats of wetting (5). A variety of liquids has been used (especially in attempts to evaluate pore size distribution). Many studies of internal surface area have used methanol (37), primarily because the heat of wetting with this material is delivered rapidly. Griffith and Hirst (38) showed, for example, that 95% of the heat was released within 10 min and 99% within 25 min.

When methanol, or many other liquids, penetrates coal, the coal begins to swell; a coking coal may show a volume increase of as much as 10% in methanol vapor at 80 mm Hg (5). The swelling process is endothermic. The interaction between methanol and the coal surface is exothermic. The resulting difference, which is invariably exothermic, is the heat of wetting. It has been demonstrated that methanol sorption is reversible, but that small irreversible changes in structure of the coal gel result from its sorption–desorption (39). Maggs (40) concluded (from heats of wetting and quantities absorbed—both with methanol) that the heat of wetting amounted to 1 calorie per 10 m² of internal surface. Bond and Spencer (*34*) studied the wetting of carbon blacks of known surface area. They showed that each of 16 liquids with molecular volumes below 300 cm³/mole liberated about 0.93 calories per 10 m² of surface; maximum deviations from this value were ±25%, but most values were within ±10%. From heats of wetting by methanol, Griffith and Hirst (*38*) found values ranging from 10 to 200 m²/g for surface areas of a wide variety of coals. Surface areas were relatively low (near 80 m²/g) in the coals with 75-80% carbon, rose steeply (to near 200 m²/g) at about 82% carbon, declined steeply to a low (10-25 m²/g) at 90% carbon, and increased to about 70 m²/g near 95% carbon. The range from 75 to 80% carbon is represented by only two coals; values for this range may be atypical.

Such measurements have in the past been considered absolute values, but it is difficult to assess their real accuracy. The implicit assumption that all pore surfaces are reached may be incorrect. If so, this fact and the endothermicity of the swelling could cause the results to be low. However, methanol is known to be imbibed by coal, and this process almost certainly creates additional (new) surface. Moreover, methanol reacts exothermically with some functional groups (e.g., phenols) in coal (*41,42*). Both these effects would tend to make the values high.

2. Adsorption of Gases and Vapors

By this well-known method (*43*) the quantity (V) of gas or vapor sorbed at constant temperature is determined at different relative pressures (p/p_0) and the surface area is evaluated from the value of v at which the entire available surface is covered by a monolayer of sorbate and from the cross-sectional area of the sorbate molecule in its adsorbed state (*44*). Early application of this method to coal using argon and nitrogen at 77-90 K gave values (0.1-10 m²/g) that were far smaller than those obtained from heats of wetting (*5,45-47*). Maggs (*48*) suggested (1) that the pores of coal closed by thermal contraction at low temperatures or (2) that diffusion of nitrogen or argon into the very fine pores required a considerable energy of activation essentially unavailable at these very low temperatures. Zwietering and van Krevelen (*49*) showed (by measuring densities of coal in He at 77-373 K) that the first explanation was certainly incorrect.

Zwietering and van Krevelen (*49*) studied adsorption of N_2 and methane at much higher temperatures (0-50°C). They concluded that such measurements, which yielded surface areas of about 90 m²/g with considerable uncertainty, could give values in reasonable agreement with those from heats of wetting. Since these studies had also shown that the surface area of coal was due to two separate pore systems (see the subsequent section), Zwietering and van Krevelen concluded that N_2, for example, needed an appreciable activation energy to penetrate the fine pore structure and, at very low temperatures, was capable of sensing only the macropore system.

Walker and Geller (*50*) showed a value of 175 m^2/g for surface area of an anthracite when measured with CO_2 at $-78°C$, while the same coal gave only 11 m^2/g with N_2 at $-196°C$. Marsh and O'Hair (*51*) showed that adsorption of CO_2 and of N_2O gave essentially identical surface areas for several coals. Bond and Spencer (*34*) used neon at 25°C and assumed that a unit surface of coal absorbed neon as did a unit surface of a (known surface area) carbon black. They found surface areas smaller by three- or fourfold for low-rank (83–85% C) coals than those obtained from heats of wetting, but the two methods gave essentially identical values for anthracites (92–94% carbon).

As of the early 1960s (*5*), several authorities (*34,50,52,53*) had apparently concluded that the true value for coal surface areas lay between those from heats of wetting and those from low-temperature gas absorption. At present, there seems to be a tendency to view the surface areas determined from CO_2 adsorption as the absolute surface area of coal (*54–56*).

Recent studies (*54*) have extended such measurements (with N_2 and CO_2) to many coals. Several authors (*51,55,57–59*) have shown that the Dubinin–Polanyi equation gives surface area values very similar to those from the Brunauer–Emmett–Teller equation when applied to CO_2 or to methanol adsorption data near room temperature.

Concerted attempts to understand the fundamentals of sorption–desorption behavior of coal appear to be relatively few. One ongoing program is studying three coals in the subbituminous–bituminous range using a variety of techniques (*14,15*). This study evaluates surface areas using the Brunauer–Emmett–Teller method with N_2, CO_2, and H_2O after a variety of outgassing methods, examines the kinetics of adsorption and desorption, carefully monitors the amounts and identities of outgassed species under vacuum at temperatures to 200°C, and determines both rapid and slow components of heats of wetting (with H_2O) on the outgassed materials. These studies show—in general agreement with previous authors—low apparent surface areas (3–10 m^2/g, depending upon outgassing) for N_2 at $-196°C$, high apparent surface areas (125–200 m^2/g, varying with the coal but with little dependence on outgassing) for CO_2 at -78 and 23°C and 68 m^2/g (Illinois No. 6) and 274 m^2/g (Wyodak) for apparent surface areas by H_2O adsorption at 20°C. Such sorptions appear to be dominated by dipolar and London forces, as noted by Larsen (*22*), and seem to be dependent markedly upon the molecular properties (polarizability, dipole moment, etc.) of the sorbate.

Models that are based upon a rigid porous sorbent with an associated "hard" electrostatic field (*60*) have successfully correlated sorption of gases or crystalline inorganic materials (*61*). However, and as might be expected, such models do not correlate with coal sorption data. Some success (*62*) has been obtained from a model that views the coal surface as a yielding "soft" electronic structure (an extensive π bond network with polar functional groups more or less at random) in which image forces can be induced by the sorbate molecule. This is not unlike a model applied with some success to adsorption on metals. (60). At present, it seems possible that an understanding of the sorption of gases by coals may be realizable from such models.

For the present, however, it is clear that techniques using some sorption–desorption equilibria measure something related to the chemical properties of the sorbate as well as to the absolute surface area of coal. Values obtained using CO_2, N_2O, and Xe seem generally to be in good agreement, and (see below) such values usually agree reasonably well with small-angle X-ray scattering (SAXS) in the few cases where real comparisons are possible. A systematic study, obtaining kinetic as well as equilibrium data, using several gases (N_2, CH_4, CH_3OH, H_2O, CO, CO_2, N_2O, Xe) as well as SAXS on carefully pretreated coals and solid coal products, should give valuable insights regarding the nature of reagent interaction with the coal surface. The Brunauer–Emmett–Teller and related surface area methods are simple and easy to apply. The values obtained with appropriate reagents give valuable information about the surface area available to specific reagents, and when Xe or CO_2 is used, give values that must at least approximate the real area in pores of diameter greater than about 4 Å.

3. Small-Angle X-Ray Scattering

Interactions of coals with the surface-area "probe" (if any are important) can, of course, be avoided by use of purely physical probes. One of these with promise for determination of internal surface area of coal is the study of small-angle X-ray scattering (SAXS) by coals.

The several studies of internal surface area of coals by SAXS (58,63–66) seem to be based upon Porod's (67) analysis of the theory of small-angle X-ray scattering in two phase (full and empty) systems. According to Porod, the scattered intensity at relatively high angles (outer part of the scattering curve) depends directly on the total interphase surface area of matter in the sample, the square of the electron density difference between the scatterers and the air, and inversely upon the fourth power of the scattering vector. It should be noted that this method is capable of measuring total porosity, including sealed as well as open pores.

Kroger and Mues (63) obtained reasonable agreement for specific surfaces of macerals by SAXS and by heats of immersion in methanol. Durif (64) applied SAXS to determine specific surface of hard coals and their carbonization products; Chiche et al. (65) applied SAXS and several adsorption techniques to coals and their carbonization products. These authors described a number of uncertainties in the SAXS method when applied to natural vitrains and chars obtained at temperatures below 1000°C and did not consider the data obtained on these low-temperature items to be particularly reliable. Marsh and O'Hair (*51*) compared values obtained by adsorption of CO_2, neon, and N_2O on coals of various ranks with the values from SAXS previously obtained (*65*) on low-temperature chars. They concluded that agreement between the adsorption methods and SAXS had not been proved. Spitzer and Ulicky (*58*) used SAXS and adsorption of methanol vapor at 20°C with 20 Czechoslovakian coals. They found that among lignites and brown coals (carbon contents 64.5 to ~ 70%) surface areas via methanol adsorption were higher (but generally not more than 30% higher) than those from SAXS. For coals of higher rank (76–91% C), however, SAXS gave higher values (frequently two- to fourfold higher) than did methanol adsorption. Spitzer and Ulicky (*58*) also applied SAXS, methanol adsorption at 20°C, and CO_2 adsorption at −78°C to two samples of American coal (a St. Nicholas with 92.5% C and an Illinois No. 6 with 78.3% C). As indicated in Table I, values obtained by SAXS agree reasonably well with those of CO_2 adsorption at 25°C for both coals. The value from methanol adsorption at 20°C on the Illinois No. 6 coal is also in agreement.

Lin *et al.* (*66*) have also examined Illinois No. 6 coal with a sophisticated SAXS assembly. They found a surface area of 140 ± 20 m^2/g.

4. Summary

Relative surface areas and surface areas available to specific sorbates can be determined with confidence from sorption–desorption equilibria with gases and vapors on coal. The absolute surface areas, however, are still somewhat uncertain. One may be reasonably confident that they fall from values of 150–250 m^2/g among lignites and subbituminous coals to 50–100 m^2/g for high bituminous coals and rise again to values above 100 m^2/g among anthracites. A considerable measure of agreement for at least some coals appears to exist among SAXS and sorption measurements with CO_2 at 25°C.

TABLE I

Surface Areas of Two American Coals by Several Methods

Method	Surface area of coal (m^2/g)	
	Illinois No. 6	St. Nicholas
CH_3OH adsorption at 20°C (*58*)	175	122
CO_2 adsorption at −78°C (*58*)	136	165
CO_2 adsorption at 25°C[a]	160	234
SAXS (*58*)	179	275

[a] These values were obtained by P. L. Walker, Jr., of Pennsylvania State University, who supplied the samples for the study (*58*).

Although its precision will have inherent limitations even with the most modern instrumentation, it may be that SAXS has the best chance of development into an absolute method if, indeed, one is needed.

C. Pore Size Distribution in Coal

Zwieterling and van Krevelen (49) established that coal contains a macro- and micropore system. By studying the penetration of mercury into coal as a function of pressure up to 1100 atm, they were able (after correction for the compressibility of coal) to establish the existence of pores with radii in the range from 30,000 to about 250 Å. They observed essentially no pores in the range from 250 Å to 60 Å (the smallest that their technique could have established). Volume in the > 250 Å pores was found to comprise only about 65% of the total pore volume of the sample. The average diameter of the micropore system was clearly much smaller than 120 Å. From consideration of a detailed study of macro- and micropore volumes and internal surface versus coal rank (53), van Krevelen (5) proposed that, as an approximation,

$$S_{internal} \ (cm^2/g) = 1.8 \times 10^7 \ V_{micro} \ (cm^3/g)$$

If so, and if the micropores are assumed to be cylindrical, their average diameter must be about 20 Å.

Within recent years improvements in sample preparation and in high-voltage transmission electron microscopes (TEM) have apparently permitted visual observation of such micropores. Lin et al. (66), using equipment and techniques previously described (68), have found that 2 mm thicknesses of vitrinite from Illinois No. 6 coal show layers of finely divided (70–120 Å diameter) minerals that are primarily kaolinite, micropores (average diameter between 20 and 30 Å), and mesopores (diameters between 200 and 500 Å with mean diameter near 250 Å). Harris and Yust (68) had previously shown that a Kentucky splint-type coal, with macerals consisting almost entirely of exinite and inertinite, contained mesopores (20–500 Å diameter) with few, if any, micropores.

As indicated above, SAXS can give the total internal surface area directly. However, determination of the size distribution requires several assumptions, among which is the assumption that the scatterers are spherical. Several mathematical methods for winning size distributions from SAXS data have been described and applied (69). Lin et al. (66), by use of these assumptions, arrive at values of about 30 Å as the diameter of the micropores, 100 Å as the mean diameter of the mineral matter, and 220 Å as the mean diameter of the mesopore system in Illinois No. 6 coal. They found 140 ± 20 m²/g as the surface area of the micropores, 3 ± 1 m²/g for the microminerals, and 10 ± 3 m²/g for the mesopore system.

For a long time there has been evidence that smaller pores—or smaller constrictions in these pores—exist in coal macerals. It is known that the heat of wetting of any coal decreases as the molecular volume of the wetting liquid increases (5). Accordingly, by measuring heats of wetting (or apparent densities) with various liquids and by plotting the first derivative against the molar volumes of the liquid it has been shown (34) that the critical dimensions of the micropores are distributed around two modal values corresponding to the molecular diameter of liquids having molecular volumes of about 50 and 150 cm²/mole. These correspond to critical pore diameters of about 5 and 8 Å. Anderson et al. (52) seemed to find pore restrictions of about 5 Å diameter.

The foregoing all suggests that the internal surface of coal is largely associated with a micropore system in which 30 Å (or smaller) diameter pores are constricted or linked by 5–8 Å passages. Gan et al. (54) have shown that most of the internal free volume in coal of low rank (< 75% C) is due to macro- and transition (meso-) pores, but that microporosity becomes dominant as the rank increases. Spitzer et al. (59), however, seem to find characteristic diameters of 12–14 Å for the micropores in a wide spectrum of Czechoslovakian coals.

IV. EFFECTS OF HEAT ON COAL STRUCTURE

All coal conversion technologies and virtually all end uses of coal require the application of heat. The structural changes upon heating accordingly influence essentially all aspects of coal technology.

A. Changes in Pore Structure on Heating

Changes in the internal structure of coal during carbonization have been studied extensively though less so than changes with coal rank. The several studies, collectively, have used essentially all the techniques mentioned above, and, in a qualitative sense, seem to be in general agreement.

Franklin (70), who relied on density measurements with several displacement fluids, showed that true density increased with increasing carbonization temperature in the high-temperature range and that, for most coals, porosity increases as carbonization temperature is raised to 500–600°C (although accessibility of the pores to larger molecules decreases).

Cannon et al. (71), who measured the effect of carbonization on heats of wetting of a series of coals, found an increase to a maximum at about 300°C, followed by some fluctuations but generally similar areas through about 600°C and then by a steady decrease to very low values at 1000°C. Anthracites showed relatively little effect below 600°C, although the decrease at higher temperatures was observed. Bond and Spencer (*34*) obtained generally similar results. In studies of sorption–desorption equilibria, they concluded that the essential capillary structure changes little up to 600°C, although molecular sieve properties become more marked, and internal volume increases above 400°C. Above 600°C the internal volume continues to increase, but accessibility decreases markedly and only He, Ne, and H_2 are capable of reaching the considerable internal surface in reasonable times after carbonization at 1100°C.

Chiche *et al.* (*65*), who used sorption–desorption equilibria with He, N_2, CH_3OH, and H_2O, as well as SAXS to study carbonization of a series of coals, reached similar conclusions. They concluded that H_2O adsorption primarily measured the number of surface oxygen groups. From the SAXS measurements they concluded that noncaking coals of whatever rank retained a considerable internal surface area, even to very high (3000°C) carbonization temperatures, while coking coals and other readily graphitizable coals tended to lose internal surface area (and, apparently, to eliminate fine pores rapidly) upon carbonization above 1500°C. In all cases, internal surface, as measured by CH_3OH and H_2O adsorption, had reached low values upon carbonization at 1000°C. Comparison of total porosity by SAXS with helium density values on high temperature cokes seemed to show a substantial fraction of closed porosity after carbonization at temperatures above 1500°C.

B. Devolatilization

All coals release volatile matter when heated. The quantities evolved depend upon coal rank, upon the ultimate temperature to which the coal is heated, and upon the system pressure. The mix of species evolved at any given temperature is also a function of coal rank as well as the temperature, the system pressure, and the previous temperature history (including heating rate) of the sample. All industrial nations that produce and consume large quantities of coal have adopted standard procedures for the determination of volatile matter, specifying sample size, degree of grinding, type of crucible, heating rate, and ultimate temperature (*5*). The values obtained are of great technological importance, but they throw relatively little light upon the physical or chemical structure of coal.

Devolatilization of coal upon heating at a slow and uniform rate (as, for example, is approximately the case in a coke oven) includes a continuum of processes in which the mix of products and their rates of evolution show only gradual changes. It is possible, however, to distinguish three more or less distinct regimes. The onset of devolatilization is generally taken to be about

350°C, but it is clear that relatively small quantities of many species are evolved at much lower temperatures. At very low temperatures (below 200°C) and especially under reduced pressure, loosely bound H_2O and CO_2 are evolved early, followed by more strongly bound molecules of these species. Recent studies with mass spectrometric analysis of species devolatilized under vacuum show that hydrocarbons (C_2, C_3, and C_4) appear above 150°C, accompanied at above 180°C by detectable quantities of aromatics, and sulfur-containing species (15). Many products can be distilled under vacuum at 150°C from bituminous coal (8-14). Above 350°C primary (rapid) devolatilization occurs with a mix of molecules, including low molecular weight gases and chemically complex liquid and semisolid tars, as the products. If the temperature is increased at a constant rate, the primary devolatilization rate passes through maximum; the temperature at which devolatilization is a maximum increases as the rate of heating is increased (5). The rate of devolatilization decreases with increasing coal rank. Exinite shows the highest devolatilization rate, followed by vitrinite and micrinite, but the differences largely disappear at coal carbon contents above 90%. This primary devolatilization is essentially complete at 550°C. Loss in weight during the primary devolatilization proceeds as a first-order reaction (with respect to the quantity remaining) with an activation energy near 50 kcal (5, 72, 73). If so, one might conclude that decomposition rates are controlled by rupture of covalent bonds (including C—C bonds). This conclusion, based on thermogravimetric measurements, has been challenged (74) on the basis that the rate of weight loss depends upon rate of diffusion of volatile matter rather than upon rate of its formation. Initial rates of devolatilization are far higher than allowed by first-order rate equations that fit the data for subsequent devolatilization; these initial rates show activation energies near 5 kcal/mole (74). Secondary devolatilization, occurring largely in the 600–800°C range, produces H_2 and CH_4 as the major products (5). When the fractional loss of a given element is plotted against total fractional weight loss, it appears that the fractional loss of hydrogen is higher than that of carbon over all coal ranks. Fractional loss of oxygen (always as oxygen compounds, not as O_2) is very high in lignites (5) and low-rank coals (70). However, the oxygen is much more tightly bound in higher rank coals, and increasingly higher temperatures (550°C for coal with 90% carbon and 915°C for 94% carbon) are necessary to produce appreciable devolatilization of the oxygen (70).

It is obvious that devolatilization is a complex process that involves—and proceeds simultaneously with—coal decomposition. Berkowitz (74) would postulate a rapid decomposition of coal (by bond scission) upon heating to create a considerable pressure gradient of volatile matter in the coal pore system with devolatilization occurring by diffusion over a considerable time span.

C. Plastic Behavior

Some coals—referred to as caking or agglutinating coals—pass, when heated, through a transient plastic stage, where they soften, swell, and resolidify into a somewhat distended coke. If the plastic properties are such as to make the coals suitable for conversion into high-strength metallurgical cokes, they are referred to as coking coals. The best coking coals tend to be medium-volatile bituminous coals at about 88% C and with about 29% volatiles. The temperature interval over which these changes occur coincides roughly with the interval of primary devolatilization.

The inertinite group of macerals generally shows no plasticity on heating, although exceptions may be noted among massive cretaceous micrinites (sometimes termed macrinites). Vitrinites and exinites seldom show plastic behavior if their carbon content is below 80.5% or above 91%. Exinites that show plastic behavior show high fluidities, while vitrinites of similar rank become less fluid (5). In general, plastic properties become evident in coals when the content of volatile matter exceeds 12–15%, become more pronounced as the volatile content increases to 30–35%, become less pronounced at higher volatile content, and disappear when the coal contains about 40% volatiles. In the same general way, the plastic properties are correlated (inversely) with coal porosity (32, 75). The most plastic coals (at 88–89% carbon) are also the least porous. Exceptions are found to these generalizations, as to virtually all others concerning coal. Loison (75) suggests that neither volatile content nor porosity provides a better correlation with plastic properties than does coal rank.

This sequence of phenomena (softening, swelling, and resolidification), which is obviously interrelated with decomposition and devolatilization, can be considerably affected by heating rate as well as system pressure. For example, all coals can be devolatilized without swelling if the heating rate is sufficiently small (76). Reduced pressure during pyrolysis reduces plastic capability and can completely destroy it in weakly caking coals (75), whereas mild pyrolysis under pressure increases fluidity in subsequent plastometer tests (23, 77). All aspects of plasticity are diminished by sufficiently fine grinding, by preheating in an inert atmosphere, and, more drastically, by oxidation. Plastic properties can be enhanced by mild hydrogenation or by removal of a substantial fraction of the contained mineral matter (23).

Assessments of caking propensities of coals are made by standardized and empirical tests that measure (1) free-swelling indices upon rapid heating (78), (2) maximum dilation of the coal upon slow heating at a controlled rate (79), or (3) determination of the mechanical strength (the Roga index) of the resulting coke button (80). These tests, along with the Gray-King index (81), are widely used for laboratory screening of coals for particular end uses. The four tests have been shown to correlate well, although correlations of free-swelling index with the others are somewhat more complex (*5*).

For many years, fluidity of the plastic coal has been measured with a constant torque plastometer incorporating a rabble-armed stirrer rotating in a compacted (5 g) charge of coal heated at 3°C/min (*82,83*). Variable torque plastometers are also used (*75,84–86*). In these the retort or the stirrer is rotated at constant speed and the required torque is measured. Such plastometers permit the coal softening point and the coke resolidification point to be determined with good precision. When plastic coals are heated at a constant 3°C/min, the fluidity increases (and may remain nearly constant for a time) and then decreases upon further heating. Since the evolution of gas seriously interferes with physical interpretation of fluidity measurements, such values are, of necessity, empirical. It is, however, possible to express plasticity in terms of an apparent viscosity by comparing the results for coal with those of model substances such as pitches (*5,75*). Such apparent viscosity depends upon the foam structure within the experimental capsule as well as upon the true coal plasticity and has, as a consequence, no simple physical significance. Good coking coals are plastic over a range of about 80°C and show minimum apparent viscosities of the order 10^4–10^5 P (*83*). Kirov and Stephens (*25*) have criticized the Giesler plastometer as furnishing misleading information during early stages of coal plasticity; they have described a viscometer that they consider capable of correct absolute values for fluid coal with viscosities below 10^7 P. Minimum viscosities for the coals of that study (*25*) seem generally to be in the 10^6 P range.

In general, the duration of the plastic stage seems to be longest for vitrinites with about 88% C and 29% volatile matter. It has been noted that in standard tests of good coking coals, the temperature of maximum fluidity corresponds closely with those of maximum swelling and of maximum rate of devolatilization (*5*).

It has long been known that certain coals after preheating (to about 400°C) and cooling give a higher yield upon solvent extraction than do nonpreheated coals (*87–89*). This phenomenon is generally referred to as thermosolvolysis, a term applied by van Krevelen (*5*). Illingsworth (*90*) showed that the yield from coking coals passes through a maximum at a definite preheating temperature. Chloroform has been shown to be a very suitable solvent for preheated coal (*91*) and that extract yield is little affected by prolonged heating at 350°C, but the yield decreases rapidly with time of heating at 420°C (*92*). Such information has long suggested that the material extracted by chloroform plays a role in the softening and melting of coking coals.

Kirov and Stephens (*25*) [see also Squires (*29*)] performed parallel thermogravimetric, viscometric, calorimetric, and chloroform extraction experiments on Australian coals. They seem to have observed an endothermic reaction between 250 and 330°C that led to increased coal acidity and a second reaction between 340 and 400°C that led to an increase in phenoiic oxygen. The temperature ranges of these endotherms (and particularly of the second) correspond well with a marked increase in chloroform solubility and with marked diminution in viscosity.

Kirov and Stephens (25) emphasize that these reactions appear to involve only a tiny fraction of the chemical bonds in the coal, and have *speculated* that the first may involve breaking of hydrogen bonds while the second may be a splitting of ether bonds. At preheating temperatures above about 400°C, the quantity of chloroform extract from coking coals is known to show an additional sharp increase (25,26,93).

The *speculation* that rupture of ether bonds in small numbers occurs with the onset of increased fluidity seems to receive some support from studies of Lazarov and Angelova (94,95). On the other hand, however, there is evidence (96-98) that the supression of fluidity by oxidation of coal is associated with the creation of ether linkages.

The great preponderance of evidence, as previously described, seems to show that when samples of coal—including samples of strongly caking coal—are brought to increasing temperatures as high as 500–600°C and cooled rapidly, their surface areas and microporosity change only slightly. This strongly suggests that the fine structure of coal is little modified by fusion (92). This fact—as well as the effect of system pressure on coal plasticity—would seem to virtually rule out the suggestion by Audibert (99) that coal undergoes true complete melting along with chemical decomposition. It may be conceivable, but, as suggested by Dryden and Griffith (100), it seems extraordinarily improbable that fissure formation on cooling or the bubble structure resulting from coal decomposition results generally in microporosity similar to that of the original coal. Several authors (5, 75) have considered that only a fraction (the coking principle or bitumen) of the coal can fuse, but that that fraction is capable of fluidizing the entire mass. These older theories seem to assume that the coal substance is a physical mixture of different chemical components. This is hardly tenable since pure "fusible" macerals have been shown to be completely homogeneous under the electron microscope (75,101). Many authors (92,102-104) consider the softening of coal to result from the pyrolytic formation of fusible compounds referred to as "thermobitumens." It has been assumed (102,105,106) that the vitrinite is transformed by thermal depolymerization into a fluid mass (a "metaplast"), and that the metaplast is subsequently decomposed (by cracking) into volatile primary gas and semicoke. These two reactions—depolymerization and cracking—overlap in time so that undecomposed solid coal, fluid metaplast, primary gas in the process of escaping through the metaplast, and solid semicoke all exist simultaneously. The semicoke subsequently decomposes to secondary gas and coke, but this reaction occurs at temperatures above the plastic range. Chermin and van Krevelen (106) assumed that each of the transformations were first order and developed a mathematical model of the overall process (5, 75). Upon the further assumption that fluidity is proportional to metaplast concentration, many experimentally observed features are correlated in this model. It is, however, important to realize that the metaplast is merely a useful mathematical construct that represents no real chemical knowledge. It is not, for example, known whether the material that causes softening exists in the coal or whether it is a pyrolysis product. Moreover, it is not certain whether the plastic fraction acts as a dispersion agent or as a lubricant for the still solid portion of the coal.

A more modern view of coal plasticity, presented thus far only in qualitative terms, offers a synthesis of several older views with some recent information (23). It appears to follow from the explicit recognition by Wiser (27) of the relationship between coal pyrolysis and coal liquefaction, and to build upon the suggestion of Brown and Waters (26) that low molecular weight, hydrogen-rich compounds (the bitumens) serve as the initiators of plastic behavior. These materials, which may begin to melt at temperatures below 200°C, solvate [or alternatively lubricate—as suggested by Berkowitz (74)] the coal micelles as they become thermally loosened. Most important, however, they serve as mobile hydrogen donors to stabilize free radicals that are produced thermally before general coal plasticity begins. As the temperature increases, general mobility of the micelles begins, pyrolytic bond rupture becomes significant, additional low molecular weight vehicle forms, and free radicals can contact donor hydrogen in thermobitumen molecules. Porosity associated with the coal micelles is largely preserved. Thermal decomposition (increasing with temperature) yields volatile materials that escape from the coal particles by pressure-driven flow; the viscous plastic mass tends to retain the gases and swelling of the mass occurs. As the temperature is increased, the formation of free radicals exceeds the capacity of the coal to cap them by hydrogen-donor reactions; the free radicals combine or attack available large molecules, mesocoke formation increases, and plasticity is lost. According to this mechanism, plasticity in coal would seem to require the presence of at least a minimum quantity of low molecular weight bitumen. It has long been known that removal (by extraction with chloroform or benzene) of these bitumens removes the plastic properties of coal. However, they are not sufficient to explain the process; Berkowitz (107) has called attention to the fact that during isothermal pyrolysis the chloroform solubles disappear before maximum fluidity is reached. However, other features are also necessary. The requirements include the presence of lamellae-bridging structures capable of thermal rupture, a supply (both in the bitumen and the balance of the coal) of hydroaromatic hydrogen, and an intrinsic potential for micellular and lamellae mobility not associated with chemical bond rupture (23).

This model appears to explain, in a qualitative way, a large number of observations of coal plasticity, including its enhancement or its diminution by various pretreatments. It would ascribe the nonplasticity of low-rank coals to their high oxygen contents and their consequent formation of strong cross-linkages from reactions at temperatures in the range below the coal softening point along with high unproductive consumption of donatable hydrogen by oxygen-associated reactions. Anthracites would be nonplastic because of the scarcity of bitumen and the relatively high extent of initial cross-linking.

It is, however, certain that many aspects of coal pyrolysis remain unexplained. As an example, if one were needed, it is known that the process of coke formation from the plasticized and decomposing fluid involves formation of nematic (threadlike) liquid crystals (108-110). It may be that the ability of prime coking coals to produce metallurgical grade coke is due to an optimum in capability for, and extent of, liquid crystal formation (108). As far as is known, no one has included the properties of liquid crystals in any model of coal pyrolysis and coke formation.

REFERENCES

1. Stopes, M. C. (1919). *Proc. R. Soc. London, Ser. B* **90**, 470.
2. "International Handbook of Coal Petrology" (1963). 2nd ed. CNRS, Paris (addendum to 2nd ed., 1971).
3. Mackowsky, M.-T. (1971). *Fortschr. Geol. Rheinl. Westfalin* **19**, 173.
4. Mackowsky, M.-T. (1975). *Microsc. Acta* **77**(2), 114.
5. van Krevelen, D. W. (1961). "Coal." Elsevier, Amsterdam.
6. Vahrman, M. (1970). *Fuel* **49**, 5.
7. Spence, J. A., and Vahrman, M. (1970). *Fuel* **49**, 395.
8. Rahman, M., and Vahrman, M. (1971). *Fuel* **50**, 318.
9. Palmer, T. J., and Vahrman, M. (1972). *Fuel* **51**, 14.
10. Palmer, T. J., Vahrman, M. (1972). *Fuel* **51**, 22.
11. Watts, R. H., and Vahrman, M. (1972). *Fuel* **51**, 130.
12. Watts, R. H., and Vahrman, M. (1972). *Fuel* **51**, 235.
13. Hayatsu, R., Winans, R. W., Scott, R. G., Moore, L. P., and Studier, M. H. (1977). *Fuel* **57**, 541.
14. Fuller, E. L., Jr. (1978). Personal communication, based on Documents ORNL/MIT-264, December 20, 1977; ORNL/MIT-266, March 1, 1978; ORNL/MIT-270, March 28, 1978.
15. Fuller, E. L. *et al.* (1979). "Annual Report of ORNL Chemistry Division for 1978," ORNL-5485, pp. 13-14. Oak Ridge Natl. Lab., Oak Ridge, Tennessee.
16. Hirsch, P. B. (1950). *Proc. Conf. Sci. Use Coal, 1950* pp. A29-33.
17. Tschamler, H., and deRuiter, E. (1963). *In* "Chemistry of Coal Utilization" (H. H. Lowry, ed.), p. 35. Wiley, New York.
18. Krulen, D. J. W. (1948). "Elements of Coal Chemistry." Rotterdam.
19. Dryden, I. G. C. (1952). *Fuel* **30**, 39.

20. Sternberg, H. (1976). "Storch Award Address," Fuel Div. Prepr., ACS Natl. Meet. Am. Chem. Soc., New York.
21. Morgans, W. T. A., and Terry, N. B. (1958). *Fuel* **37**, 201.
22. Larsen, J. W. (1978). *Proc. Coal Chem. Workshop, 1978* CONF-780372.
23. Neavel, R. C. (1975). *Proc. Coal Agglomeration Conversion Symp., 1975* p. 37.
24. Whitehurst, D. D., Farcasiu, M., Mitchell, T. O., and Dickert, J. J., Jr. (1977). *Annu. Rep. Res. Proj.* EPRI AF-480, 410-1.
25. Kirov, N. Y., and Stephens, J. N. (1967). "Physical Aspects of Coal Carbonization." Department of Fuel Technology, University of New South Wales, Sydney, Australia.
26. Brown, H. R., and Waters, P. L. (1966). *Fuel* **45**, 17.
27. Wiser, W. (1968). *Fuel* **47**, 475.
28. Neavel, R. C. (1976). *Fuel* **55**, 237.
29. Squires, A. M. (1978). *Appl. Energy* **4**, 161.
30. Bangham, D. H. (1943). *Annu. Rep. Chem. Soc.* **40**, 29.
31. Krunin, P. O. (1963). *Bull.—Ohio State Univ. Eng. Exp. Stn.* 195.
32. King, J. G., and Wilkins, E. T. (1944). *Proc. Conf. Ultra-fine Struct. Coals Cokes, 1944,* p. 46.
33. Franklin, R. E. (1949). *Trans. Faraday Soc.* **45**, 274.
34. Bond, R. L., and Spencer, D. N. T. (1958). *Ind. Carbon Graphite, Paper Conf., 1957* p. 231.
35. Kotlensky, W. V., and Walker, P. L., Jr. (1960). *Proc. Carbon Conf., 4th, 1959* p. 423.
36. Danforth, J. D., and DeVries, T. (1939). *J. Am. Chem. Soc.* **61**, 873.
37. Bangham, D. H., and Razouk, R. I. (1938). *Proc. R. Soc. London, Ser. A* **166**, 572.
38. Griffith, M., and Hirst, W. (1944). *Proc. Conf. Ultra-fine Struct. Coals Cokes, 1944,* p. 80.
39. Fugassi, P., Hudson, R., and Ostapchenko, G. (1958). *Fuel* **37**, 39.
40. Maggs, F. A. P. (1943). *J. Inst. Fuel* **17**, 49.
41. Kini, K. A., Nandi, S. P., Sharma, J. N., Igengarana, M. S., and Lahiri, A. (1956). *Fuel* **35**, 71.
42. Marsh, H. (1965). *Fuel* **44**, 253.
43. Brunauer, S., Emmett, P. H., and Teller, E. (1938). *J. Am. Chem. Soc.* **60**, 309.
44. Gregg, S. J. (1944). *Proc. Conf. Ultra-fine Struct. Coals Cokes, 1944* p. 110.
45. Leroy Malherbe, P. (1951). *Fuel* **30**, 97.
46. Lecky, J. A., Hall, W. K., and Anderson, R. B. (1951). *Nature (London)* **168**, 124.
47. Zwietering, P., Oele, A. P., and van Krevelen, D. W. (1951). *Fuel* **30**, 203.
48. Maggs, F. A. P. (1952). *Nature (London)* **169**, 793.
49. Zwietering, P., and van Krevelen, D. W. (1954). *Fuel* **33**, 331.
50. Walker, P. L., Jr., and Geller, I. (1956). *Nature (London)* **178**, 1001.
51. Marsh, H., and O'Hair, T. F. (1966). *Fuel* **45**, 301.
52. Anderson, R. B., Hall, W. K., Lecky, J. A., and Stein, K. C. (1956). *J. Phys. Chem.* **60**, 1548.
53. Sevenster, P. G. (1954). *J. S. Afr. Chem. Inst.* **7**, 41.
54. Gan, H., Nandi, S. P., and Walker, P. L., Jr. (1972). *Fuel* **51**, 272.
55. Walker, P. L., Jr., and Patel, R. L. (1970). *Fuel* **49**, 91.
56. Anderson, R. B., Bayer, J., and Hotet, L. J. E. (1965). *Fuel* **50**, 345.
57. Marsh, H., and Siemieniewska, T. (1965). *Fuel* **44**, 355.
58. Spitzer, Z., and Ulicky, L. (1976). *Fuel* **55**, 21.
59. Spitzer, Z., Biba, V., Bohac, F., and Malkova, E. (1977). *Fuel* **56**, 313.
60. Steele, W. A. (1974). "The Interactions of Gases with Solid Surfaces," Chapter 2. Pergamon, Oxford.

61. Fuller, E. L., Jr., and Agron, P. A. (1976). "Reactions of Atmospheric Vapors with Lunar Soils," ORNL-5129. Oak Ridge Natl. Lab., Oak Ridge, Tennessee.
62. Fuller, E. L., Jr. (1981). *Adv. Chem. Ser.* (in press).
63. Kroger, C., and Mues, G. (1961). *Brennst.-Chem.* **42**, 77.
64. Durif, S. (1963). *J. Chem. Phys.* **60**, 816.
65. Chiche, P., Durif, S., and Pregermain, S. (1965). *Fuel* **44**, 5.
66. Lin, J. S., Hendricks, R. W., Harris, L. A., and Yust, C. S. *J. Appl. Crystalallogr.* **11**, 621 (1978); Hendricks, R. W., *ibid.* p. 15.
67. Porod, G. (1951). *Kolloid-Z.* **124**, 83; **125**, 51, 109 (1952).
68. Harris, L. A., and Yust, C. S. (1976). *Fuel* **55**, 233.
69. A useful review is given in J. Schelton and R. W. Hendricks, *J. Appl. Crystallogr.* (1978).
70. Franklin, R. E. (1949). *Trans. Faraday Soc.* **45**, 668.
71. Cannon, C. G., Griffith, M., and Hirst, W. (1944). *Proc. Conf. Ultra-fine Struct. Coals Cokes, 1944,* p. 131.
72. Boyer, A. F. (1952). *C. R. Congr. Ind. Gaz.* **69**, 653.
73. van Krevelen, D. W., van Heerden, C., and Huntjens, F. J. (1951). *Fuel* **30**, 253.
74. Berkowitz, N. (1960). *Fuel* **39**, 47.
75. Loison, R., Peytavy, A., Boyer, A. F., and Grillot, R. (1963). *In* "Chemistry of Coal Utilization" (H. H. Lowry, ed.). Wiley, New York.
76. Dulhunty, J. A., and Harrison, B. L. (1953). *Fuel* **32**, 441.
77. Skylar, M. G., and Shustikov, V. I. (1962). *Khim. Tekhnol. Topl. Masel* **7**, 39.
78. See, for example, *Am. Soc. Test. Mater., Book ASTM Stand.* D271-48.
79. Audibert, E., and Delmas, L. (1927). *Rev. Ind. Miner.* **7**, 1.
80. Roga, B., and Wnekowska, L. (1951). *Biull. Inst. Weglowego, Komun.* No. 101.
81. *Fuel Res. Surv. Pap.* No. 44 (1940); No. 53 (1946).
82. Giesler, K. (1934). *Glueckauf* **70**, 178.
83. Soth, G. C., and Russell, C. C. (1943). *Proc., Am. Soc. Test. Mater.* **43**, 1176.
84. Davis, J. D. (1931). *Ind. Eng. Chem., Anal. Ed.* **3**, 43.
85. Echterhoff, G. (1954). *Glueckauf* **90**, 510.
86. Boyer, A. F., and Lahouste, J. (1954). *Rev. Ind. Miner.* **35**, 1107.
87. Burgess, M. J., and Wheeler, R. V. (1911). *J. Chem. Soc.* **99**, 649.
88. Jones, D. T., and Wheeler, R. V. (1915). *J. Chem. Soc.* **107**, 1318.
89. Bone, W. A., and Sarjant, R. J. (1920). *Proc. R. Soc. London, Ser. A* **96**, 119.
90. Illingsworth, S. R. (1922). *Fuel* **1**, 213.
91. Dryden, I. G. C., and Pankhurst, K. S. (1955). *Fuel* **34**, 363.
92. Brown, J. K., Dryden, I. G. C., Dunerein, D. H., Joy, W. K., and Pankhurst, K. S. (1958). *J. Inst. Fuel* **31**, 259.
93. den Hertog, W., and Berkowitz, N. (1958). *Fuel* **37**, 358.
94. Angelova, G., and Lazarov, L. (1968). *Fuel* **47**, 333.
95. Lazarov, L. (1968). *Fuel* **47**, 342.
96. Wachowska, H., Nandi, B. N., and Montgomery, D. S. (1974). *Fuel* **53**, 212.
97. Ignasiak, B. S., Szladow, A. J., and Montgomery, D. S. (1974). **53**, 12.
98. Ignasiak, B. S., Szladow, A. J., and Berkowitz, N. (1974). *Fuel* **53**, 229.
99. Audibert, E. (1926). *Fuel* **5**, 229.
100. Dryden, I. G. C., and Griffith, M. (1954). *Br. Coal Util. Res. Assoc., Mon. Bull.* **18**, 62.
101. Pregermain, S., and Guillemot, G. (1960). *Int. Conf. Electron. Microsc., Proc., 4th, 1958.*
102. van Krevelen, D. W., Huntjens, F. J., and Dormans, H. N. M. (1956). *Fuel* **35**, 462.
103. Fitzgerald, D., and van Krevelen, D. W. (1959). *Fuel* **38**, 17.
104. Oele, A. P. (1952). *Brennst.-Chem.* **33**, 231.

105. Fitzgerald, D. (1956). *Trans. Faraday Soc.* **52**, 362.
106. Chermin, A. G., and van Krevelen, D. W. (1957). *Fuel* **36**, 85.
107. Berkowitz, N. (1967). *Proc. Mon. Bull. Symp. Sci. Technol. Coal, 1967,* p. 149.
108. Marsh, H. (1973). *Fuel* **52**, 206.
109. Marsh, H., Dachille, F., Iley, M., Walker, P. L., Jr., and Whang, P. W. (1973). *Fuel* **52**, 253.
110. Marsh, H., Hermon, G., and Cornford, C. (1974). *Fuel* **53**, 168.

Magnetic Resonance Studies of Coal

H. L. RETCOFSKY

United States Department of Energy
Pittsburgh Energy Technology Center
Pittsburgh, Pennsylvania

I.	Introduction	43
II.	The Aromaticity of Coal	44
	A. Methods for Determining Coal Aromaticities	44
	B. Aromaticity Changes during Vitrinization	45
	C. Structural Changes during Coal Liquefaction	47
III.	Free Radicals in Coal	50
	A. Vitrains	50
	B. Fusains	64
	C. Selected Macerals	66
	D. Comparisons with Other Carbonaceous Materials	67
IV.	Preliminary ESR and NMR Investigations of Chinese Coals:	
	A Cautionary Note on Coal Structure Studies	69
V.	ESR Studies of Respiratory-Size Coal Particles	73
	A. Experimental	73
	B. Anderson Sampler Fractions of Raw Coal	76
	C. Respirable Dust from Coal Workers' Personal Samplers	77
	D. Respirable Dust from Pneumoconiosal Lung Tissue	79
VI.	Summary	80
	References	80

I. INTRODUCTION

Electron spin resonance (ESR) and nuclear magnetic resonance (NMR) spectrometries have become extremely valuable tools in coal research, both for probing the chemical structure of coal itself and for elucidating the mechanisms involved in its conversion to environmentally acceptable fuels.

The techniques have also provided insight into the origin of coal and its subsequent metamorphism.

Uebersfeld *et al.* (1954) and Ingram *et al.* (1954) were the first to apply ESR to coal, whereas the first published report of NMR measurements on coal was authored by Newman *et al.* (1955). For each of the two techniques, reports of applications to coal appeared just nine years after successful observations of the respective resonances in bulk matter (Zavoisky, 1945; Bloch *et al.,* 1946; Purcell *et al.,* 1946). Early ESR studies of coal have been reviewed by van Krevelen (1961), by Tschamler and DeRuiter in the treatise by Lowry (1963), and by Ladner and Wheatley (1965). The first two of these three references also contain extensive coverage of early NMR studies of coal. Several chapters describing more recent applications of NMR spectrometry in coal research can be found in the second of Karr's volumes (Karr, 1978). The newly revised supplement of the Lowry treatise (Elliot, 1981) contains comprehensive and up-to-date reviews of both techniques in coal research.

The NMR and ESR work described here was undertaken by the Pittsburgh Energy Technology Center for a variety of reasons. Of primary concern was the need for a better understanding of the chemical structure of coal, particularly the nature of the aromatic units and the naturally occurring free radicals. This knowledge of coal structure, coupled with other information, allows one to deduce changes in chemical structure that occur (1) during the coalification process and (2) during the conversion of coal to liquid products. In addition, a limited ESR study of respirable coal dust was carried out as part of a United States Bureau of Mines effort to obtain information relating to the cause(s) of coal workers' pneumoconiosis.

II. THE AROMATICITY OF COAL

Few concerns over the chemical structure of coal have proved more controversial and have provoked more heated debate than the question of its carbon aromaticity f_a. Retcofsky (1978), for example, traced literature f_a values for an 82.6% C coal over the period 1955–1976 and found them to range from a low of less than 0.5 to a high of nearly 0.9. This is clearly a nontrivial difference and suggests that the classical view of coal as a highly aromatic material has not been without its critics.

A. Methods for Determining Coal Aromaticities

Schuyer and van Krevelen (1954) and Dryden and Griffith (1955) developed methods for estimating f_a values of coals from heat of combustion data. The methods, however, required a knowledge of the hydrogen aromaticity, a number perhaps equally as elusive as its carbon counterpart. Dormans *et al.*

(1957) published an extensive list of carbon aromaticities for macerals from coals whose carbon contents ranged from 70.5 to 96.0%. The values were obtained by the graphical densimetric method, which makes use of the (true) densities of coals and a number of *a priori* assumptions of the distribution of oxygen, nitrogen, and sulfur functional groups. Ultraviolet (UV)–visible spectrophotometry was used by Friedel (1957) and by DeRuiter and Tschamler (1958) in coal aromaticity studies, and it triggered perhaps the first controversy over aromatic versus diamond-like structures in coal. Van Krevelen *et al.* (1959), using velocity of sound measurements, reported carbon aromaticities for six coals. The second major controversy over the high aromaticity of coal occurred in 1974 (Chakrabartty and Kretschmer, 1974) and was based upon the analysis of products resulting from hypochlorite oxidation of coal. These studies represent only a few of the many early attempts to obtain reliable f_a values for coals.

The most recent—and probably the most useful—technique for determining coal aromaticities is cross-polarization carbon-13 NMR, developed by Pines *et al.* (1972). It has been applied to coal by a number of investigators, including VanderHart and Retcofsky (1976a,b), Bartuska *et al.* (1976), Miknis *et al.* (1980), and Taki *et al.* (1980). A major problem in the technique is the overall effectiveness of the cross polarization between the protons and carbons in the sample. Should the effect not be uniform, the resulting f_a values would be suspect. The work of VanderHart and Retcofsky (1976a,b) suggests the technique to be quantitative, whereas Taki *et al.* (1980) report that not all the carbons in the coal are being observed.

Retcofsky and co-workers have demonstrated[1] the excellent agreement between f_a values obtained by the cross-polarization technique and values obtained by conventional high-resolution ^{13}C NMR methods, the latter measurements being done in a manner that ensures quantitative integrity. The materials studied were all derived from coal and were soluble in solvents suitable for high-resolution studies. Cross-polarization measurements were made after removal of the solvent. The soluble fractions included coal extracts, coal-derived oils, and coal-derived asphaltenes. The results, depicted graphically in Fig. 1, increase our confidence in the aromaticity values obtained by the cross-polarization technique and provide added support for the validity of the aromaticity values for coal vitrains presented in the next section.

B. Aromaticity Changes during Vitrinization

Since most United States coals are rich in vitrain, the chemical structure of the vitrain (or vitrinite) component is typically considered representative of that of the whole coal. Thus, vitrinization is considered to be synonymous

---

[1] H. L. Retcofsky, T. A. Link, and D. L. VanderHart, unpublished results.

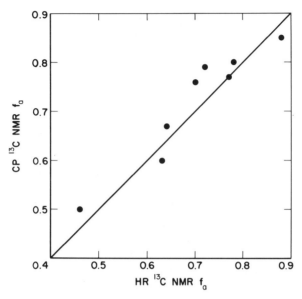

Fig. 1.   Comparison of experimental carbon aromaticities for coal-derived materials.

with coalification, although more detailed studies would include other processes such as fusinization. Cross-polarization ^{13}C NMR carbon aromaticity
values for vitrains from nine United States coals ranging in rank from peat to
anthracite are shown in Fig. 2. The results support the classical view that
coals, as a whole, are highly aromatic materials and that, at least to a first approximation, the aromaticity of coal increases with increasing rank or carbon
content.

Figure 3 summarizes the change in aromaticity with coal rank for solvent
extracts of coal and compares the results with those on whole coals described
above. The f_a values for the pyridine extracts were estimated from high-
resolution proton NMR spectra taken in pyridine-d_5 solution. The Brown and
Ladner (1960) method as modified by Retcofsky (1977) was used to convert
the hydrogen distribution into a carbon skeleton. The f_a values for the carbon
disulfide extracts were obtained from solution spectra via conventional high-
resolution ^{13}C NMR. The extract data are taken from earlier publications
(Retcofsky and Friedel, 1970; Retcofsky, 1977). The plots show that the
aromaticities of the extracts behave similarly to those of the parent coals, that
is, each set of extracts shows an increase in aromaticity as vitrinization of the
original coals progresses. The pyridine extracts have aromaticities closer in
value to those of the whole coals, as would be expected since they represent a
larger portion of the whole coals than the carbon disulfide extracts. For each

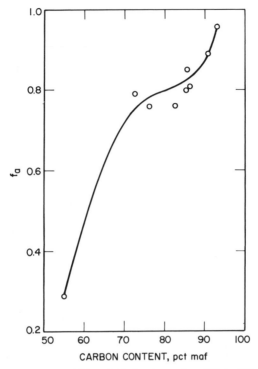

Fig. 2.   Carbon aromaticities of vitrains from selected United States coals.

set of data, however, the extracts have aromaticities lower than those of the parent coals.

C.   Structural Changes during Coal Liquefaction

Once confidence is established in the carbon aromaticity values, the data can be extended to other aspects of coal constitution and, in particular, can be used to probe the changes in chemical structures that occur during coal lique-faction. Information into the latter can be deduced by the generation of mean structural parameters for process coals, their oil products, and their "im-mediate" products, defined operationally as asphaltenes and preasphaltenes. The separation scheme used to obtain the oil, asphaltene, and preasphaltene fractions in the present work was that of Schweighardt and Thames (1978). The mean structural units of interest were the carbon aromaticity f_a; the degree of aromatic ring substitution σ; and the H/C atomic ratio for the hypothetically unsubstituted polynuclear condensed aromatic ring system

48   H. L. Retcofsky

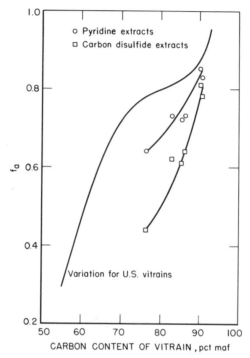

Fig. 3. Carbon aromaticities of solvent extracts of United States coal vitrains.

H_{aru}/C_{aro} . The latter is a measure of the number of condensed aromatic rings per mean structural unit.

Carbon aromaticities for the oils and asphaltenes were determined by high-resolution ^{13}C NMR, whereas the Brown and Ladner treatment of high-resolution proton NMR data was used to estimate the carbon aromaticity of the preasphaltene fraction. The carbon aromaticity of the process coal was determined by cross-polarization ^{13}C NMR. The cross-polarization technique was used to confirm the carbon aromaticities of the oil, asphaltene, and preasphaltene fractions.

The degree of aromatic ring substitution and average number of condensed rings per mean structural unit were estimated using the equations of Brown and Ladner (1960) and slight variations thereof. The Brown and Ladner-type calculations require, in addition to the carbon aromaticity, knowledge of the elemental composition of the material, its hydrogen aromaticity, and phenolic and/or aryl ether oxygen content. The hydrogen aromaticity for the soluble fractions was obtained by high-resolution proton NMR spectrometry, whereas that of the process coal was estimated by combined ^{1}H NMR infrared (IR) measurements (Retcofsky, 1977). Phenolic OH groups for the soluble materials were estimated from IR measurements (D. H. Finseth, unpublished results, 1980); those for the whole coal were interpolated from data published previously by Friedman *et al.* (1961) and based on studies of trimethylsilyl ether derivatization of whole coals.

The coal-derived materials examined were subfractions of the centrifuged liquid product from a coal liquefaction run in the Pittsburgh Energy Technology Center 400 lb/day process development unit (Akhtar *et al.,* 1975). The changes in mean structural units occurring as the process coal is converted to a pentane-soluble oil through the preasphaltene and asphaltene intermediary stages are shown in Fig. 4. The plot of f_a (Fig. 4a) indicates an increase in aromaticity, 0.76 to 0.84, in going from coal to preasphaltene. One factor contributing to this observed change in aromaticity is the loss of light hydrocarbon gases. Cleavage of aliphatic groups from the coal matrix, resulting in gas production, would lead to some increase in the aromaticity of the remaining coal fragments. Also, disproportionation accompanied by hydrogen transfer would create more highly aromatic material. Such aromatized residues would be expected to have a low solubility when compared to the asphaltene and oil fractions, which appear to contain more highly

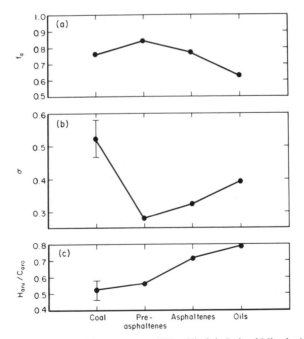

Fig. 4. NMR mean structural parameters of West Virginia Ireland Mine hvAb coal and its liquefaction products: (a) f_a, (b) σ, and (c) H_{aru}/C_{aro}.

hydrogenated material. The sharp decrease in the degree of aromatic ring substitution (Fig. 4b) is consistent with the hypothesis that the primary lique-faction step involves free radical intermediates; however, it could also result from aromatization, cyclization, or other bond cleavage mechanisms.

In going from preasphaltene to asphaltene to oil, the aromaticities and average size of the polynuclear condensed aromatic ring system decrease, whereas the degree of aromatic ring substitution exhibits a gradual increase (Fig. 4a, c, and b, respectively). These observations are undoubtedly due to both an increasing alkyl substitution and an increasing content of hydro-aromatic material.

III. FREE RADICALS IN COAL

A considerable body of evidence suggests very strongly that the ESR signals arising from coal are associated with organic free radical structures (and not mineral matter or organic charge transfer complexes) and that these free radicals are strongly aromatic. For low-rank coals, the unpaired electrons are partially localized on heteroatoms. As coalification increases, the radicals become more "hydrocarbon-like," resulting ultimately in graphite-like struc-tures. Recent studies of 64 vitrains from coal (Retcofsky *et al.,* 1978; 1981) support this view. In the following section, we expand on these and related studies. In addition to investigations of vitrinization, the ESR behavior of fusains from 29 coals, discussed briefly in an earlier report (Retcofsky *et al.,* 1978), is examined in more detail. The ESR data for the vitrains and fusains investigated are given in Tables I and II, respectively.

A. Vitrains

As indicated previously, coalification, at least with respect to United States coals, is most closely associated with vitrinization; thus, the terms will be con-sidered synonymous in this section, as was the case for Section II. For simplicity, the discussion of the ESR characteristics of vitrains will be divided into three sections, each dealing with one of the three commonly reported ESR parameters—the intensity of the ESR signal, its spectral linewidth, and its *g* value.

1. ESR Intensities

A number of researchers, including, but not limited to, Austen *et al.* (1958), Smidt and van Krevelen (1959), Toyoda and Honda (1966), Retcofsky *et al.* (1968, 1975, 1978, 1981), and Petrakis and Grandy (1978), have observed the near-exponential increase in the free radical contents of coals with increasing carbon content, at least up to carbon contents of ~ 94%. In this work, intensity data were obtained for 35 of the 64 vitrains listed in Table I. A least squares treatment of the ESR intensity–carbon content relationship, reported graphically in a previous publication (Retcofsky *et al.*, 1978), results in the following regression equation:

$$\ln I = 7.3 \times 10^{-2}(\% \text{ C}) + 38.5 \tag{1}$$

The standard error of estimate, however, is a disappointingly high 3.6×10^{18} for the free radical contents, which cover only the range 0.16×10^{19} to 5.3×10^{19}. The data for the 95.7% C coal was excluded from the statistical treatment because of its very low value (see discussion below). This data point was unfortunately misplotted in Fig. 1 of the paper by Retcofsky *et al.* (1978); the proper value is 5×10^{17} free spins/g (Table I) and not 5×10^{18} as depicted in the reference.

The explanation for the dependence of the free radical content has been adequately summarized by Tschamler and deRuiter (Lowry, 1963, p. 81):

> It is assumed that the radicals arise through chemical reactions or degradations and are stabilized by resonance in the aromatic system. It might be expected that such stabilization would be greater the greater the average size of the aromatic system. This view is in accord with X-ray measurements which show that the average size of the aromatic lamellae increases only slowly between 80 and 90% carbon content, but grows rapidly above 90% C. The decrease in radical concentration beyond 95% C is ascribed to the rapidly increasing electrical conductivity which permits electron pairing.

Based on the above explanation, a correlation between the free radical contents and carbon aromaticities of coals is anticipated. Increased confidence in the f_a values determined by the cross-polarization ^{13}C NMR technique (see Section II) prompted the plot of Fig. 5. Figure 5 shows that the free radical contents of coals increase with increasing aromaticity, although not in the linear manner suggested earlier (Yen and Erdman, 1962). The observed decrease in free radical contents of coals after chemical reduction (Reggel *et al.*, 1961) also supports the views expressed by Tschamler and deRuiter.

2. ESR Spectral Linewidths

Retcofsky *et al.* (1975) presented a detailed study of the ESR linewidths of coals and concluded that unresolved electron–nuclear hyperfine interactions contribute significantly to the observed linewidths in accordance with earlier reports (Austen *et al.*, 1966; Toyoda and Honda, 1966; Retcofsky *et al.*, 1968). This explanation is based principally on the smooth variation of linewidth with hydrogen content. Hydrogen is the most abundant element in coal having a nonzero magnetic moment and the most likely to interact with the free radical electrons in coal.

Electron spin resonance linewidths for vitrains from 64 coals are given in

TABLE I ESR Parameters and Other Information for Vitrains from Selected Coals

	Source			Ultimate analysis (maf)[1]					Electron spin resonance data		
Rank of coal[2]	Mine	Seam	Location	C (%)	H (%)	O (%)	N (%)	S (%)	Free spin conc. (10^{19} spins/g)	g Value	Linewidth (G)
Peat		Hawk Island Swamp	Manitowoc Co., Wis.	58.3	5.2	33.5	2.0	1.0	0.16	2.0041_8	5.9
Peat			Ireland	nd	nd	nd	nd	nd	nd	2.0035_9	4.5
Lignite	Zap	Beulah–Zap	Mercer Co., N. Dak.	67.8	5.6	25.1	0.9	0.6	0.46	2.0040_1	6.6
Subbituminous	Wyodan (strip)	Smith–Roland	Campbell Co., Wyo.	70.2	6.0	21.4	1.1	1.3	1.2	2.0038_1	7.3
Subbituminous	Big Horn	Dietz	Sheridan Co., Wyo.	70.8	5.1	22.5	0.6	1.0	nd	2.0038_1	8.0
Lignite	Beulah	Beulah–Zap	Mercer Co., N. Dak.	72.6	5.1	21.0	0.6	0.7	3.0	2.0042_3	7.4
		Tempoku	Japan	72.7	5.1	20.1	1.6	0.5	nd	2.0039_3	7.4
Lignite	Baukol-Noonan	Noonan	Divide Co., N. Dak.	74.2	4.5	19.4	1.3	0.6	nd	2.0039_6	7.1
Lignite		Rasa	Yugoslavia	75.0	5.7	6.4	1.7	11.2	nd	2.0038_0	6.8
Lignite		Sharigh	Pakistan	75.8	6.1	12.3	1.7	4.1	nd	2.0037_7	7.5
Bituminous	Mecca	Illinois No. 6	Henry Co., Ill.	76.1	5.5	15.3	1.0	2.1	1.2	2.0032_0	7.4
Bituminous	Harmattan	No. 7	Illinois	76.1	5.0	nd	nd	nd	nd	2.0032_0	7.6
subC	D. O. Clark	No. 7½	Sweetwater Co., Wyo.	76.1	4.6	17.0	1.6	0.7	2.3	2.0033_4	6.2
subA		Adaville	Lincoln Co., Wyo.	76.3	5.2	16.5	1.3	0.7	1.8	2.0038_3	6.5
Bituminous	Rainbow No. 7	No. 11	Sweetwater Co., Wyo.	76.5	5.7	15.0	1.9	0.9	1.9	2.0032_0	7.2
Bituminous	Liberty	Liberty	Carbon Co., Utah	77.7	6.2	13.7	1.7	0.8	1.1	2.0033_5	8.0
		Taiheiyo	Japan	77.8	6.0	14.9	1.1	0.2	nd	2.0035_9	8.0
		Kashima	Japan	78.1	5.9	12.5	0.8	2.7	nd	2.0033_6	8.2
hvCb	(Herrin Bed)	Illinois No. 6	St. Clair Co., Ill.	78.1	5.6	11.7	1.9	2.7	nd	2.0031_8	7.7
hvCb	Green Diamond	Illinois No. 6	St. Clair Co., Ill.	78.1	4.9	13.7	1.3	2.0	2.4	2.0032_4	7.2
hvBb	River King	Illinois No. 6	St. Clair Co., Ill.	78.4	5.6	14.1	1.1	0.8	2.7	2.0033_8	8.6
hvCb	Harmattan	Illinois No. 6	Vermillion Co., Ill.	78.6	5.7	12.6	1.4	1.7	2.2	2.0033_8	7.7
Bituminous	Orient No. 3	Illinois No. 6	Jefferson Co., Ill.	79.1	5.3	12.5	1.9	1.2	1.2	2.0031_5	6.9
hvCb	Banner	Colchester No. 2	Fulton Co., Ill.	79.2	5.8	11.2	1.3	2.5	1.9	2.0034_4	7.7
Bituminous	Pleasant View	Kentucky No. 9	Hopkins Co., Ky.	79.3	6.0	11.1	1.6	2.0	1.6	2.0030_4	7.4
		Takamatsu	Japan	79.8	5.5	12.8	1.0	0.9	nd	2.0033_8	7.8
Bituminous	DeKoven	Kentucky No. 9	Union Co., Ky.	80.7	6.0	9.9	1.3	2.1	1.4	2.0029_0	7.7
		Bibai	Japan	81.1	6.0	11.0	1.6	0.3	nd	2.0032_3	8.1
		Akabira	Japan	83.4	6.2	8.4	1.7	0.3	nd	2.0030_6	7.8
hvAb	Experimental	Pittsburgh	Allegheny Co., Pa.	83.5	5.7	8.1	1.7	1.0	1.9	2.0028_2	7.0
Bituminous	Majestic	Pond Creek	Pike Co., Ky.	83.7	5.6	8.5	1.4	0.8	0.6	2.0027_9	7.2
Bituminous		Zollverein	Germany	83.9	5.2	8.7	1.6	0.6	2.8	2.0029_3	5.9

Rank²	Mine	Coal bed	Location	C	H					Density	
hvAb	Miike		Japan	84.5	6.1	7.1	1.2	1.1	nd	2.0030_8	7.4
hvAb	Elk Creek No. 1	Powellton (A)	Wyoming Co., W.Va.	85.1	5.3	7.1	1.5	1.0	1.7	2.0030_6	7.2
	No. 52	Lower Banner	Russell Co., W.Va.	86.1	5.5	5.6	1.8	1.0	1.9	2.0026_9	5.5
		Yūbari	Japan	86.2	6.3	5.3	1.9	0.3	nd	2.0028_4	7.2
Bituminous	Kopperston	Eagle	Wyoming Co., W.Va.	86.2	5.6	5.7	1.5	1.0	2.2	2.0027_4	6.1
		Hashima	Japan	87.2	5.8	5.4	1.3	0.3	nd	2.0028_9	7.1
mvb	Dip		Indiana Co., Pa.	87.7	4.9	nd	nd	0.8	3.3	2.0027_3	6.6
sa	Trevorton		Northumberland Co., Pa.	88.5	3.8	5.3	1.5	0.7	nd	2.0029_3	5.2
mvb	Dutch Creek	Mesa Verdi	Pitkin Co., Colo.	89.1	5.2	3.1	2.1	0.5	3.6	2.0029_8	6.9
lvb	Dawes	Lower Hartshorne	LeFlore Co., Okla.	89.5	5.0	3.0	1.8	0.7	3.1	2.0028_9	6.1
lvb	Virginia Pocahontas No. 3	Pocahontas No. 3	Buchanan Co., Va.	90.0	4.4	1.1	0.7	3.8	3.5	2.0027_0	6.0
Bituminous	Steinman No. 10	Upper Kittanning	Cambria Co., Pa.	90.1	5.0	2.6	1.5	0.9	4.0	2.0028_0	6.4
lvb		Pocahontas No. 4	McDowell Co., W.Va.	90.4	4.6	3.1	1.4	0.5	3.7	2.0026_8	5.3
sa	Trevorton	Pocahontas No. 3	Northumberland Co., Pa.	90.6	4.0	3.0	1.5	1.0	4.2	2.0028_2	3.4
lvb	Itmann		West Virginia	90.7	4.8	2.5	1.3	0.7	nd	2.0026_7	5.4
an		Canmore	Canada (Alberta)	91.5	4.4	1.6	1.9	0.6	nd	2.0027_5	5.8
sa		Buck Mountain	Schuylkill Co., Pa.	92.4	4.4	0.6	nd	nd	4.6	2.0028_6	5.1
an	Trevorton		Northumberland Co., Pa.	92.5	3.9	1.6	1.5	0.5	5.2	2.0030_2	5.9
an		Dorrance	Luzerne Co., Pa.	92.8	2.7	2.9	1.0	0.6	5.3	2.0028_6	0.87
an		Omine	Japan	93.2	3.3	1.2	1.7	0.6	nd	2.0028_0	4.0
an	Huber		Luzerne Co., Pa.	93.4	2.6	2.3	1.0	0.9	nd	2.0028_1	2.1
an			Iron Co., Mich.	93.4	0.8	4.3	1.2	0.3	nd	2.002_9	63
an			Providence Co., R. I.	94.1	1.1	3.0	0.5	1.3	nd	2.0027_2	2.2
an	Pallaska		Peru (Chimbote)	94.7	1.6	2.2	0.6	0.9	nd	2.0028_5	0.70
an		Wanamie No. 19	Luzerne Co., Pa.	95.2	2.4	0.9	1.0	0.5	4.5	2.0028_2	0.65
an			Providence Co., R. I.	95.2	0.9	1.8	0.3	1.8	nd	2.0028_8	5.5
ma			Rhode Island	95.7	0.5	3.2	0.1	0.5	0.050	2.002_6	14
an			Antarctica	95.9	0.5	3.6	0.0	0.0	nd	2.010_6	22
ma			Newport Co., R. I.	97.4	0.3	1.0	0.1	1.2	nd	2.004_4	10
ma			Providence Co., R. I.	97.7	0.3	1.7	0.1	0.2	nd	2.006_8	19
ma			Newport Co., R. I.	97.9	0.2	1.7	0.2	0.0	nd	2.003_9	20
ma			Austria (Leoben)	98.2	0.2	1.3	0.2	0.1	nd	2.0031_8	1.6

¹ Moisture ash-free.

² Rank designations, wherever possible, are those specified in ASTM (1981): ma—metaanthracite, an—anthracite, sa—semianthracite, lvb—low volatile bituminous, mvb—medium volatile bituminous, hvAb—high volatile A bituminous, hvBb—high volatile B bituminous, hvCb—high volatile C bituminous, sub A—subbituminous A, sub C—subbituminous C.

TABLE II

ESR Parameters and Other Information for Fusains from Selected Coals

	Source			Ultimate analysis (maf)					Electron spin resonance data		
Rank of coal	Mine	Seam	Location	C (%)	H (%)	O (%)	N (%)	S (%)	Free spin conc. (10^{19} spins/g)	g Value	Linewidth (G)
Peat	Kankakee		Manitowoc Co., Wis.	nd	nd	nd	nd	nd	nd	2.0034_0	4.8
Lignite	Velva		Ward Co., N. Dak.	58.2	2.4	nd	nd	nd	nd	2.0028_9	1.2
Brown coal	Rhenish		King Co., Wash.	64.2	4.9	28.7	nd	nd	nd	2.0031_4	5.0
Brown coal	Fortuna		Cologne, Germany	nd	nd	26.1	nd	nd	0.037	2.0026_7	0.37
Brown coal	Velva		Ward Co., N. Dak.	nd	nd	24.7	nd	nd	nd	2.0031_9	5.8
Lignite	Ville		Cologne, Germany	67.5	4.6	26.1	nd	nd	nd	2.0031_7	5.7
Brown coal	Zap	Zap-Beulah	Mercer Co., N. Dak.	68.2	4.1	20.5	0.7	1.1	2.9	2.0032_4	5.8
Lignite	Wyodak		Campbell Co., Wyo.	69.8	4.1	19.8	nd	nd	0.087	2.0028_8	0.38
subC	Beulah		Beulah Co., N. Dak.	72.6	5.0	18.9	0.8	0.8	nd	2.0034_5	6.6
Lignite	Harmattan	No. 7	Illinois	74.4	4.1	nd	nd	nd	3.0	2.0029_9	3.0
hvCb		Illinois No. 7	Vermillion Co., Ill.	76.3	3.2	10.6	0.3	7.7	2.3	2.0030_9	4.5
subB	Eagle		Weld Co., Colo.	78.5	2.8	nd	nd	nd	nd	2.0026_2	0.68
subA		Rock Springs No. 7½	Sweetwater Co., Wyo.	78.8	2.6	12.7	nd	nd	2.3	2.0027_4	0.59
Subbituminous		Wapumun No. 3	Alberta, Canada	nd	nd	nd	nd	nd	1.6	2.0030_0	4.4
Subbituminous		Wapumun	Alberta, Canada	82.1	3.6	nd	nd	nd	2.6	2.0029_1	3.4
hvBb	River King	Illinois No. 6	Illinois	82.8	4.7	3.9	nd	nd	0.48	2.0027_4	0.95
hvCb	Green Diamond	Illinois No. 6	St. Clair Co., Ill.	83.7	3.6	11.1	0.5	1.1	0.72	2.0028_4	0.93
hvCb	Banner	Colchester No. 2	Fulton Co., Ill.	85.5	3.7	8.6	nd	nd	1.6	2.0028_7	0.96
mvb	Marian	Sewell	Wyoming Co., W. Va.	90.2	2.9	4.3	0.2	0.7	0.39	2.0028_1	0.56
hvCb	(Herrin Bed)	Illinois No. 6	St. Clair Co., Ill.	90.4	3.5	5.2	nd	nd	nd	2.0027_3	0.67
lvb	Buckeye No. 3	Pocahontas No. 3	Wyoming Co., W. Va.	91.2	3.5	4.7	0.3	0.6	2.2	2.0026_5	0.43
hvAb	Experimental	Pittsburgh	Allegheny Co., Pa.	91.4	2.8	4.9	nd	nd	1.6	2.0026_5	0.77
sa	Trevorton	Buck Mountain	Northumberland Co., Pa.	91.8	4.4	3.8	nd	nd	0.63	2.0027_5	0.56
an		Wanamie No. 19	Schuylkill Co., Pa.	92.5	3.2	2.7	nd	nd	0.65	2.0027_2	0.57
an			Luzerne Co., Pa.	92.9	2.8	4.3	nd	nd	2.8	2.0027_4	0.36
mvb	Dip		Indiana Co., Pa.	93.5	2.5	3.2	0.3	0.5	1.0	2.0027_0	0.43
lvb	Dawes	Hartshorne	LeFlore Co., Okla.	93.7	2.9	2.8	nd	nd	1.0	2.0026_9	0.49

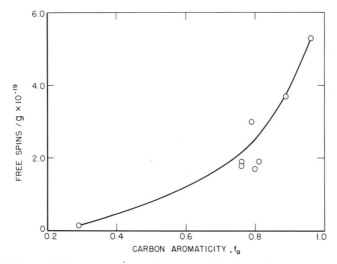

Fig. 5. Free radical content versus carbon aromaticity for vitrains from selected United States coals.

Table I. Typically, the linewidths fall within the range 0.6–8.6 G, the exceptions being six high-rank samples having linewidths of 10, 14, 19, 20, 22, and 63 G. The latter vitrains, each of which contains less than 1.0% hydrogen, have unusual electrical properties and are discussed in a later section.

Statistical treatment of the linewidth data, excluding those samples having % H < 1.0 and, of course, those of unknown hydrogen content, leads to the relationship

$$\Delta H = 1.5\,(\%\,H) - 1.2 \tag{2}$$

Equation (2), based on a least squares analysis of 55 samples, has a standard error of estimate of 0.8, which is roughly comparable to the precision of the linewidth measurement (Table III). Equation (2) differs only slightly from an earlier one based on fewer samples (Retcofsky *et al.*, 1975).

The proton-broadening hypothesis is also supported by a comparative ESR study of coals and dehydrogenated coals (Retcofsky *et al.*, 1975). For vitrains from ten coals, the linewidth of each dehydrogenated sample was considerably less than that of the original coal. The percentages of decreases in linewidths of the samples upon dehydrogenation were found to parallel closely the hydrogen evolution data.

Other effects, in addition to proton–electron interactions, may also influence the ESR linewidths of coal. One such effect, electron exchange narrowing (Tschamler and deRuiter, 1963; Kwan and Yen, 1979), can result from free electron transfer from one aromatic site in coal to another. For electron

TABLE III

Precision[a] of ESR g Value and Spectral Linewidth Measurements

Identity of vitrain	C (%)	Linewidth (G)	g Value
Anthracite from Pallaska Mine (Peru)	94.7	0.70 ± 0.18	2.00285 ± 0.000026
Metaanthracite from Leoben, Austria	98.2	1.63 ± 0.19	2.00318 ± 0.000050
Anthracite from Providence Co., R. I.	94.1	2.17 ± 0.39	2.00272 ± 0.000075
Metaanthracite from Newport Co., R. I.	97.4	10.00 ± 1.70	2.00436 ± 0.00049
Metaanthracite from Providence Co., R. I.	97.7	18.80 ± 0.75	2.00684 ± 0.00089
Anthracite from Iron Co., Mich.	93.4	63.00 ± 0.80	2.00286 ± 0.00039

[a] Precision expressed as ± 1 standard deviation. All values are based on three measurements.

exchange frequencies less than the hyperfine splitting constants, the resulting band of unresolved hyperfine lines is Gaussian, whereas more rapid exchange results in a Lorentzian band. By carefully investigating the line shapes of 15 coals having carbon contents within the range 71–93%, Kwan and Yen reported that the line shapes of coals are closer to Lorentzian than Gaussian and concluded that the results of their study are consistent with exchange narrowing of the ESR resonances. Another possible contributor to observed ESR linewidths in coals is any heterogeneity (or anisotropy) in the g value (see Section III,A,3). As is so frequently the case in coal research, single simple explanations for observed behavior seldom exist.

3. ESR g Values

Electron spin resonance g values, largely ignored in early ESR studies of coals, have proven to be perhaps the most useful parameter in elucidating the immediate chemical environment of the free radical electrons in coal. Systematic studies of the ESR g values of coal have been reported by Toyoda and Honda (1966), Retcofsky *et al.* (1967, 1968, 1978, 1981), Elofson and Schulz (1970), Yen and Sprang (1970), and Petrakis and Grandy (1978). The usefulness of the g value in coal research is a direct result of the fact that g values of organic free radicals are greatest for those radicals in which the unpaired electrons are localized or partially localized on atoms such as oxygen, nitrogen, and sulfur that have high spin–orbit coupling constants. Yen and Sprang (1970) and Retcofsky *et al.* (1978, 1981) have reported good correlations between ESR g values and heteroatom contents of coals.

The ESR g values of vitrains from 52 of the coals in Table I were treated statistically to further explore the dependence of g values on heteroatom contents. Coals having hydrogen contents less than 1% were excluded, as in the linewidth discussion. The resulting 14 regression equations are given in Table IV. The standard error of estimate was used to compare the relative "goodness of fit" of the equations. The dependent variable in each case is $g - g_e$, where g_e refers to the g value of the free electron (2.00232). Only linear and multiple linear regression analyses were considered. The purpose of the exercise was to determine the relative importance of oxygen-, nitrogen-, and sulfur-containing radicals in coals. As indicated above, highly metamorphosed coals having hydrogen contents of less than 1% are excluded from the calculations.

In the first set of calculations [Table IV, Eqs. (A)–(C)], linear regressions were carried out for equations of the type $g - g_e = f(\Sigma\ X\xi_X)$, where X is the atom fraction of atom X present in the coal and ξ_X is its spin–orbit coupling constant. Spin–orbit coupling constants used were those estimated by McClure (1952) for oxygen (152 cm^{-1}), sulfur (382 cm^{-1}), and nitrogen (70 cm^{-1}). It should be noted that the analytical data in Table I are on an maf basis and that no differentiation between sulfur forms in the various coals is made. It is also important to point out that for the vitrains considered, oxygen contents range from 0.6 to 33.5, whereas nitrogen contents cover the much smaller range of 0.5–2.1%. The sulfur contents, however, can be divided into two groups, one set consisting of coals having sulfur contents between 0.3 and 4.1% and a second set consisting of the two coals that are exceptionally high in sulfur content.

Treatment of the oxygen data alone results in a regression coefficient of 7.0×10^{-5} and a standard error of estimate of 2.13×10^{-4} [Table IV, Eq. (A)], the latter being considerably larger than the expected precision of the g value measurements (Table III). Addition of a sulfur term leaves the regression coefficient essentially unchanged, but reduces the standard error of estimate by approximately 12% [Table IV, Eq. (B)]. The addition of a third term for nitrogen does nothing to improve the fit. These results suggest that oxygen and sulfur atoms play important roles in the free radical structures in coal. The importance of nitrogen, however, cannot be deduced from the information at hand. Equations (A')–(C') (Table IV) were obtained by repeating the calculations using the data for all samples except the two high-sulfur coals. The standard error of estimate of Eq. (A') is less than that of Eq. (A) but very nearly the same as that for Eq. (B), which supports the conclusion drawn from consideration of Eqs. (A)–(C).

Equations (D)–(G) and (D')–(G') are of the form $g - g_e = f(O, N, S)$. Note that Eqs. (D) and (D') are identical to Eqs. (A) and (A'), respectively. The use of multiple linear regression analyses has certain advantages over the

TABLE IV

Results for Linear Regression Analyses Showing Relationship between Electron Spin Resonance g Values and Heteroatom Contents of Vitrains from Selected Coals

	Regression equation	Standard error of estimate	Comments
(A)	$g - g_a = 7.0 \times 10^{-5}(O\xi_O) + 3.5 \times 10^{-4}$	2.13×10^{-4}	52 Samples
(B)	$g - g_e = 7.1 \times 10^{-5}(O\xi_O + S\xi_S) + 2.6 \times 10^{-4}$	1.88×10^{-4}	Vitrains having %H $<$ 1.0 excluded
(C)	$g - g_e = 7.1 \times 10^{-5}(O\xi_O + S\xi_S + N\xi_N) + 2.2 \times 10^{-4}$	1.89×10^{-4}	
(A')	$g - g_e = 7.1 \times 10^{-5}(O\xi_O) + 3.2 \times 10^{-4}$	1.69×10^{-4}	50 Samples
(B')	$g - g_e = 6.9 \times 10^{-5}(O\xi_O + S\xi_S) + 2.6 \times 10^{-4}$	1.84×10^{-4}	Vitrains having %H $<$ 1.0 excluded
(C')	$g - g_e = 6.9 \times 10^{-5}(O\xi_O + S\xi_S + N\xi_N) + 2.2 \times 10^{-4}$	1.86×10^{-4}	Vitrains having %S $>$ 4.0 excluded
(D)	$g - g_e = 1.1 \times 10^{-2}(O) + 3.5 \times 10^{-4}$	2.13×10^{-4}	52 Samples
(E)	$g - g_e = 1.1 \times 10^{-2}(O) + 2.6 \times 10^{-2}(S) + 2.6 \times 10^{-4}$	1.89×10^{-4}	Vitrains having %H $<$ 1.0 excluded
(F)	$g - g_e = 1.1 \times 10^{-2}(O) - 1.1 \times 10^{-2}(N) + 4.4 \times 10^{-4}$	2.14×10^{-4}	
(G)	$g - g_e = 1.1 \times 10^{-2}(O) + 2.5 \times 10^{-2}(S) - 9.3 \times 10^{-3}(N) + 3.4 \times 10^{-4}$	1.90×10^{-4}	
(D')	$g - g_e = 1.1 \times 10^{-2}(O) + 3.2 \times 10^{-4}$	1.69×10^{-4}	50 Samples
(E')	$g - g_e = 1.1 \times 10^{-2}(O) - 1.7 \times 10^{-2}(S) + 3.6 \times 10^{-4}$	1.68×10^{-4}	Vitrains having %H $<$ 1.0 excluded
(F')	$g - g_e = 1.1 \times 10^{-2}(O) - 2.0 \times 10^{-2}(N) + 4.8 \times 10^{-4}$	1.66×10^{-4}	Vitrains having %S $>$ 4.0 excluded
(G')	$g - g_e = 1.1 \times 10^{-2}(O) - 2.8 \times 10^{-2}(S) - 2.7 \times 10^{-2}(N) + 6.1 \times 10^{-4}$	1.60×10^{-4}	

simplified treatment of the data that resulted in Eqs. (A)–(C) and (A′)–(C′). In particular, regression coefficients for each of the three atom terms will result; thus, predetermined weighting terms such as spin–orbit coupling constants are not required. Regression equations (D)–(G) were obtained using data for all 52 samples; data for the two high-sulfur coals were deleted in the statistical treatment leading to Eqs. (D′)–(G′).

Addition of a sulfur term to the g value–oxygen content equation [Table IV, Eq. (E)] results in an improved standard error of estimate, suggesting again that sulfur species as well as oxygen species are important in the free radical structures in coals. Furthermore, the ratio of regression coefficients for the sulfur and oxygen terms, that is, $(2.6/1.1) \times 10^{-2}$, is for all practical purposes identical to the ratio of their spin–orbit coupling constants, 382/152. Addition of a nitrogen term to the oxygen equation does not improve the fit of the data [Table IV, Eq. (F)] and results in a negative regression coefficient. The same statement applies to Eq. (G), which results from the addition of a nitrogen term to the oxygen and sulfur equation. Elimination of the data for the two high-sulfur coals results in negative terms for both sulfur and nitrogen [Table IV, Eqs. (E′)–(G′)]. The results require a slight modification of our g value interpretation given above based on regression equations (A)–(C) and (A′)–(C′). Consideration of the ESR g values leads to the general conclusion that the free radical electrons in coals are delocalized over aromatic structures, with some partial localization on oxygen atoms in all except the most highly metamorphosed coals. For coals having high sulfur contents, some localization onto sulfur atoms occurs. The importance of nitrogen cannot be defined.

Another interesting aspect of the observed g value relationships is the intercept of the regression lines. The best value, $g - g_e = 2.6 \times 10^{-4}$, is given by regression equation (E) and corresponds to $g = 2.00258$. Stone (1963), using his general theory of the g value in polyatomic systems, proposed that $g - g_e = b + \lambda C$ for aromatic radicals. In the equation, b and c are experimentally determined constants and λ is a numerical factor obtained by solving the appropriate secular equation (in the Hückel molecular orbital approximation). Stone deduced from experimental g values for semiquinone radicals (Blois et al., 1961) that the g values for all neutral alternant hydrocarbon radicals and other radicals for which $\lambda = 0$ should be 2.00257. A slightly higher value, 2.002638, was proposed shortly thereafter from a comprehensive study of aromatic hydrocarbon free radicals (Segal et al., 1965). The intercept for vitrains, 2.00258, is quite close to the projected values of Stone and Segal and lends added support for the free radical interpretation of the ESR spectra of coals.

The uniqueness of the two high-sulfur coals is worthy of discussion at this point. Characterization of sulfur groups in both the Sharigh sample (Roy,

1956; Iyengar *et al.,* 1960) and the Rasa sample (Kreulen, 1952) have been reported. Approximately 90% of the sulfur in Sharigh lignite is organic sulfur, most of which occurs in SH groups associated with aromatic structures. The Rasa sample, also rich in organic sulfur, is thought to have most of its organic sulfur in ring structures. Free radical species having such structures would be expected to possess relatively high g values, as observed for these two coals.

It should be noted that the g value dependence on heteroatom content as shown for coals has considerable precedence in studies of organic free radicals of known structure (Fig. 6). For example, the g values of semiquinone anion

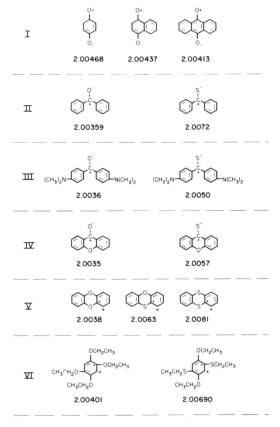

Fig. 6. ESR g values for oxygen- and sulfur-containing radicals. (I) Semiquinone anion radicals; (II) anion radicals from benzophenone and benzothione; (III) anion radicals from the electrochemical reduction of Michler's ketone and thione; (IV) anion radicals from reduction of 9-xanthone and 9-xanthione; (V) cation radicals produced from dibenzo-*p*-dioxin, phenoxanthin, and thianthrene; (VI) cation radicals from 1,2,4,5-tetraethoxybenzene and 1,4-diethoxy-2,5-diethylthiobenzene.

radicals (Blois *et al.*, 1961) illustrate the lowering of the *g* value as the unpaired electron becomes more delocalized over aromatic rings and thus spends less time on oxygen atoms (Fig. 6, I). The higher *g* values observed for sulfur-containing radicals when compared with their oxygen analogs are exemplified by comparing *g* values of the (1) anion radicals from benzophenone and ben-zothione (Blois *et al.*, 1960; Schulz and Elofson, 1966) (Fig. 6, II); (2) anion radicals from the electrochemical reduction of Michler's ketone and thione (Schulz and Elofson, 1966) (Fig. 6, III); (3) anion radicals from reduction of 9-xanthone and 9-xanthione (Schulz and Elofson, 1966) (Fig. 6, IV); and (4) cation radicals from 1,2,4,5-tetraethoxybenzene and 1,4-diethoxy-2,5-diethylthiobenzene (Sullivan, 1973) (Fig. 6, VI). The gradual *g* value change by progressively replacing oxygen with sulfur can be vividly seen by compar-ing data (Fig. 6, V) for cation radicals produced from dibenzo-*p*-dioxin (Yang and Pohland, 1972), phenoxanthin (Schmidt *et al.*, 1964), and thianthrene (Shine *et al.*, 1964).

The importance of heteroatoms in the free radical structures in coal is also reflected by decreases in ESR *g* values after the coals are heat treated at mild temperatures. Heat treatment of coals, particularly coals of low and medium rank, is known to result in the evolution of heteroatom-containing gases such as CO_2, H_2O, and SO_2 (Sharkey *et al.*, 1963; Groom, 1969).

Figure 7 depicts the decrease in *g* values of vitrain and fusain samples separated from a lignite and an hvAb coal during heat treatment at 350°C. The lignite vitrain shows the largest *g* value change, as would be anticipated on the basis of its high oxygen content and its high content of volatile matter.

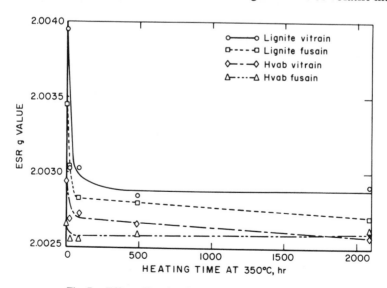

Fig. 7.   Effect of heating time on ESR *g* values of macerals.

Only small changes are observed when heat treatment is extended beyond 24 hr. Based on these preliminary results, a series of vitrains was selected for further study. Samples of vitrains from two lignites, and from a sub-bituminous and two bituminous coals were heat treated at 350°C for 24 hr. The total gases evolved during this period were collected and analyzed by mass spectrometry. The high-sulfur Rasa coal discussed previously was one of the two lignites investigated.

The decrease in g value and the amount of heteroatom gases evolved for each sample is summarized in Table V. As would be expected, the total quantity of gases evolved closely paralleled the ranks of the coals investigated. The amount of heteroatom gases evolved and the g value change behave in a similar fashion. The high-sulfur Rasa coal, in contrast to the others, evolves large quantities of H_2S and SO_2. The loss of sulfur and oxygen gases and the accompanying decrease in g value supports the view presented above for heteroatom participation in free radical structures in coal.

Electron spin resonance g values of coals before and after dehydrogenation were also determined to elucidate further the nature of the free radicals in coal. The samples were from the investigation of Reggel *et al.* (1968) and were the same ones discussed above with respect to ESR spectral linewidth correlations. The g values of the original and dehydrogenated coals are shown as a function of the carbon contents of the original coals in Fig. 8. Figure 8 shows

TABLE V

Gas Evolution and Change in ESR g Value for Selected Coals Heat-Treated at 350°C for 24 hr

Coal	Beulah–Zap	Adaville	Rasa	Pittsburgh	Kentucky No. 9
Carbon content (%, maf)	67.8	76.3	75.0	83.5	80.0
Total gas evolution (mm — ml/g coal)	133,000	62,000	48,000	22,000	29,000
Sulfur- and oxygen-containing gases (μg gas/g coal)					
CO_2	96,000	42,000	8900	5600	650
CO	23,000	12,000	—	2500	2400
H_2O	76,000	30,000	15,000	10,000	14,000
H_2S	1400	1300	22,000	380	2100
SO_2	—	2300	8100	1800	830
COS	—	110	820	trace	82
Total S + O gases (μg gas/g coal):	196,000	87,700	54,800	20,300	20,000
Δg	0.0009	0.0011	0.0007	0.0003	0.0004

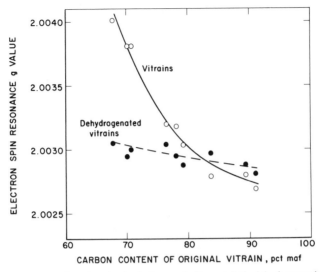

Fig. 8. ESR g values for vitrains before and after catalytic dehydrogenation.

that the g value change is most pronounced for the low-rank coals and that for these coals, the g value of the dehydrogenated vitrain is lower than that for the original coal. This behavior can be best explained on the basis of heteroatoms involved. Since the dehydrogenation probably involves aromatization, the g values may simply be reduced owing to the greater delocalization of the unpaired electrons, as is the case of the decrease in g value in the semiquinone series shown in Fig. 6, I. Another factor is the evolution of heteroatom gases, particularly CO_2, during the dehydrogenation of subbituminous coals and lignite. Irrespective of which of these two causes predominate, the expected result is a decrease in g value, as is observed. Dehydrogenation of the higher rank coals results in much smaller g value changes. For the three samples involved, dehydrogenation leads to higher g values in contrast to the results for the lower rank samples. No rationale for this behavior is immediately obvious.

Absent from the above discussions of g values and spectral linewidths is any consideration of the six coals that exhibit unusually large spectral linewidths. These vitrains, each of which contains less than 1.0% hydrogen, are classified as anthracites or metaanthracites. The X-ray diffraction studies of Mentser *et al.* (1962) have shown these and other coals similar in rank to be highly graphitic. As Wagoner (1960) and Singer and Wagoner (1963) have reported, graphite and partially graphitized materials exhibit large g value anisotropies. The presence of anisotropic g values is the most likely explanation for the large linewidths observed in the ESR linewidths of very high-rank coals. The g value anisotropies of 1.8×10^{-4} for Huber Mine anthracite (Retcofsky *et al.*,

1978) and 4×10^{-5} for St. Nicholas anthracite (present work) support this view. These two anthracites do not exhibit the very large linewidths (> 10 G) observed for the highly graphitized coals; the electrical properties of the latter coals thus far have prevented measurements of their g value anisotropy. The observed anisotropies for the Huber and St. Nicholas samples are much smaller than those reported for single-crystal graphite (Wagoner, 1960), a high-temperature (2800°C) graphitized carbon film (Toyoda et al., 1972), and acenaphthylene char heat treated at 3000°C (Singer and Cherry, 1969) (see Table VI). The sign of the g value anisotropy for the two anthracites is opposite both to that observed for chars heat treated to low and medium temperatures (Singer and Cherry, 1969) and to that predicted from Stone's (1964) theoretical treatment of neutral odd-alternant radicals.

B. Fusains

The reasonably well-understood ESR correlations for vitrains from coal are not reflected in fusains. This undoubtedly results from the "extraordinary wide variation in properties and composition" of fusains (Marshall, 1954) as well as from the factors that lead to the origin of fusain and its subsequent metamorphism. The specific heat studies of Terres et al. (1956) suggested that fusains were formed by a thermal process, presumably in forest fires, the average temperature being between 700 and 1000°C. Schopf (1948) reported that fusinization of plant remains has an early inception and progresses rapidly in the peat and lignitic stages, after which the metamorphic changes are almost imperceptible. The ESR studies of Austen et al. (1966) suggest strongly that the origin of fusinitic macerals commonly involved exposure to fairly high temperatures, while those of Retcofsky et al. (1978) suggest that fusinization, in *some* cases, may continue into the bituminous stages of coalification.

Electron spin resonance g values of vitrains and fusains from the present study (Tables I and II) are plotted against their carbon contents in Fig. 9. To reiterate, the g value–carbon content plot for vitrains can be interpreted as follows. The high g values for vitrains from peat and lignites suggest the presence of aromatic radicals with some partial localization of the unpaired electrons on heteroatoms, principally but not exclusively oxygen. As coalification progresses, the g values decrease, suggesting that the radicals become more "hydrocarbon-like." During the final stages of coalification, the g values become quite large, as expected for continued condensation of these aromatic rings into graphite-like structures. The corresponding plot for fusains supports a dual theory of metamorphic changes, that is, some fusains appear to be completely metamorphosed before or shortly after incorporation into the peat bed, while others undergo continued metamorphism,

TABLE VI

ESR g Value Anisotropies for Selected Materials

	$g_{\parallel}$[a]	$g_{\perp}$[a]	$\dfrac{g_{\parallel} - g_{\perp}}{\times 10^4}$
Anthracitic coals			
Huber Mine	2.00295	2.00277	1.8
Dorrance Mine	2.00284	2.00280	0.4
Graphite single crystal[b]	2.0496	2.0026	470
Graphitized carbon film (2800°C)[c]	∼2.0267	∼2.0063	∼200
Acenaphthylene chars[d]			
700°C	2.002368	2.002743	−3.75
1000–1100°C			∼0
3000°C			Limit 470
			($T \to 3000°C$)
Odd-alternant aromatic[e]	2.00238	2.00266	−2.8
Hydrocarbon radicals			
(calculated values)			

[a] The g values presented are all reasonably well represented by the expression

$$g = g_{\perp} + A \cos^2 \theta$$

where θ is the angle between the static magnetic field and the C axis in the case of graphite and chars, and the axis perpendicular to the bedding plane in the case of coals.

[b] Wagoner (1960).

[c] Toyoda et al. (1972).

[d] Singer and Cherry (1969).

[e] Stone (1964).

although the metamorphic changes are not as pronounced as in the case of vitrains. Curiously enough, we have seen no evidence supporting graphitization as the final stage of fusinization.

Interpretation of the ESR linewidths of fusains also requires a dual explanation. The data, plotted as a function of hydrogen content in Fig. 10, appear to fall into two groups. Sixteen of the samples having known hydrogen contents exhibit very narrow ESR resonances (< 2 G) that show little variation with hydrogen content. These are thought to be samples that have undergone rapid metamorphism. Seven samples, however, show a linewidth-hydrogen content variation quite similar to that observed for vitrains; that is, except for one point, they fall within the two lines labeled 2A, which are spaced at intervals equal to ±2 standard error of estimates from the regression line for *vitrains*. Linear regression analysis of the 16 "well-metamorphosed" samples leads to the equation

$$\Delta H = 0.02(\% H) + 0.58 \tag{3}$$

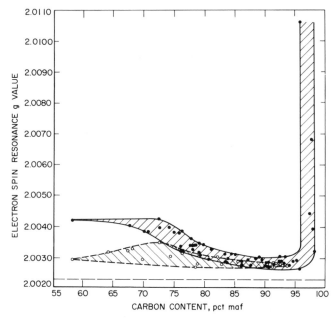

Fig. 9. ESR *g* values as a function of carbon contents for vitrains (●) and fusains (○) from coal.

having a very small, 0.25 standard error of estimate. The seven samples that closely follow the vitrinization regression line are best fitted to the equation

$$\Delta H = 1.2\,(\%\,H) - 0.12 \tag{4}$$

Note that Eq. (4) is quite similar to the corresponding vitrain equation [Eq. (2)].

Owing to the lack of elemental analysis data for the fusains, reliable conclusions based on *g* value correlations cannot be made (Fig. 11). For the seven samples shown in Fig. 11, half appear to fall within the vitrinization bands represented by the dotted lines. The plots suggest that heteroatom involvement is not as important as was found for vitrains.

C. Selected Macerals

Electron spin resonance studies of high purity macerals have been reported by Austen *et al.* (1966). For the few samples studied, fusinites (five samples) and the single micrinite gave the narrowest lines, 1.4–2.0 G. The two exinites exhibited linewidths comparable to nonanthracitic vitrinites. No *g* values were determined. The ESR linewidths and *g* values were measured for

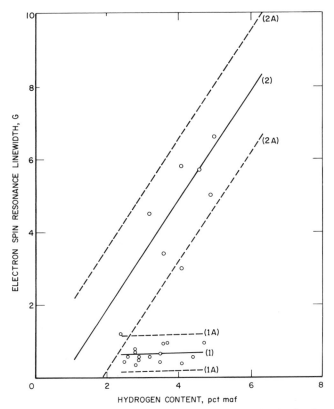

Fig. 10. ESR linewidths as a function of hydrogen contents for fusains from coals. (1) Regression line for fusains having linewidths < 2 G; (1A) regression line (1) ± 2 standard error of estimate; (2) regression line for vitrains; (2A) regression line (2) ± 2 standard error of estimate.

vitrinite, exinite, and micrinite from Zollverein (German) coal during the present study. The g values—2.0029_3, 2.0028_6, and 2.0028 for the vitrinite, exinite, and micrinite, respectively—are all quite low, indicating little hetero-atom involvement. The corresponding linewidths are 5.9, 5.0, and 1.4 G.

D. Comparisons with Other Carbonaceous Materials

In Sections III,A and III,B, it was shown that vitrains and fusains exhibit distinctly different ESR properties. In this section, we expand our study to include other carbonaceous materials. Because of the limitations on data available to us, the following discussion is confined to g values, the g values being correlated with the weight percent oxygen and sulfur in the samples.

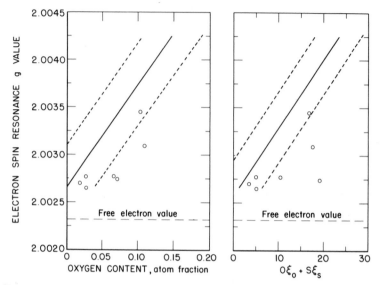

Fig. 11. Functional dependences of the ESR g values of fusains from selected coals.

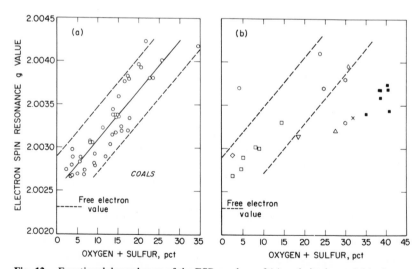

Fig. 12. Functional dependences of the ESR g values of (a) coal vitrains and (b) other carbonaceous materials. ($\square$) Pyridine extracts of coal; ($\diamondsuit$) carbon disulfide extract of coal; ($\triangledown$) 200°C cotton char; ($\triangle$) 200°C pine sawdust char; (hexagon) kerogens; ($\lozenge$) wood near coal; ($\blacksquare$) humic acids; ($\times$) wood in volcanic tuff.

The regression line for vitrains from coal (Fig. 12a) serves as our baseline. The dotted lines in both the left and right sides of Fig. 12 are identical and are spaced at plus or minus twice the standard error of estimate from the regression line for vitrains; this will be referred to as the vitrain band.

As might be expected, g values for solvent extracts of the vitrains fall within the vitrain band, whereas those of most other carbonaceous materials do not. Although two of the kerogens fall within the vitrain band, three others do not. The g values of the kerogens appear to be reasonably independent of their heteroatom content. The seven soil humic acids all show g values characteristic of oxygen-containing radicals, the samples being quite similar in elemental composition and ESR properties. The g values observed are consistent with the explanation by others (Steelink, 1966; Atherton *et al.,* 1967) that the ESR absorption is caused by a semiquinone species. A sample of wood found in close proximity to coal lies in the vitrain band, albeit near the edge. However, low-temperature chars of wood and cotton used in artificial coalification studies (Friedel *et al.,* 1970) do not. A highly carbonized wood fragment found in volcanic tuff also does not fall in the vitrain band, but surprisingly enough has both a high g value and a high content of heteroatoms.

IV. PRELIMINARY ESR AND NMR INVESTIGATIONS OF CHINESE COALS: A CAUTIONARY NOTE ON COAL STRUCTURE STUDIES

Many of the relationships shown in the previous sections, for example, the change in carbon aromaticity with increasing coal carbon content and the dependence of ESR g values on heteroatom contents of coals, tend to provide a sense of security in what is thought to be an understanding of coal structure. Such a "comfortable" feeling can cause one to be unprepared for extension of his work to additional coals. Three coal samples from the People's Republic of China were recently received by this laboratory. The results of NMR and ESR measurements on the samples prompted the preparation of this cautionary note section.

The elemental composition and magnetic resonance data for the three coals are summarized in Table VII. The most unusual features of the coals appear to be their relatively low carbon aromaticities and, in the case of the Shang-Shi sample, its high content of organic sulfur.

The cross-polarization ^{13}C NMR spectrum of Ming-Shan coal is reproduced in Fig. 13. The nearly equal intensities of the aromatic and aliphatic resonances are clearly evident, resulting in an f_a value of 0.57. In Fig. 14, the plot of carbon aromaticities versus carbon content for United States vitrains (Fig. 2) is used as a basis for comparison of Chinese and United States coals.

TABLE VII

Elemental Analyses, Sulfur Forms, and Magnetic Resonance Data for Three Chinese Coals

	Ming-Shan	Chang-Guang	Shang-Shi
Ultimate analysis (%, maf)			
C	82.7	82.6	78.3
H	6.8	6.3	5.5
O	7.1	4.7	4.7
N	1.6	1.3	1.1
S	1.9	5.0	10.5
Sulfur forms (%)			
Sulfate	0.0	0.0	0.1
Pyrite	0.6	2.3	2.6
Organic	1.3	2.7	7.8
Carbon aromaticity, $f_a{}^a$	0.57	0.64	0.67
Electron spin resonance			
g value	2.0028_5	2.0027_7	2.0028_8
Linewidth (G)	6.0	8.6	5.7
Free spins per gram	1.4×10^{19}	1.5×10^{19}	1.2×10^{19}

[a] Determined by cross-polarization ^{13}C NMR spectrometry.

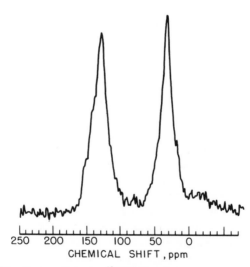

Fig. 13.   Cross-polarization ^{13}C NMR spectrum of Ming-Shan coal.

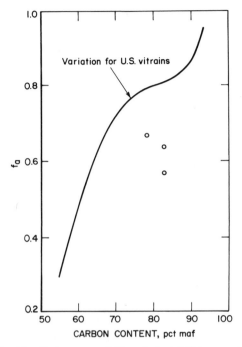

Fig. 14. Carbon aromaticities of selected Chinese coals.

The data points for the Chinese coals fall far below the curve established for United States coals and suggest fundamental differences between the two.

The ESR results were perhaps even more baffling (Fig. 15). Although the data points for the Chinese coals fall within the United States vitrain band in the plot of g values versus oxygen contents, the addition of a sulfur term produced the disturbing results shown in the plot on the right of Fig. 15. The ESR results indicate that the unpaired electrons do not interact with the sulfur despite the fact that one of the three coals has a total sulfur content of more than 10%, nearly 75% of which is in organic structures. The explanation for these totally unexpected results became evident only after the petrographic compositions of the coals were determined (Table VIII). Table VIII demonstrates, first, that in contrast to United States coals, which are highly vitrinous, the vitrinite content of the Chinese samples is quite low—much less than 50% for two of the three samples. Thus, comparison with vitrains is not justified. Second, the two coals showing the lowest aromaticities are rich in liptinite. Liptinites originate from relatively hydrogen-rich plant materials such as cutin, resins, waxes, fats, and oils and presumably are rich in aliphatic material. Thus, the low aromaticity values are accounted for in at least the Ming-Shan and Chang-Guang coals. The Shang-Shi sample has a lower car-

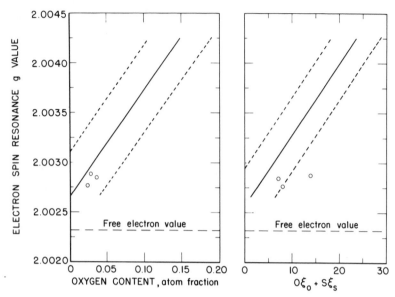

Fig. 15. Functional dependences of the ESR g values of three Chinese coals.

bon content, more vitrinite, more semifusinite, and more sporinite than the other two coals. The first and last of these would normally result in lower aromaticities, whereas the higher concentrations of vitrinite and semifusinite would normally add to the aromaticity.

With the large amounts of aliphatic material present in the Chinese coals, it is conceivable that the sulfur, although organic, is associated with aliphatic

TABLE VIII

Petrographic Composition of Chinese Coals: Percentage of Macerals

Coals	Ming-Shan	Chang-Guang	Shang-Shi
Vitrinite	32.4	14.3	53.9
Fusinite	1.3	1.5	1.7
Semifusinite	10.9	10.1	20.7
Macrinite	1.7	3.0	2.4
Micrinite	2.0	1.4	7.0
Sporinite	—	—	14.3
Exinite	2.2	—	—
Resinite	—	0.5	—
Liptinite	49.5[a]	69.2[b]	—

[a] Massive liptinite maceral, identified as a cutinite substance.
[b] Liptodetrinite substance with 8% cuticles.

structures rather than with aromatic structures, as is the case for the Sharigh and Rasa samples discussed previously. This would account for the low *g* values, since partial localization of the unpaired electrons onto aliphatic sulfur structures is unlikely.

V. ESR STUDIES OF RESPIRATORY-SIZE COAL PARTICLES

The Bureau of Mines, as part of an effort to obtain information relating to the cause(s) of coal workers' pneumoconiosis (CWP), has been engaged in a number of studies involving the detection and possible quantitative determination of specific materials in respiratory-size coal dust. Previous investigations have included the detection of organic compounds (Shultz *et al.*, 1972; Freedman and Sharkey, 1972), analysis for free silica (Freedman and Sharkey, 1972), and the determination of trace elements (Freedman and Sharkey, 1972; Kessler *et al.*, 1971). During this investigation, samples of respirable dusts and related materials were examined by ESR spectrometry. The primary purpose was to determine if any correlation exists between the fusain content of respirable coal mine dusts and the occurrence of CWP. Fusain, a lithotype found in nearly all United States coals, is thought to be an extremely friable material and has been reported to concentrate in the very fine particle sizes (International Committee on Coal Petrography, 1963). It seemed desirable to determine if fusain plays an important part in the composition of respirable dust and whether its presence is dependent upon such factors as particle size, rank, and geographical origin of the coal involved. Since several of the respirable dust samples examined were from mines that were included in the Interagency Study of Coal Workers' Pneumoconiosis (formerly the "National Study") (Morgan *et al.*, 1973), comparison of the ESR results with data relating to the prevalence of CWP in specific mines was possible.

Experimental procedures are given in more detail in this section, owing to the uniqueness of the samples and the necessary close attention that must be paid to sample preparation and spectrometer operation.

A. Experimental

1. Samples

The respirable dusts examined were obtained from personal samplers (Jacobson and Lamonica, 1969) worn by coal workers during normal working shifts in 23 mines (Table IX). A simple brushing or scraping proved suffi-

TABLE IX

Source and Rank of Coal for Respirable Dust Investigation

Rank	Mine No.	Seam	Location
hvCb	1	Kentucky No. 11	Hopkins Co., Ky.
	2	Illinois No. 6	Montgomery Co., Ill.
hvBb	3	Illinois No. 6	Franklin Co., Ill.
	4	Double Freeport	Allegheny Co., Pa.
hvAb	5	Thick Freeport	Allegheny Co., Pa.
	6	Powellton	Fayette Co., W. Va.
	7	Pittsburgh	Greene Co., Pa.
	8^a	Pittsburgh	Monongalia Co., W. Va.
	9	Sunnyside	Carbon Co., Utah
	10^b	Pittsburgh	Greene Co., Pa.
	11	Elkhorn No. 2	Pike Co., W. Va.
	12	Cedar Grove	Logan Co., W. Va.
	13	Taggart	Wise Co., Va.
	14	Pittsburgh	Washington Co., Pa.
	15	Pond Creek	Pike Co., Ky.
mvb	16^c	Lower Kittanning	Cambria Co., Pa.
lvb	17	Taggart	Wise Co., Va.
	18	Pocahontas No. 4	McDowell Co., W. Va.
	19^b	Pocahontas No. 3	Buchanan Co., Va.
	20	Pocahontas No. 4	McDowell Co., W. Va.
	21	Pocahontas Nos. 3 and 4	McDowell Co., W. Va.
Anthracite	22^a		Luzerne Co., Pa.
	23	Mammoth Vein^d	Schuylkill Co., Pa.

[a] Float dust.

[b] Two samples; removed from filters by different techniques.

[c] Float dust and respirable dust.

[d] Two samples; one collected in working area, the other in loading area.

cient to remove the bulk of the sample from the filter. In three cases, samples were removed by ultrasonic agitation.

The single sample of coal dust from pneumoconiosal lung tissue was scraped from a small section of lung that had been removed from the body of a coal miner during postmortem examination. A small amount of tissue, which appeared to be free of coal particles, was also removed from the specimen.

Sized fractions of raw coal were obtained by crushing the coal with a jaw crusher, followed by grinding in a micromill. The − 200 mesh samples were separated by sieving the crushed material. In order to obtain samples of raw coal having various particle sizes down to the micrometer range, portions of the -200 mesh coal were dispersed in a wind tunnel and then separated and collected by an Anderson sampler.

Petrographic separation of lithotypes from raw coal, needed to supplement this investigation, was carried out by removing bands of vitrain and/or lenses of fusain from selected lumps of coal. These were then crushed and, in some cases, upgraded microscopically.

All samples were evacuated to 10^{-6} torr and allowed to outgas at that reduced pressure for at least 48 hr prior to the ESR measurements. This was done primarily to prevent the objectional spectral line broadening that results from shortening of the electron relaxation times by interaction with molecular oxygen (Austen *et al.*, 1958).

2. Spectral Measurements

A Varian[2] Associates Model V–4500 ESR spectrometer equipped with a 100 kHz field modulation unit was used for all spectral measurements in this section. The nominal operating frequency of the spectrometer was 9.5 GHz.

A ruby crystal was mounted within the microwave cavity of the spectrometer for use as a secondary intensity standard. This arrangement, which permitted recording of the spectrum of both the sample and ruby standard, automatically compensates for changes in cavity Q and in other factors that may alter spectrometer sensitivity when samples are changed. A spectrum of fusain from Hartshorne coal was obtained before and after that of each sample; the number of free spins in the fusain (1.0×10^{19}) was carefully determined using a fresh sample of diphenylpicryhydrazyl for comparison.

Owing to the small amount of sample generally available, specially constructed thin-walled capillary cells were used in place of the usual 4 mm ESR sample tubes. For some of the respirable dusts, the amount of sample available was not sufficient to completely fill the "effective" length of the microwave cavity even in these very small cells. In these cases, a cavity "map," obtained by measuring spectral intensities of samples of Hartshorne fusain in capillary tubes filled to various depths with sample, was used to correct the observed spectra.

Representative spectra of respirable dust and raw coal from the same mine are reproduced in Fig. 16. Each spectrum consists of two overlapping lines—a broad component approximately 6 G wide (assignable to "vitrain-like" spins) and a narrower one approximately 1 G wide (assignable to "fusain-type" spins). The overlapping lines were deconvoluted by assuming the lineshape of the broader component to be intermediate between pure Gaussian and pure Lorentzian.

---

[2] Reference to specific equipment is made to facilitate understanding and does not imply endorsement by the United States Department of Energy.

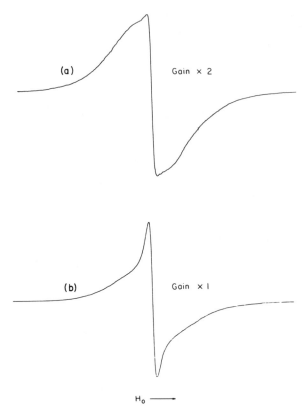

Fig. 16.   ESR spectra of (a) coal (3.3–5.5 μm) and (b) respirable dust from Mine No. 6.

B.   Anderson Sampler Fractions of Raw Coal

In order to determine the concentration of fusain-type spins in raw coal as a function of particle size in the respirable range ($< 10\,\mu$m), Anderson sampler fractions of coals having particle sizes in the following ranges: 5.5–9.2, 3.3–5.5, 2.0–3.3, 1.0–2.0 μm, and in two cases, $< 1\,\mu$m, were examined. The results, which are summarized in Fig. 17, indicate that, at least for four of the five mines involved, the concentration of fusain-type spins of the coal particles in the 3.3–9.2 μm range are higher than those for the fractions containing smaller particles. This finding was contrary to expectation; the reportedly high friability of fusain suggests that it would be more prevalent in the fractions containing the smaller size particles. Although it has been reported that fusain concentrates in the very fine particles of coal, it is not clear whether this

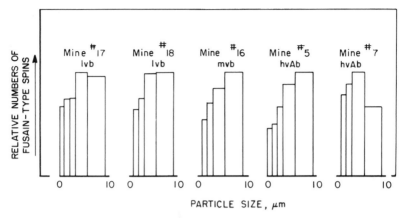

Fig. 17.   Relative numbers of fusain-type spins as a function of particle size for coals from selected mines.

statement was meant to refer to particles as small as a few micrometers in diameter.

C.   Respirable Dust from Coal Workers' Personal Samplers

For a few mines, samples of both respirable dust removed from personal samplers and raw coal were available. Comparisons of the intensities of the narrow components of the ESR spectra again were used to determine relative numbers of fusain-type spins. The raw coal samples were either -200 mesh or the 3.3–5.5 μm Anderson sampler fractions; for coals from three mines, both were available.

The respirable dust samples examined, with the exception of the two anthracites, were collected in bituminous mines and represent all ranks of bituminous coals (Table IX). Samples from two or more mines within the same seam were available for five seams: Taggart, Pocahontas Nos. 3 and 4, Pittsburgh, Illinois No. 6, and the Mammoth (anthracite) vein. Twelve of the dusts were from mines that had been included in the Interagency Study of Coal Workers' Pneumoconiosis (Morgan *et al.,* 1973). The above information regarding the respirable dust samples permitted the ESR data to be examined for possible correlations with coal rank, seam, and the known prevalences of CWP in certain mines. Before proceeding, it must be emphasized that the number of samples examined was not large, and in most cases, only a single sample from any particular mine was examined.

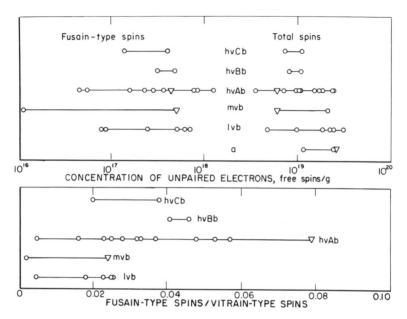

Fig. 18. ESR intensity data for respirable dusts from coals of different ranks: (○) respirable dust; (▽) float dust.

The results of the attempted correlation with coal rank are summarized graphically in Fig. 18. Electron spin resonance spectral intensities are given for the fusain (narrow) component of the spectrum and also for the sum of the fusain and vitrain (broad) components; the data are expressed in units of unpaired electrons per gram of sample. It should be noted that the ESR intensities depend upon the amount of coal in the dust and will be reduced by material such as rock dust, shale, or other mineral matter that is undoubtedly present in most of the dust samples. In order to present the data in a manner independent of the presence of such material, the ratio of fusain spins to vitrain spins was computed for each sample. Of the various ranks of coals included in this study, hvAb coals were represented by the largest number of respirable dust samples. The range of fusain spins (4×10^{16}–1.3×10^{18} g^{-1}) for the hvAb coal mine dusts overlaps considerably with the range for dusts from mines of other rank (Fig. 18). The results indicate that dusts from different mines having the same rank coal may have considerably different concentrations of fusain spins. Furthermore, Fig. 18 suggests that no correlation exists between the concentration of fusain spins and coal rank. In addition, no correlation of the ratio of fusain to vitrain spins with coal rank is apparent.

Treatment of the data also showed that dusts from different mines within

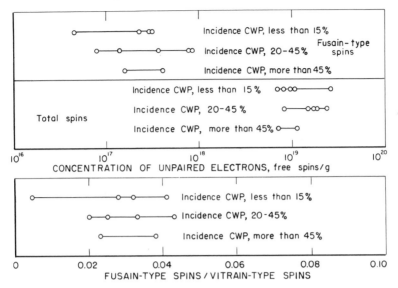

Fig. 19.   ESR intensity data for respirable dust samples from mines included in the inter-agency study of coal workers' pneumoconiosis.

the same coal seam can also exhibit a wide variety of spin concentrations; as was the case with coal rank, no apparent correlation of the ESR data with coal seam is evident (plot not shown). The results of attempted correlations of the ESR results with the prevalence of CWP are summarized in Fig. 19. It is easily seen that the ESR intensity data from mines having low incidence of CWP overlap considerably with data from mines having medium and high incidences. Again no correlation is apparent.

D. Respirable Dust from Pneumoconiosal Lung Tissue

During the course of this investigation, a section of lung tissue that had been removed from the body of a coal worker during postmortem examination was made available. Coal particles imbedded in the lung tissue were removed and examined by ESR. A spectrum of a quantity of dust particles removed from the lung tissue is shown in Fig. 20; the spectrum compares quite favorably with those of Fig. 16 except for the slight modulation broadening of the narrow component. Interference from the lung tissue from which the particles were removed was found to be negligible (Fig. 20b). For comparison, a spectrum of fusain from Hartshorne coal (Fig. 20c) is also shown. The position and shape of the fusain resonance is essentially identical with that of the narrow component found in the spectrum of the dust particles.

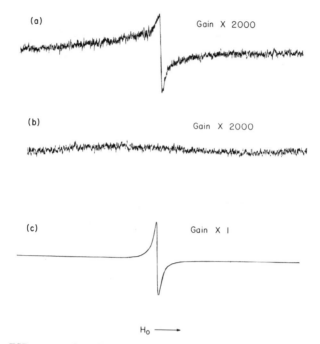

Fig. 20. ESR spectra of (a) dust particles removed from pneumoconiosal lung tissue; (b) lung tissue free of dust particles; (c) fusain from Hartshorne coal.

VI. SUMMARY

The material discussed in this chapter is illustrative of the many uses of magnetic resonance spectrometry in coal research. Electron spin resonance and nuclear magnetic resonance provide valuable data that have led and undoubtedly will continue to lead to a better understanding of coal structure and how this structure is altered during the natural processes involved in coalification and during man-made processes for converting coal into more suitable fuels. On the other hand, ESR studies of respiratory-size mine dust appear to give little additional insight into our understanding of coal workers' pneumoconiosis.

REFERENCES

Akhtar, S., Mazzocco, N. J., Weintraub, M., and Yavorsky, P. M. (1975). *Energy Commun.* **1**, 21–36.
ASTM (1981). "Standard Classification of Coals by Rank," ASTM Standard D 388-77, 1980 Annual Book of ASTM Standards, Part 26, Gaseous Fuels: Coal and Coke; Atmospheric Analysis, pp. 223–227. American Society for Testing and Materials, Philadelphia.

Atherton, N. M., Cranwell, P. A., Floyd, A. J., and Haworth, R. D. (1967). *Tetrahedron* **23**, 1653–1667.

Austen, D. E. G., Ingram, D. J. E., and Tapley, J. G. (1958). *Trans. Faraday Soc.* **54**, 400–408.

Austen, D. E. G., Ingram, D. J. E., Given, P. H., Binder, C. R., and Hill, L. W. (1966). *In* "Coal Science" (P. H. Given, ed.), pp. 344–362. Am. Chem. Soc., Washington, D.C.

Bartuska, V. J., Maciel, G. E., Schaefer, J., and Stejskal, E. O. (1976). *Prepr. Coal Chem. Workshop, 1976* pp. 220–228.

Bloch, F., Hansen, W. W., and Packard, M. (1946). *Phys. Rev.* **69**, 127.

Blois, M. S., Jr., Brown, H. W., and Maling, J. E. (1960). *Arch. Sci.* **13**, 243–255.

Blois, M. S., Jr., Brown, H. W., and Maling, J. E. (1961). *In* "Symposium of Free Radicals in Biological Systems" (M. S. Blois, Jr., ed.), pp. 121–131. Academic Press, New York.

Brown, J. K., and Ladner, W. R. (1960). *Fuel* **39**, 87–96.

Chakrabartty, S. K., and Kretschmer, H. O. (1974). *Fuel* **53**, 132–135.

DeRuiter, E., and Tschamler, H. (1958). *Brennst.-Chem.* **39**, 362–363.

Dormans, H. N. M., Huntjens, F. J., and van Krevelen, D. W. (1957). *Fuel* **36**, 321–339.

Dryden, I. G. C., and Griffith, M. (1955). *Fuel* **34**, S36–S47.

Elliot, M. (ed.) (1981). "Chemistry of Coal Utilization," 2nd Suppl. Vol. Wiley, New York (in press).

Elofson, R. M., and Schulz, K. F. (1970). *In* "Spectrometry of Fuels" (R. A. Friedel, ed.), pp. 202–207. Plenum, New York.

Freedman, R. W., and Sharkey, A. G., Jr. (1972). *Ann. N.Y. Acad. Sci.* **200**, 7–16.

Friedel, R. A. (1957). *Nature (London)* **179**, 1237–1238.

Friedel, R. A., Queiser, J. A., and Retcofsky, H. L. (1970). *J. Phys. Chem.* **74**, 908–912.

Friedman, S., Kaufman, M. L., Steiner, W. A., and Wender, I. (1961). *Fuel* **40**, 33–46.

Groom, P. S. (1969). *Fuel* **48**, 161–169.

Ingram, D. J. E., Tapley, J. G., Jackson, R., Bond, R. L., and Murnahgan, A. R. (1954). *Nature (London)* **174**, 797–798.

International Committee on Coal Petrography (1963). 2nd ed. CNRS, Paris.

Iyengar, M. S., Guha, S., Beri, M. L., and Lahiri, A. (1960). *Proc. Symp. Nat. Coal, 1959* pp. 206–214.

Jacobson, M., and Lamonica, J. A. (1969). *Tech. Prog. Rep.—U.S., Bur. Mines* **TPR 17**.

Karr, C., Jr. (1978). "Analytical Methods for Coal and Coal Products," Vol. 2. Academic Press, New York.

Kessler, T., Sharkey, A. G., Jr., and Friedel, R. A. (1971). *Tech. Prog. Rep.—U.S., Bur. Mines* **TPR 42**.

Kreulen, D. J. W. (1952). *Fuel* **31**, 462–467.

Kwan, C. L., and Yen, T. F. (1979). *Anal. Chem.* **51**, 1225–1229.

Ladner, W. R., and Wheatley, R. (1965). *Br. Coal Util. Res. Assoc., Mon. Bull.* **29**, 201–231.

McClure, D. S. (1952). *J. Chem. Phys.* **20**, 682–686.

Marshall, C. E. (1954). *Fuel* **33**, 134–144.

Mentser, M., O'Donnell, H. J., and Ergun, S. (1962). *Fuel* **41**, 153–161.

Miknis, F. P., Sullivan, M., Bartuska, V. J., and Maciel, G. E. (1981). *Org. Geochem.* **3**, 19–28.

Morgan, W. K. C., Burgess, D. B., Jacobson, G., O'Brien, R. J., Pendergrass, E. P., Reger, R. B., and Shoub, E. P. (1973). *Arch. Environ. Health* **27**, 221–226.

Newman, P. C., Pratt, L., and Richards, R. E. (1955). *Nature (London)* **175**, 645.

Petrakis, L., and Grandy, D. W. (1978). *Anal. Chem.* **50** (2), 303–308.

Pines, A., Gibby, M. G., and Waugh, J. S. (1972). *J. Chem. Phys.* **56**, 1776–1777.

Purcell, E. M., Torrey, H. C., and Pound, R. V. (1946). *Phys. Rev.* **69**, 37–38.

Reggel, L., Raymond, R., Steiner, W. A., Friedel, R. A., and Wender, I. (1961). *Fuel* **40**, 339–356.

Reggel, L., Wender, I., and Raymond, R. (1968). *Fuel* **47**, 373–389.

Retcofsky, H. L. (1977). *Appl. Spectrosc.* **31**, 116–120.

Retcofsky, H. L. (1978). *Proc. Sci. Probl. Coal Util., 1977,* pp. 79–97.

Retcofsky, H. L., and Friedel, R. A. (1970). *In* "Spectrometry of Fuels" (R. A. Friedel, ed.), pp. 70–89. Plenum, New York.

Retcofsky, H. L., Stark, J. M., and Friedel, R. A. (1967). *Chem. Ind. (London)* pp. 1327–1328.

Retcofsky, H. L., Stark, J. M., and Friedel, R. A. (1968). *Anal. Chem.* **40**, 1699–1704.

Retcofsky, H. L., Thompson, G. P., Raymond, R., and Friedel, R. A. (1975). *Fuel* **54**, 126–128.

Retcofsky, H. L., Thompson, G. P., Hough, M., and Friedel, R. A. (1978). *In* "Organic Chemistry of Coal" (J. W. Larsen, ed.), pp. 142–155. Am. Chem. Soc., Washington, D.C.

Retcofsky, H. L., Hough, M. R., Maguire, M. M., and Clarkson, R. B. (1981). *In* "Coal Structure" (M. L. Gorbaty and K. Ouchi, eds.), pp. 37–58. Am. Chem. Soc., Washington, D.C. (in press).

Roy, M. M. (1956). *Naturwissenschaften* **43**, 497–498.

Schmidt, U., Kabitzke, K., and Markau, K. (1964). *Chem. Ber.* **97**, 498–502.

Schopf, J. M. (1948). *Econ. Geol.* **43**, 207–225.

Schultz, J. L., Friedel, R. A., and Sharkey, A. G., Jr. (1972). *Tech. Prog. Rep.—U.S., Bur. Mines* **TPR61**.

Schuyer, J., and van Krevelen, D. W. (1954). *Fuel* **33**, 348–354.

Schweighardt, F. K., and Thames, B. M. (1978). *Anal. Chem.* **50**, 1381–1382.

Segal, B. G., Kaplan, M., and Fraenkel, G. K. (1965). *J. Chem. Phys.* **43**, 4191–4200.

Sharkey, A. G., Jr., Shultz, J. L., and Friedel, R. A. (1963). *Rep. Invest.—U.S., Bur. Mines* **RI–6318**.

Shine, H. J., Das, C. F., and Small, R. J. (1964). *J. Org. Chem.* **29**, 21–25.

Shultz, K. F., and Elofson, R. M. (1966). *Pap., 49th Conf., Can. Inst. Chem.*

Singer, L. S., and Cherry, A. R. (1969). *Summ. Pap., 9th Conf. Carbon* p. 49.

Singer, L. S., and Wagoner, G. (1963). *Proc. Conf. Carbon, 5th, 1961* Vol. 2, pp. 65–71.

Smidt, J., and van Krevelen, D. W. (1959). *Fuel* **38**, 355–368.

Steelink, C. (1966). *In* "Coal Science" (P. H. Given, ed.) pp. 80–90. Am. Chem. Soc., Washington, D.C.

Stone, A. J. (1963). *Mol. Phys.* **6**, 509–515.

Stone, A. J. (1964). *Mol. Phys.* **7**, 311–316.

Sullivan, P. D. (1973). *J. Phys. Chem.* **77**, 1853–1859.

Taki, T., Sogabe, T., Murphy, P. D., and Gerstein, B. C. (1980). *Fuel* (in press).

Terres, E., Dahne, H., Nandi, B., Scheidel, C., and Trappe, K. (1956). *Brennst.-Chem.* **37**, 269–277.

Toyoda, S., and Honda, H. (1966). *Carbon* **3**, 527–531.

Toyoda, S., Yamakawa, T., Kobayashi, K., and Yamada, Y. (1972). *Carbon* **10**, 646–647.

Tschamler, H., and DeRuiter, E. (1963). *In* "Chemistry of Coal Utilization" (H. H. Lowry, ed.), Suppl. Vol., pp. 35–118. Wiley, New York.

Uebersfeld, J., Etienne, A., and Combrisson, J. (1954). *Nature (London)* **174**, 614.

VanderHart, D. L., and Retcofsky, H. L. (1976a). *Fuel* **55**, 202–204.

VanderHart, D. L., and Retcofsky, H. L. (1976b). *Prepr. Coal Chem. Workshop, 1976.* pp. 202–218.

van Krevelen, D. W. (1961). "Coal," pp. 382–399. Elsevier, Amsterdam.

van Krevelen, D. W., Cherman, H. A. G., and Schuyer, J. (1959). *Fuel* **38**, 483–488.

Wagoner, G. (1960). *Proc. Conf. Carbon, 4th,* pp. 197–206. Pergamon, New York.

Yang, G. C., and Pohland, A. E. (1972). *J. Phys. Chem.* **76**, 1504–1505.

Yen, T. F., and Erdman, J. G. (1962). *Anal. Chem.* **34**, 694–700.

Yen, T. F., and Sprang, S. R. (1970). *Prepr. Pap.—Am. Chem. Soc., Div. Petrol. Chem.* **15**(3), A65–A76.

Zavoisky, E. (1945). *J. Phys. (Moscow)* **9**, 211–216.

Molecular Structure of Coal[1]

IEA Coal Research
London, England

I.	Introduction	84
	A. Structural Parameters	85
	B. The Literature to 1973	86
II.	Molecular Weight of Coal	87
	A. Depolymerization	87
	B. Reductive Alkylation	88
	C. Acylation	94
	D. Base/Alcohol Hydrolysis	95
	E. Nonchemical Methods	96
	F. Comments	97
III.	Carbon Aromaticity	98
	A. ^{13}C NMR Studies of Solid Coals	99
	B. Infrared Spectrometry	104
	C. Sodium Hypochlorite Oxidation	104
	D. Fluorination	105
	E. Differences between the Macerals	106
	F. Comments	107
IV.	Aromatic Ring Structures	108
	A. ^{13}C NMR Studies	108
	B. Oxidative Degradation	110
	C. NaOH/Alcohol Hydrolysis	113
	D. Solvation Studies	114
	E. Comments	115
V.	Aliphatic Structures	115
VI.	Low Molecular Weight Compounds	120
	A. Flash Heating	122
	B. Supercritical Gas Extracts	123

[1] This article was previously published by IEA Coal Research, London, as Report No. ICTIS/TR 08.

83

VII.   Free Radicals   126
VIII.   Functional Groups and Heteroatoms   132
        A. Oxygen   132
        B. Sulfur   134
        C. Nitrogen   136
IX.   Structural Changes with Rank   136
        A. Molecular Weight   136
        B. Carbon Aromaticity   137
        C. Aromatic Ring Structures   138
        D. Low Molecular Weight Compounds   139
        E. Free Radicals   139
        F. Anthracites   139
        G. Brown Coal/Lignite   141
X.   Macromolecular Skeletal Structures   142
        A. Polyamantane Structures   143
        B. Aromatic/Hydroaromatic Structures   146
        C. Molecular Sieve Structures   150
        D. Structural Models   152
        E. Comments   153
XI.   Conclusions   154
        References   155

I. INTRODUCTION

Coal is a sedimentary rock composed principally of two basic classes of materials: inorganic crystalline *minerals* and organic carbonaceous *macerals.* The macerals form the combustible part of the coal and are, in turn, divided into three maceral groups—vitrinite, exinite (or liptinite), and inertinite. The macerals are of different origin and differ in chemical composition. The best characterized macerals are the vitrinites. Vitrinites and coals are mixtures of macromolecules; in fact, these mixtures are very complex, and each macromolecule is considered to be composed of *structural units* or *constituent molecules,* which are those units of the macromolecules which survive mild chemical operations on coal.

The problem of ascertaining the molecular structure of the organic part of coal is that coal is not structurally dependent on a single molecule but on a complex mixture of molecules which varies according to the type of coal. Coal is chemically heterogeneous.

Coal is considered to have been formed from the peat deposits produced in swamps through the accumulation of plant substances. These substances contained distinctive organic chemical structures such as celluloses and lignins. The plant substances were subjected to various processes such as compaction by subsequent sediments and heating. These conditions of pressure and temperature account for the *coalification* of plant matter. The relative magnitude of severity of the conditions leads to increasing *coal rank*. For most purposes the various coals can be arranged in rank order:

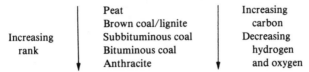

	Peat	Increasing
Increasing	Brown coal/lignite	carbon
rank	Subbituminous coal	Decreasing
	Bituminous coal	hydrogen
	Anthracite	and oxygen

Many physical and chemical properties vary with rank, but this variation is not a simple progression. Neavel (1979) has suggested that the concept of rank should be clarified and a scientific classification be developed. This would involve basing the classification on the fundamental properties of vitrinite with a knowledge of the element concentrations and molecular structure configuration. It is the latter requirement that will be examined in this review—the molecular structure configuration of coal. Aspects that will not be examined are the elemental analysis of coal—the percentage composition in terms of carbon, hydrogen, oxygen, and other elements in the organic part of the coal, and the pore structure, which may be considered to be a reflection of the underlying molecular structure.

A. Structural Parameters

The concept of coal structure is difficult to define as the macromolecules of coal are not composed of repeating monomeric units (as are, for example, the proteins). The problem is one of assigning a structure to a mixture. However, coal can be described in terms of structural parameters. These may include:

1. size distribution of the macromolecules,
2. degree of cross-linking,
3. type of cross-links,
4. carbon aromaticity,
5. average size of condensed aromatic units,
6. number of hydroxyl groups, and
7. scissile bridging structure.

The above list is neither exhaustive nor necessarily correct. It is based on the assumption that coal has a polymeric character—that it does, in fact, consist of macromolecules.

Much of the basic information concerning early ideas on the structure of coal can be found in van Krevelen's *Coal* (1961). It is not the intention of this review to repeat van Krevelen, but the importance of this book is inescapable (as a glance at the citation lists of many of the articles reviewed here will show). The conclusions of van Krevelen can be used as a useful starting point for this review:

1. Coal is a substance of high molecular weight and nonuniform structure.

2. Coal as a whole is strongly aromatic; its aromaticity increases more or less steadily with rank, reaching its maximum value of unity at about 94% carbon in vitrinites.

3. Coal has a polymeric structure.

4. The average structural unit in coals from the lignite stage to the low-volatile bituminous stage contains about 20 C atoms and about 4–5 rings. The aromaticity in this range increases from 0.70 to 0.90. In the anthracite stage the size of clusters, and hence the number of rings per cluster, increases rapidly.

These views on the structure of coal were reasonable in 1961 and may remain so, with modifications, today. But, rather than accepting these views completely, they should be considered as test questions against which we may assess more recent evidence from the literature. The earlier views were often based on assumptions and interpretations which, although largely reasonable, can be tested using modern techniques and reevaluated. Thus, our understanding of the structure of coal, and what we mean by it, can be reassessed.

B.   The Literature to 1973

The literature up to 1973 has been reviewed in several excellent publications: van Krevelen (1961) has been mentioned, but other good sources include Lowry's *Chemistry of Coal Utilization* (1963) [a new edition was published in 1981 (Elliott, 1981)], especially the chapter by Wender *et al.* in the new edition. Other useful sources include reviews by Ignasiak *et al.* (1975) and by the Information Division of Oak Ridge National Laboratory (Ensminger, 1977). A very comprehensive review was prepared by Tingey and Morrey (1973). They noted that if it is assumed that coal is completely made up of condensed benzene rings, the average number of rings for 70–83% carbon coal is two, increasing to between three and five for 90% carbon. These figures were based on total carbon rather than fixed carbon, but they represent a reduction of the four- to five-ring structure postulated by van Krevelen (1961), at least for lower rank coals.

Given the existence of the above-mentioned sources, this review will be limited largely to work performed from 1973 onwards. This date limit is also justified by the renaissance in the study of the fundamental chemistry of coal in recent years. Ample evidence for this comes from the annual workshops on coal chemistry, held in the United States since 1975, and the expansion of the journal *Fuel* from a quarterly to a monthly journal. Bartle *et al.* (1979d) have reviewed the analytical techniques employed in the analysis of coal-derived products. Their review covers the same period as that reported here and should be a useful source of information on the theory and practice of many of the analytical techniques referred to herein.

II. MOLECULAR WEIGHT OF COAL

There is a real problem in understanding what is meant by the "molecular weight" of coal. One can consider the molecular weight (MW) of each macromolecule in coal or the molecular weight of the fragments between cross-links in the coal structure (the constituent molecules). There is not a great deal in the published literature that is directly relevant to the molecular weight distribution in solid coal. This was remarked upon by Larsen in 1975, who also pointed out that it is hardly possible to claim any real understanding of a coal whose molecular weight distribution is not known. This point is particularly important since most studies of the molecular weight of coal provide the number average of coal fragments rather than the relative distribution of those fragments within the coal. Most methods of determining molecular weights of coals depend on converting the largely insoluble coal into soluble substances under mild conditions to ensure that the coal molecules remain unchanged as far as possible. The main problem is to avoid the cracking reactions which take place above 350°C. The following sections describe various methods of determining the molecular weight of coal.

A. Depolymerization

Depolymerization of coal is the technique which, it is claimed, solubilizes coal by cleaving methylene bridges in the coal. A review of the chemistry involved is provided by Larsen and Kuemmerle (1976) and Heredy (1979), but, simply put, the methylene chains joining aromatic groups can be cleaved at the ring and the methylene group can then alkylate another aromatic substrate:

$$Ar-(CH_2)_n-Ar' + PhOH \overset{HA}{\rightarrow} ArH + HOPh(CH_2)_nAr'$$
$$HOPh(CH_2)_nAr' + PhOH \overset{HA}{\rightarrow} HOPh(CH_2)_nPhOH + Ar'H$$

This technique is often called Heredy–Neuworth depolymerization and can be accomplished by using phenol as solvent and boron trifluoride as catalyst. Larsen *et al.* (1978) have used this technique to obtain the MW distributions of coals and note that the number average molecular weights (M_N) of the material soluble in pyridine after depolymerization can be affected by the presence of colloidal material. The reported M_N values have often been in the region of 400, but after centrifugation to remove the colloidal material M_N rises to 1000.

The molecular weight distribution of the true pyridine-soluble part of the depolymerized coal after centrifugation is shown in Table I. The coal was a Bruceton coal and the pyridine-soluble material accounts for 37% of the coal, 29% being soluble in benzene–ethanol. This paper has been followed by further results from Larsen and Choudhury (1979), which are similar to those in the results of 1978. It appears that low MW materials are not the chief products of the Heredy–Neuworth depolymerization of bituminous coals. The material extracted contains much colloidal and high MW material. Reliable conclusions based on instrumental measurements on these solutions are not possible. It is pointed out that, in the reaction conditions employed, AR—CH$_2$ linkages should be cleaved to yield predominantly low MW products; this is not observed. This leads to the possibilities that either these linkages are not present in coal to any great extent or that, if present, they are largely unreactive—either intrinsically or because access to them is restricted.

B. Reductive Alkylation

The chemistry of this reaction is also reviewed in the paper by Larsen and Kuemmerle (1976). The technique was developed by Sternberg et al. (1971; Sternberg and Delle Donne, 1974). Coal is treated with an alkali metal in tetrahydrofuran (THF) in the presence of a small amount of naphthalene, the

TABLE I

Molecular Weight Distribution of the True Pyridine
Solubles after Centrifugation[a]

wt. %	Cumulative wt. %	M_N
<0.5		
14.7	14.7	>3000
20.3	35.0	>3000
15.2	50.2	>3000
15.5	65.7	2960
12.8	78.5	2440
7.0	85.5	880
5.9	91.4	560
4.3	95.7	560
2.7	98.4	370
1.6	100.0	370

[a] From Larsen et al. (1978).

product being the "coal polyanion," which is readily alkylated by alkyl halides. Ether bridges are also cleaved under these conditions and, in addition, carbon–carbon bond cleavage (Alemany *et al.,* 1979) although it was originally expected that the reaction would not cleave coal molecules. Sternberg examined the MW distribution of an ethylated Pocahontas coal (90% C) using gel permeation chromatography (GPC) and vapor pressure osmometry (VPO) and obtained a MW distribution. The number average MW was 3300, and most of the coal had a MW close to this value. The distribution is fairly symmetrical and there is little material above 20,000 (Fig. 1).

Sun and Burk (1975) reported a method of obtaining the molecular weight distribution using a modified Sternberg alkylation procedure. They obtained a yield of 88% of the total carbon in the coal. A sample was separated on a prep-GPC column into fractions of narrow size range. In this sample 91% of the coal was found in 15 fractions out of a total of 40, and some of these had their molecular weights determined. Analysis by GPC showed that there was a narrow range of molecular sizes present in each fraction, and it was thus assumed that the weight average molecular weight of each fraction was close to the number average. Therefore, the molecular weights measured by vapor phase osmometry were considered to be representative of the true MW of each fraction. The results are presented in Fig. 2 and show a distribution ranging from about 200 to about 200,000. About 60% of the alkylated coal had a MW of less than 1000. These results for a Southern Illinois coal contrast sharply with Sternberg's; the distribution is unsymmetrical, and there is a greater pro-

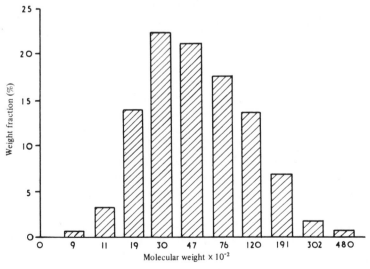

Fig. 1. Molecular weight distribution of ethylated Pocahontas coal (Sternberg *et al.,* 1971).

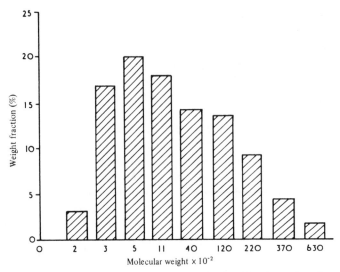

Fig. 2. Molecular weight distribution of ethylated Illinois coal (redrawn from Sun and Burk, 1975).

portion of low MW material in the Illinois coal (60%) and also of high MW material. Sun and Burk noted that only a small amount of chain cleavage appeared to have taken place.

Other studies on coals using the reductive alkylation technique include those by Ignasiak and Gawlak (1977) and Wachowska (1979). Ignasiak and Gawlak investigated high-rank vitrinite from medium-volatile bituminous Balmer coal from British Columbia. They noted that reactions involving alkali metals as electron donors attack carbon–oxygen bonds while leaving carbon–carbon bonds essentially intact. The vitrinite was reductively alkylated using potassium in THF in the presence of naphthalene and sodium in liquid ammonia, followed by the addition of an alkyl iodide. The number average molecular weight was determined by GPC and VPO of benzene and pyridine extracts. There were differences in the results from the two extracts, but these declined with increased size of the introduced alkyl groups—the differences being ascribed to associative forces between the coal clusters. They concluded that the macromolecular network of the vitrinite from the Balmer coal was composed of relatively small units, of number average MW of 670, interconnected by ether linkages. The clusters could be divided into 20 wt. % of number average MW below 500 and the remainder of relatively low uncorrected MW of about 2000. These figures do not agree with those of Sun and Burk (1975), but the coals were different in each case. The study of Balmer coal was followed by a study of sulfur-rich Rasa (Yugoslav) lignite, which indicated that about 80% of Rasa lignite is composed of uniform and relatively small clusters of MW 500, the clusters being connected by sulfur (and probably oxygen) bridges (Ignasiak *et al.*, 1978b).

Recently, studies have been carried out by Wachowska (1979), who studied (Polish?) coals of different rank ranging from 78.2% C to an anthracite with 92.6% C. The molecular weights of the benzene-extractable fractions were determined. The extractability of the alkylated coals increased markedly with increasing chain length of the alkyl substituent. The benzene extractability of the alkylated coals increased with rank (% C) for the bituminous coals but anthracite was only slightly extractable. These results are in agreement with those of Hodek *et al.* (1977), who observed a similar increase using olefins and HF/BF$_3$, as shown in Fig. 3, and Sternberg and Delle Donne (1974), who noted that anthracite cannot be successfully alkylated. Wachowska interprets these results in terms of a highly condensed structure for anthracite and the loss of lower MW material of the lower rank coals caused by washing with 70% ethanol. The number average MWs increase with increasing chain lengths of the alkyl substituent and the corresponding increasing extractability in benzene. This result disagrees with that of Ignasiak and Gawlak (1977), but experimental and procedural differences might account for this. Ignoring the results for anthracite, Wachowska's paper indicates that the MWs range from 500–800 for the lowest-rank coal to 1300–2000 for the highest rank coal (Table II). Wachowska notes that these figures characterize only the higher MW material and that the results indicate that the reduction process involves degradation of the coal by cleavage of ether links. Wachowska *et al.* (1979) used similar procedures on British Columbia Balmer 10 coal and on fusinite from an Illinois coal. The MWs of the benzene-soluble fractions were determined: the vitrinite from the Balmer coal had a MW of 1120, oxidized vitrinite

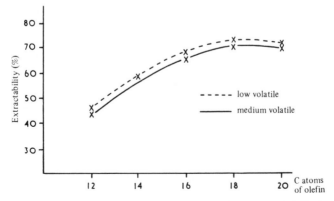

Fig. 3. Extractability of alkylated coal with increasing chain length of olefin (Hodek *et al.*, 1977).

TABLE II

Characteristics of Alkylation Products for Different Rank Coals[a]

C in orig. coal (%)	Alkyl introduced	Alkylated product			Benzene extract	
		Atomic H/C	No. of alkyl groups per 100 C atoms	Benzene extractability (wt. %)	No. av. mol. wt.[b] determined	No. av. mol. wt.[b] corrected
78.2	CH_3-	0.91	10.5	38.6	606	546
	C_2H_5-	0.99	9.6	44.5	655	549
	C_4H_9-	1.04	7.2	48.2	1028	801
	$C_8H_{17}-$	1.17	6.3	49.5	1210	810
81.1	CH_3-	0.94	9.6	34.4	786	712
	C_2H_5-	0.99	8.0	44.5	826	707
	C_4H_9-	1.04	6.2	52.4	1253	1003
	$C_8H_{17}-$	1.16	5.4	58.1	2094	1472
86.9	CH_3-	0.85	9.7	50.3	700	633
	C_2H_5-	0.92	8.9	69.9	1059	923
	C_4H_9-	0.99	7.1	73.6	1560	1208
	$C_8H_{17}-$	1.16	6.8	78.4	2084	1342
87.9	CH_3-	0.78	9.9	46.0	839	756
	C_2H_5-	0.85	8.6	57.9	1052	890
	C_4H_9-	1.00	8.7	60.3	1861	1366
	$C_8H_{17}-$	1.14	7.4	75.0	3186	1990
92.6	CH_3-	0.30	3.2	1.2	370	358
	C_2H_5-	0.29	1.8	1.5	445	430
	C_4H_9-	0.31	1.3	2.9	472	450
	$C_8H_{17}-$	0.39	1.3	5.4	500	455

[a] From Wachowska (1979).
[b] Number average molecular weights corrected for introduced alkyl groups in alkylated products.

1130, and fusinite 910. A volatility analysis was carried out to assess the MW distribution of the coal macerals. The results suggested that while the aromatic clusters increase in size from vitrinite to semifusinite to fusinite, there is a simultaneous increase in the percentage of low MW fragments.

Originally, it was thought that the alkylation reaction did not have a complex chemistry. Other than the cleavage of ether linkages it was assumed that the increase in solubility was largely due to the molecular "layers" being pushed apart by the bulky alkyl substituents weakening hydrogen bonds. This view led to conclusions that medium-rank coals are systems of plane aromatic molecules and that there are no three-dimensionally cross-linked structures

(Lazarov and Angelova, 1976; Shapiro and Al'terman, 1977). However, it seems that alkylation does not merely cleave ether links but that other chemistry is involved. Ignasiak *et al.* (1978a) point out that the alkylation reaction involves much more complex reaction paths (and side reactions) than originally suggested by Sternberg and that, when used to solubilize a low-rank coal (77.7% C), solvent molecules can be incorporated. The characteristics of the alkylation reaction are now being studied at the University of Chicago (Alemany *et al.*, 1978, 1979; L. M. Stock, private communication, 1979), and it is believed that the reaction involves cleavage of ether bonds, elimination reactions, carbon–carbon bond cleavage, and reduction of carbonyls to semi-quinones or ketyls, but that few rearrangements take place under the basic conditions.

It is considered that the structures of the anionic products are quite closely related to the structures of the molecular fragments in coal. However, the alkylation of the anionic products is complicated by competitive electron transfer reactions, which yield alkyl radicals. Larsen and Urban (1979) note that in some fractions of alkylated coal obtained by the Sternberg procedure there can be incorporation of naphthalene, tetrahydrofuran, and ethyl. However, it is still possible to separate cleanly very complex products containing about 50% of the coal if the first fractions are ignored. They warn that this holds for Illinois No. 6 coals and warn against extrapolating their results to coals of different rank. Ignasiak *et al.* (1979) announced a method of ethylating vitrinite in liquid ammonia with sodium and potassium amides. This converts a significant portion of the vitrinite into chloroform and pyridine soluble products. It is claimed that this method does not cause any detectable hydrolytic covalent bond cleavage. Niemann and Hombach (1979) also reported a method of solubilizing coal using solvated electrons prepared by dissolution of potassium in 1,2-dimethoxyethane (DME) and triethylene glycol dimethyl ether (triglyme). The reduced coal is then solubilized by hydrolyzing with a mild proton donor (isopropyl alcohol). As yet there are no reported MW determinations for the products obtained from these newer methods of solubilizing coal.

The molecular weights of the solubilized products are generally determined by gel permeation chromatography and vapor phase osmometry. Ignasiak *et al.* (1978a) also warn against possible shortcomings with these techniques. Curves determined by GPC and their interpretation depend on the *shapes* of the solute molecules, and therefore the calibration must be handled with care. It also means that the calibration is affected by the knowledge (or lack of it) or the assumptions made about the molecular structure of the coal fragments. Both GPC and VPO can be affected by the aggregation of the solute molecules and by colloid formation (e.g., Larsen *et al.*, 1978). Molecular association may be avoided by using larger chain alkyl groups, but the differences in the trends here between the work reported by Ignasiak and Gawlak (1977) and Wachowska (1979) remain. The reliability of the MWs obtained by these methods remains questionable as long as the reaction mechanism of reductive alkylation is inadequately characterized and if the MW determination techniques are carried out without checking for possible sources of error. However, there are signs that MW determinations using VPO may be reliably obtained. Chung *et al.* (1979) have reexamined VPO theory and practice and suggest that only the "rectilinear region" in the relationship between the osmometer readout and nonvolatile solute concentration should be used to calculate molecular weights. Tests with coal-derived liquids indicated that in the rectilinear region there was no dissociation or association affecting the results.

C. Acylation

Hodek and Koelling (1973) showed that the Friedel–Crafts acylation of coals resulted in an increase in extractability which was dependent on the chain length of the inserted acyl groups. The MWs of the soluble products were measured by vapor-phase osmometry and were only slightly higher than those reported for many of the alkylation experiments (ranging from 930 to 4400 for a coking coal and from 1900 to 5200 for a dry steam coal). The MWs increased with increasing chain length of the acyl substituent. Reservations concerning the validity of results obtained by VPO were expressed.

Larsen and Kuemmerle (1976) report unpublished work on the determination of the molecular weight of acylated coals, noting that the acylation results suggested that the MW values were much higher than those reported from alkylation studies. The MWs were of the order 10^5–10^6. Hombach *et al.* (1979) reported a study of the Friedel–Crafts acylation of coal that yielded 85% solubility, and they attempted to get values for the molecular weight distribution since they considered that the average MW values can be misleading owing to the overemphasis of certain (mainly low MW) fractions. The solution obtained by the acylation with long-chain carboxylic halides was separated using ultrafiltration. The main portion separated had an MW of 10^5–10^6, which was calculated assuming that the molecules were ellipsoid in shape. The MW distribution as represented by the residual fraction (mg/mm) plotted against the pore diameter of the filter (Fig. 4) exists as two independent sets of molecular weight distributions although the lower MW portion is relatively small. It is interesting that the molecular weights determined by this technique are significantly higher than those reported using alkylation, GPC, and VPO.

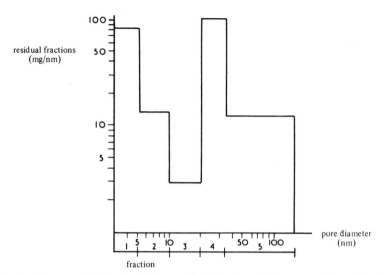

Fig. 4. Frequency distribution of the particle size in the pyridine extract of an acylated coal (Hombach *et al.,* 1979).

D. Base/Alcohol Hydrolysis

Makabe *et al.* (1978) obtained remarkable increases in the extractability of coal with pyridine by reacting coal with sodium hydroxide in ethanol. The extractability rises with the reaction temperature until 450°C, after which it decreases. Yield also increases with coal rank. The main reaction is considered to be hydrolysis, but there appears to be hydrogenation and cleavage of ether linkages among other reactions. The soluble fragments produced have been analyzed (Makabe and Ouchi, 1979; Ouchi *et al.,* 1979), and the molecular weight values have been obtained using VPO. It was found that the molecular weight of a coal "molecule" increases with rank, but that of the "unit structures" remains constant up to 82% C and then increases. The molecule appears to be the fragment obtained by the original hydrolysis, and the unit structures were calculated according to the method described by Kanda *et al.* (1978). The coal is assumed to have the structure

$$-Ar-(CH_2)_m-[AR-(CH_2)_m-]$$

where [ ] encloses the unit structure per aromatic nucleus. These are *average* units, not necessarily linked in one line or all identical. The MWs of the molecules and unit structures are shown in Fig. 5. However, as with most solubilization procedures, the degree of degradation may not be as slight as originally supposed: Winans *et al.* (1979) have examined the hydrolytic

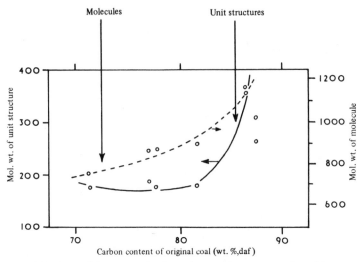

Fig. 5. Molecular weight per molecule and per unit structure of pyridine extracts of NaOH/alcohol treated coals (Makabe and Ouchi, 1979). Daf = dry ash-free.

solubilization of coal and conclude that the polycyclic aromatic rings are being reduced and oxygen heteroaromatics are destroyed. Other work on alcohol/base systems has been performed by Ross and Blessing (1979a,b). The main thrust of this research is to elucidate the reaction mechanism of base/alcohol solubilization of coal. The data obtained support a hydride transfer process. However, the MWs of the products obtained are in the 450–500 range.

E. Nonchemical Methods

Larsen and Kovac (1978) describe an interesting approach to the determination of the MW of coal. They consider that the time-dependent response of bituminous coals to a constant stress shows that they are cross-linked, three-dimensional molecules. Most of the strain is recoverable, and this cannot be explained by weak forces such as hydrogen bonds and van der Waals interactions. The statistical theory of polymer network elasticity and the Flory–Huggins theory were invoked to calculate the average molecular weight between cross-links. Some results obtained from solvent swelling studies of Japanese coals were analyzed on the basis that the solvent will be absorbed by the network until the swelling pressure is exactly balanced by the elastic force of the network. The results obtained using this technique, in the order of 1000 and 1500, are considered reasonable by the authors. Very high rank coals give very low MW values, which were expected by the authors on the grounds that the higher rank coals are likely to have more graphite-like, more highly cross-linked structures. The general MW increase with rank was ascribed to the presence of larger molecules assembled together to form the macromolecules or to fewer cross-links in the higher rank coals. The first explanation is preferred. Similar studies on Australian coals gave good agreement with the Japanese results. The results obtained from solvent swelling data were compared with other values calculated from the shear modulus of coal. However, the MW values calculated by this method are very low indeed; there is a 10^2–10^3 discrepancy when compared with the solvent swelling results. Two qualitative arguments are proposed to explain part of the discrepancy: the first based on *swollen network theory* and the second on *intermolecular obstructions*. It is pointed out, however, that the statistical model used to calculate the average MW per cross-link relies on assumptions which are probably not fulfilled in the coal structure. For example, the chains in the coal network will not obey Gaussian statistics if they are neither long nor flexible, as one might reasonably expect to be the case in coal. The chains are also unlikely to act independently. Certainly the application of an unmodified Flory–Huggins theory is not completely valid, for it is most unlikely that coal is a simple polymer network; its heterogeneity, mineral content (for example, clays might act as fillers in the polymer immobilizing any layers), and other factors indicate that a simple application of polymer theory is not justified. However, the approach is not presented as a refined method of determining MW—it is the first such described. It should not be too optimistic to expect that if some of the problems of the application of polymer theory to coals could be resolved then the method might provide a valuable adjunct to chemical methods of MW determination. Further work on the theory of non-Flory-model polymers could provide valuable information here; not only might the disturbingly low values obtained from shear modulus data be explained, but the "reasonable" values obtained from the solvent swelling data could be verified or shown to be only accidentally reasonable.

F. Comments

If the molecular weight of coal is regarded as a fundamental structural parameter of the whole coal, then it is certainly an elusive one. All the reported methods of obtaining the molecular weights of coal present values of the molecular weights of coal fragments between cross-links and not of individual or discrete coal "molecules." These fragments in many ways represent the smallest units of coal that can easily be obtained—this could be relevant when considering possible coal liquefaction processes. The chemical methods all involve breaking the cross-links in coal in some way, but there remains the problem of breaking the cross-links in a known way such that the molecular weights obtained can be related to particular coal fragments. It is difficult to disagree with Larsen and Given (1978) when they point out that reliance on number average molecular weights is dangerous and in some cases badly misleading and that techniques of obtaining weight average molecular weights should be developed (ultracentrifugation is mentioned as a possible technique). The effect of number averaging is to distort the MW by overemphasizing (usually) the low MW fragments; for example, a depolymerized coal might produce a soluble product with a number average weight between 350 and 750 when most of the products in fact have molecular weights above 2500: large numbers of small molecules dominate the number average. The use of phenol or similar solvents raises doubts that the MWs measured may be corrupted by the presence of phenol polymerization products. The problems of determining the molecular weight of coal are therefore at least twofold:

1. There is no method of breaking the cross-links in coal in a known way.

2. There are doubts about the reliability of the techniques used to measure MW, and there does not seem to be, as yet, a satisfactory method of determining weight average molecular weights.

In other words, there is a lack of certainty about what is being measured coupled with a similar lack of certainty that these measurements are correct.

III. CARBON AROMATICITY

Returning to the statements taken from van Krevelen in the introduction, it is apparent that the "classical" view of the aromaticity of coal is that coal as a whole is strongly aromatic. There are two important aromaticity values to be considered: the percentage of hydrogen atoms directly bonded to aromatic carbons and the (much more important) percentage of aromatic carbon atoms. The carbon aromaticity is usually written as f_a, and it is one of the more important structural parameters of coal. It is only in recent years that this carbon aromaticity could be measured by direct examination of the carbon in the solid coal. This has largely been due to advances in nuclear magnetic resonance (NMR) techniques. A very useful account of the application of these techniques to coal is provided by Retcofsky and Link (1978), but, simply, NMR enables the chemist to look at the molecular environment of certain atoms that are sensitive to NMR techniques, for example, ^1H and ^{13}C. Proton (^1H) NMR is largely limited to coal liquids and soluble products. The ^1H NMR data only yield carbon aromaticity values if certain assumptions are made about the aliphatic structures in coal (Brown and Ladner, 1960). These assumptions mean that the determination of the carbon aromaticity is not unambiguous. It is not unreasonable to have reservations concerning data on the structural parameters of coal that are obtained by employing assumptions about other structural parameters—however reasonable those assumptions might be. For this reason the development of improved ^{13}C NMR techniques, which examine the carbon atoms directly, is especially welcome.

A. ^{13}C NMR Studies of Solid Coals

In 1973 Retcofsky and Friedel reported a study of ^{13}C magnetic resonance in diamonds, coals, and graphite using "conventional" broad-line techniques with accompanying signal averaging of spectra. Two bituminous coals, an anthracite, and a metaanthracite were studied together with diamond and graphite. The results had no real quantitative significance, but the spectrum of Rhode Island metaanthracite (96% C) was qualitatively quite similar to that of graphite. For all the coals the position of maximum signal intensity was near the chemical shift value for liquid benzene. These findings indicated relatively high carbon aromaticity values for all the coals studied (83% C and higher) and also that the aromaticity increases with carbon content. The spectra obtained were very broad and lacked resolution, and only on the spectrum for the 83% C coal was there any indication of two components due to the presence of aromatic and aliphatic carbons. The results, however, demonstrated that ^{13}C NMR was potentially useful and gave valuable qualitative indications of the aromaticity, but there was no way of calculating reliable f_a values from them. In an aside at the beginning of the paper it is noted that Pines and his associates had recently demonstrated that *high-resolution* spectra of solids could be obtained (Pines *et al.*, 1972, 1973). This work confirmed that it was possible to resolve four functional types of carbon: simple aromatic, quaternary aromatic, oxygen-bonded aromatic, and simple aliphatic. The method, now usually called cross-polarization (CP) ^{13}C NMR, has several advantages over the conventional technique. The weak signals from the ^{13}C atoms (abundance 1.1%) are enhanced by transferring polarization from protons to carbons and then decoupling the C–H coupling by irradiation with high-power radiofrequency (rf). This enables the use of the short proton relaxation time to be used instead of the long ^{13}C relaxation time as the waiting time between experiments. These relaxation times are the decay times as excited nuclei return to their normal lower energy states. Cross-polarization ^{13}C NMR studies have been carried out on solid coals by, or on behalf of, several groups of workers.

VanderHart and Retcofsky (1976a,b) examined vitrains from Pittsburgh coal (82.6% C) and Pocahontas No. 4 coal (90.4% C). The spectra obtained are shown in Fig. 6 and consist of two overlapping resonances representing the aromatic and aliphatic carbons. The resonances are fairly broad due to aniso-

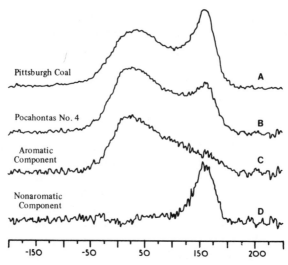

Fig. 6. CP spectra of two solid coals, A and B. C and D represent the deconvoluted spectra which were obtained by taking appropriate linear combinations of A and B (VanderHart and Retcofsky, 1976b).

tropic chemical shift effects in the solid sample and differences in the isotropic chemical shifts resulting from different chemical "environments" of the carbon atoms. The spectra were analyzed by assuming that they were two differently weighted sums of two line shape distributions. For the Pittsburgh coal the aromaticity value determined was about 75–77%. The Pocahontas coal had a higher aromaticity. The f_a values for other coals and coal-related materials were determined. The relevant spectra are shown in Fig. 7, and the f_a value of 92.8% C coal (anthracite) is nearly unity. It should be understood that the analysis of these spectra involved the use of assumptions—the most significant being that the division of the spectra into aromatic and aliphatic parts can be achieved by assuming that the aromatic component is very nearly a linearly decreasing function through the region of overlap between the two components and that there are only two components to be considered. These assumptions are recognized as such in the papers, but it is thought that the derived f_a values are still reasonable since studies with model compounds and on other carbonaceous materials showed that for a high C:O ratio (the coals studied had C:O > 13:1) the aromatic/nonaromatic distinction was valid. The CP method also makes assumptions about proton–proton and carbon–proton distances: if there are extensive regions of nonprotonated material, there would be no ^{13}C polarization and hence no signal contribution. This was tested for by comparing the CP spectra with conventional spectra generated by ^{13}C relaxation. The spectra obtained from Pocahontas No. 4 coal showed

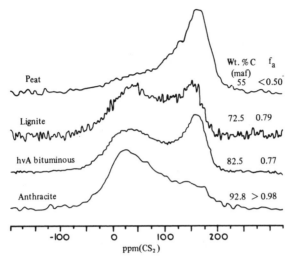

Fig. 7. CP spectra of coal and fuel related materials (VanderHart and Retcofsky, 1976b). Maf = moisture ash-free.

essentially no differences between the two. Therefore, these studies indicated that bituminous and anthracitic coals are indeed highly aromatic substances and that CP ^{13}C NMR could be used with confidence to determine the carbon aromaticity.

Pines and Wemmer (1978) report that high-resolution NMR is practical for solid materials and that the method developed by Pines *et al.* (1973) had reached the stage where analysis of some functional groups in coal is possible. The computer simulation of a coal sample spectrum, which is generated by considering the line shapes due to aliphatic carbon (26%), ether carbon (13%), aromatic carbon (53%), and polycondensed aromatic carbon (8%), is illustrated in Fig. 8. This method has been applied to samples provided by Mobil (Whitehurst, 1978; Whitehurst *et al.,* 1977). Whitehurst *et al.* (1977) provided samples and the CP NMR spectra were obtained by A. Pines at University of California, Berkeley. It is significant that the instrument was calibrated using a series of model compounds. The aromaticity values obtained were quite accurate. The typical error limits for coals were judged to be about ±5%. Typical values for the aromatic carbon content are given in Table III. The range of f_a values from about 40–50% for subbituminous coals to over 90% for anthracitic coals is noteworthy. Whitehurst (1978) points out that this range of values means that the fairly common belief that coals are highly aromatic must be modified. The Mobil studies also showed that there was relatively little correlation between the f_a values and the H/C ratio.

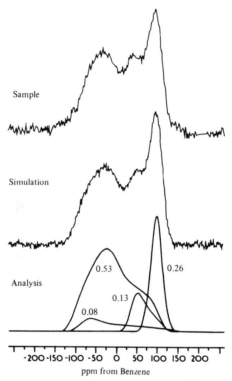

Fig. 8. Comparison of observed and computer simulated CP ^{13}C NMR spectra of the "black acids" obtained from coal + NaOCl (see Mayo and Kirshen, 1979) (Pines and Wemmer, 1978).

TABLE III

Aromatic Content of Coals[a]

Coal	% Aromatic C[b] (%)
Wyodak subbituminous	42
Ill. No. 6 hvA bituminous	61
Pittsburgh No. 8 hvC bituminous	66
Pennsylvania anthracite	100

[a] From Whitehurst *et al.* (1977).
[b] ± 5%.

Bartuska *et al.* (1976, 1977) have taken the CP technique one step further by using *magic-angle spinning,* which is a method of reducing the line broadening due to dipolar $^{13}C/^1H$ interactions. The magnitude of this broadening is governed by the factor $1-3 \cos^2 \theta$, which vanishes to zero when θ is 54.7°. Therefore, if the solid sample is rotated rapidly about an axis which is at 54.7° with the direction of the static magnetic field, there is an increase in resolution of the spectra due to the elimination of the anisotropic chemical shift effects. Magic-angle spinning does *not* remove dispersion of isotropic chemical shifts. Magic-angle spinning studies on anthracite and lignite gave f_a values of 99% (min) and 72%, respectively. It was noted that if one analyzes the spectra without spinning assuming an arbitrary straight-line continuation of the distinguishable part of the aromatic region, then the aromaticity values are reduced to 95% and 65%, respectively. Nuclear magnetic resonance spectra obtained for anthracite and lignite both with and without spinning are shown in Fig. 9. Spinning, it was suggested, is especially valuable for coals with high aromaticity but less important when the aliphatic carbon content is high. The possibility of errors introduced by not detecting carbons isolated from protons, that is, proton-free regions of about 10 nm in diameter, was noted but was not considered likely. Work by this group is still continuing and is regularly reported (see Bartuska *et al.,* 1978a; Maciel *et al.,* 1979; Miknis *et al.,* 1979). The 1979 papers present f_a values for ten solid coals, including a Texas lignite, three Wyoming subbituminous coals, an Illinois bituminous, two eastern United States bituminous coals, and an anthracite. The f_a values obtained ranged from 59% for the Wyodak coal to 95% for anthracite.

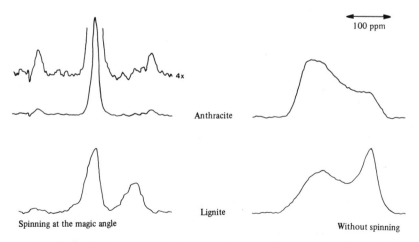

Fig. 9. CP ^{13}C NMR spectra of coals taken with and without spinning (Bartuska *et al.,* 1976).

Metallurgical coke gave a value of 100%. A plot of the f_a values against the atomic H/C ratio was obtained. This showed a fairly smooth curve in contrast to the results reported by Whitehurst (1978).

Recently Zilm et al. (1979) have reported the use of a high-speed magic-angle spinner in the study of the CP ^{13}C nmr spectra of solid coals, but no f_a values were reported.

All the investigators appear to be aware of the potential sources of error in CP NMR (e.g., VanderHart and Retcofsky, 1976b). Thus, f_a might be overestimated if the aromatic carbon band includes olefinic carbons and carbonyl carbons and underestimated if there is incomplete polarization of all the carbons. However, the reservation that extended regions of non-protonated carbons cannot be "seen" by CP NMR suggests that the applicability of the technique to very high-rank coals might be questionable. If extensive areas of graphite-like carbons exist in the very high-rank coals it is possibly immaterial that the results may not reflect all the carbons because there will be so few nonaromatic carbons present to affect the derived aromaticity values. VanderHart and Retcofsky (1976b) were unable to obtain a spectrum for semiconducting anthracite since it detuned the rf coil of their NMR spectrometer.

B. Infrared Spectrometry

Retcofsky (1977) reported studies in which infrared (IR) spectrometry was used to complement NMR spectral data for coal extracts in order to estimate the aromaticity of whole coals. Proton NMR data from solvent extracts of coal were used to calibrate the out-of-plane hydrogen bending vibrations per aromatic hydrogen. This was compared with IR spectra of vitrain and with pyridine and carbon disulfide extracts. The pyridine extract was chosen for calibration purposes since it represented a larger part of the whole coals. The aromaticities of the whole coals were then estimated using the aromaticity equation of Brown and Ladner (1960). The values obtained range from about 0.66 for 76% C to 0.70 for 86% C. It is suggested that these values are probably too low, for the calibration curves are only representative of part of the coal. This method of estimating the aromaticity of solid coals has been overtaken by the development of CP NMR, but the values obtained provide a reasonable cross check of the CP NMR values.

C. Sodium Hypochlorite Oxidation

Chakrabartty and Kretschmer (1974) presented some surprising f_a values obtained from studies on the oxidation of coal by sodium hypochlorite. These studies will be examined in greater detail later as they are of more interest to the understanding of the aliphatic structures present in coal and the carbon skeleton. However, it should be noted that this work was severely criticized by several other investigators (e.g., Mayo, 1975; Aczel *et al.*, 1975, 1979; Ghosh *et al.*, 1975; Landolt, 1975). The basic premise was that hypohalite is a highly selective reagent which discriminates between aromatic and aliphatic carbon such that no aromatic carbon can be oxidized. It was found that, under the reaction conditions used, hypohalite oxidation is substantially independent of coal rank up to about 90% C (anthracites were not oxidized). Up to 90% C the ratio of aromatic to aliphatic carbon was calculated to be approximately 2:3, indicating an aromaticity of about 40%—a very low figure indeed. The abundance of sp^3 (saturated aliphatic) carbons is shown in Table IV. It was argued that hitherto the assignment of a largely aromatic structure was based on presumption and that the hypohalite oxidation results challenged this "classical" view of a predominately aromatic structure. Recent studies by Mayo and Kirshen (1978b, 1979) under more carefully controlled conditions (pH controlled at 11 with repeated additions of small quantities of hypochlorite to an extremely finely divided coal) have resulted in a major product, which is a mixture of black acids that contain a high proportion of aliphatic carbon. The original studies had used an excess of hypochlorite and the pH was not controlled. The higher MW material in the coal, representing 15-20%, was not characterized in these studies, making the f_a values obtained open to doubt. However, Chakrabartty and Kretschmer argued that they could not be far different from the "real" values.

There are still uncertainties about the mechanism of the coal/hypochlorite reaction (Chakrabartty, 1978) and until those uncertainties are resolved the f_a values obtained by this method should be viewed with caution. Certainly the original suggestion of 40% aromaticity for all coals up to 99% C is not in agreement with the results obtained by CP NMR.

D. Fluorination

Huston and others (1976) report studies on the reaction of elemental fluorine with bituminous coal at room temperature, followed by a mass spectrometry investigation of the fluorinated coal. Fluorination can be considered

TABLE IV

Abundance of sp^3 Carbon in Coal[a]

C in coal (%)	76.1	80.5	83.1	85.0	86.4	90.2
Csp^3/C total (%)	68	70	68	63	71	59

[a] From Chakrabartty and Kretschmer (1974).

as equivalent to hydrogenation under milder conditions of temperature and pressure. The reaction is, however, strongly exothermic and the pressure has to be carefully controlled to prevent ignition. The fluorine reacts with aromatic compounds as well as aliphatic ones but the aromatic compounds react by addition as well as replacement. Hydrogen replacement results in the formation of HF but addition results in the loss of fluorine. Therefore, if the reaction proceeds to completion, gravimetric methods could be used to determine f_a . For example, consider cyclohexane and benzene with fluorine:

$$C_6H_{12} + 12F_2 \rightarrow C_6F_{12} + 12HF$$
$$C_6H_6 + 9F_2 \rightarrow C_6F_{12} + 6HF$$

It was found that when coal was fluorinated, there was a net loss of fluorine greater than could be accounted for by hydrogen replacement alone. Fluorination of adamantane and phenanthrene was also carried out, and a procedure based on the reaction stoichiometry was used to check the purely gravimetric calculations. The average aromaticity of the bituminous coal (77.9% C) examined was 68.6%. It was suggested that this figure should be considered a minimum because the fluorine may not have fully saturated all of the coal's aromaticity. Errors could also be introduced by the presence of nonprotonated carbons, which may not be fluorinated and hence will not be "seen." The f_a values may be affected by olefinic carbons being counted as aromatic, thus increasing the apparent f_a, or replaceable hydrogen in functional groups decreasing f_a (Studier *et al.,* 1978). The fluorination method is more akin to proton NMR than ^{13}C NMR since it is the hydrogen in coal which is directly studied rather than the carbon. However, the stoichiometric studies showed that the method gives at least a qualitatively correct impression of the aromaticity because the coal behaved much more like phenanthrene than adamantane. The conclusion reached was that, for the coal studied, 70% of the carbon was aromatic, 10% was nonaromatic carbon in hydroaromatic nuclei, and 20% was aliphatic carbon in the interconnecting "web." The value of 80% cyclic carbon was inferred from mass spectrometry studies.

E. Differences between the Macerals

Retcofsky and VanderHart (1978) have studied CP ^{13}C NMR spectra of the individual macerals in coal. The NMR spectra obtained are shown in Fig. 10. The macerals had been separated from a West Virginia hvA bituminous coal (85.8% C). The calculated f_a values were vitrinite 85%, exinite 66%, micrinite 85%, and fusinite 93–96%. These results were in agreement with earlier work, which had indicated that aromaticity increases in the order: exinite < vitrinite/micrinite < fusinite. Exinite is thought to be derived from the nonwoody tissue of plants, whereas the others are from the woody and cortical tissues—

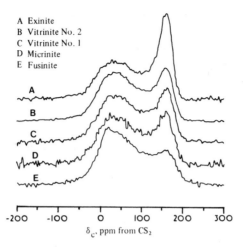

A Exinite
B Vitrinite No. 2
C Vitrinite No. 1
D Micrinite
E Fusinite

δ_C, ppm from CS_2

Fig. 10. ^{13}C–^{1}H cross-polarization nuclear magnetic resonance spectra of macerals from Hernshaw hvAb coal (Retcofsky and VanderHart, 1978).

this may offer some explanation for the lower value of exinite aromaticity. The higher aromaticity of fusinite is qualitatively substantiated by the results of Wachowska *et al.* (1979) referred to in Section II,B on molecular weight. These results suggested that fusinite possesses larger aromatic clusters than vitrinite.

F. Comments

The measurement of the carbon aromaticity of solid coals appears to be in a healthier state than the determination of molecular weights. However, some of these results do not bear out van Krevelen's statement that the aromaticity in the range of lignite to low-volatile bituminous increases from 70 to 90%. It would appear that whereas the higher value may be correct, the lower figure must be lowered still further to account for the low-rank coals. The range of aromaticity values obtained by Whitehurst *et al.* (1977) and by Maciel *et al.* (1979) force agreement, if these results are accepted, that the aromaticity of coal varies with rank over quite a wide range of values.

Certainly there are possible sources of error in the CP NMR results, but, in general, these are understood and can be tested for, for example, by comparing CP and non-CP spectra. The main doubt concerning the applicability of CP NMR is that the model studies used for calibration may not be directly applicable to a material as complex and difficult as coal. There may also be problems associated with the presence of free radicals in coal and hydrocarbons which give practically no ^{13}C NMR signal (L. L. Anderson, private communication, 1979).

IV. AROMATIC RING STRUCTURES

In Section III on carbon aromaticity it was shown that the limits for this value may be as low as 40% for low-rank coals. This figure is still representative of a large fraction of the carbon in coal, and the generally accepted higher f_a values for higher rank coals demonstrate that the bulk of carbon in coal is accounted for by the presence of aromatic ring structures. These structures can range from simple benzene rings to large condensed aromatic systems (and perhaps, though it is unlikely, larger single aromatic rings). One of the commonest conceptions of the overall structure is that of various condensed aromatic units connected by linking groups. The evidence for the f_a values considered with the elemental carbon/hydrogen ratio leads to the conclusion that such structures are indeed present in coals. It was noted in the introduction that Tingey and Morrey (1973) had postulated that for 70–83% C coal the average number of condensed benzene rings in the coal was 2, increasing to between 3 and 5 for 90% C, and at 95% C the number could exceed 40 and the rings would be tightly packed. The assumptions behind these figures were that coal is completely made up of condensed benzene rings and that all the peripheral carbon atoms are bonded to hydrogen. These are obvious oversimplifications, given the f_a values obtained for coal. These same f_a data provide a valuable source of evidence for estimating the degree of condensation of the aromatic rings in coal.

A. ^{13}C NMR Studies

Retcofsky and VanderHart (1978) used the f_a data obtained from the CP NMR spectra of four macerals to calculate the sizes of the aromatic ring systems. The assumptions involved were that the nonaromatic carbons should be predominately methylene ($-CH_2-$), and that the oxygen and half the nonaromatic carbons are directly bonded to aromatic rings. The atomic H/C ratio for the hypothetical unsubstituted aromatic nuclei (H_{aru}/C_{ar}) can be calculated by using the equation

$$H_{aru}/C_{ar} = (H - 3C_{ali}/2 + O)/C_{ar}$$

where the subscripts "ar" and "ali" denote atoms in aromatic and aliphatic groupings, respectively. The value obtained for vitrinite was 0.68, indicating a mean structural unit consisting of three to four ring condensed aromatic systems. The authors believe that the assumptions used are reasonable; they are based on the assumptions used in the interpretation of proton f_a values for various coal products (Brown and Ladner, 1960), and these derived f_a values had been found to be in accordance with the values directly obtained from ^{13}C NMR (Retcofsky et al., 1977).

TABLE V

Structural Parameters for Iowa and Virginia Vitrains[a]

Sample	C (%)	H (%)	C_{fa} (%)	H_{fa} (%)	H_{ar}/C_{ar}
Virginia "Pocahontas No. 4" vitrain	90.3	4.43	86	77	0.53
Iowa "Star" vitrain	77.0	6.04	71	31	0.40

[a] From Gerstein et al. (1979).

More recently, Gerstein et al. (1979), using high-resolution ^{13}C and ^{1}H solid-state NMR, have provided possible values for the average aromatic ring size in vitrains from Iowa (77.0% C) and Virginia (90.3% C) coals. They obtained f_a values for both carbon and hydrogen from the NMR data and calculated the H_{ar}/C_{ar} values. The results are given in Table V. It is pointed out that the lower H_{ar}/C_{ar} for the lower rank coal may be surprising, since one would expect larger ring systems in the older coal. However, Fig. 11 is illustrative of the average aromatic ring size as a function of H_{ar}/C_{ar} and *connectivity*. The connectivity is important because the average ring size depends greatly on the number of side chains connected to the ring: on the degree of ring substitution. The ring sizes preferred by the authors were that the Virginia vitrain has an average ring size of no greater than three and the Iowa vitrain has a size no greater than two. These inferences are strongly biased by the belief that the average ring size in higher rank coals should be greater than that in lower rank coals. Therefore, there is agreement with van Krevelen that the structural units increase in size with rank, but the values chosen are lower than the four to five rings suggested by earlier studies.

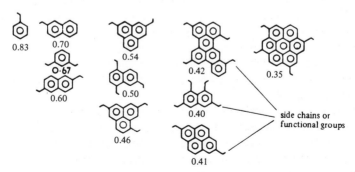

Fig. 11. Average aromatic ring size as a function of H_{ar}/C_{ar} and connectivity (Gerstein et al., 1979).

B. Oxidative Degradation

The number of aromatic rings in the unit structures in coal can be estimated if it is possible to degrade the aliphatic portion of coal while leaving the aromatic structure intact. Oxidation of coal is one approach which has been investigated, but unfortunately many common oxidizing agents are fairly drastic and their use results in the degradation of both aromatic and aliphatic components such that the only identifiable aromatic compounds are benzene carboxylic acids.

The sodium hypochlorite studies of Chakrabartty and Kretschmer (1972, 1974) and Chakrabartty and Berkowitz (1974) yielded these derivatives, and, believing that NaOCl was a selective oxidizing agent, it was concluded that coal is largely aliphatic and that benzene rings constituted the majority of the aromatic portion. However, the nonspecificity of NaOCl under the conditions used was demonstrated (e.g., Mayo, 1975; Ghosh et al., 1975; Landolt, 1975), and it seems almost certain that aromatic rings were cleaved generating carbon dioxide. An oxidizing agent that appears to be reasonably selective is sodium dichromate $(Na_2Cr_2O_7)$; this oxidizes substituted polycyclic aromatic compounds with a minimum of ring degradation.

1. Sodium Dichromate Oxidation

Hayatsu et al. (1975) used sodium dichromate to oxidize coal selectively. The coal was heated at 250°C in 0.4 M sodium dichromate and the aromatic acids produced were isolated. The Illinois bituminous coal used was completely solubilized and the resulting acids weighed 53% of the original weight of coal. The methyl esters of the acids were identified by time-of-flight mass spectrometry. Thirty-five aromatic acids were positively identified, including several heterocyclic compounds containing oxygen and sulfur in the rings. Several units containing nitrogen in the rings were isolated from the products of photochemical oxidative degradation of coal, but these units could not be isolated from the dichromate product due to some experimental difficulties. The observed units were very complex; there were up to five carboxylic groups on benzene and up to four on polycyclic aromatics. These groups are considered to be the residues of the aliphatic or alicyclic linkages between the aromatic units. Some of the aromatic units identified by mass spectrometry are shown in Fig. 12.

Chakrabartty and Berkowitz (1976a) claimed that the $Na_2Cr_2O_7$ results were not indicative of the aromatic species indigenous to coal and that, on the contrary, 65% of the total carbon existed in nonaromatic species. This letter to Nature is immediately followed by the reply by Hayatsu et al. (1976) countering this claim by pointing out that the aromatic units obtained did not

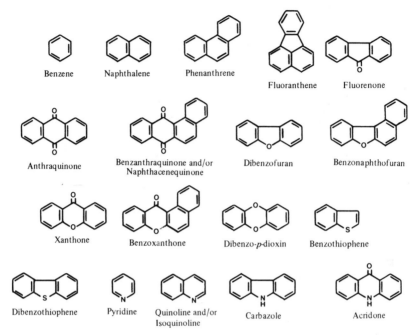

Fig. 12. Aromatic units indigenous to coal (Hayatsu *et al.*, 1975). Reprinted by permission from *Nature (London)* **257,** 379 (1975). Copyright © 1975 Macmillan Journals Ltd.

represent the full aromatic content of the coal and that the results were perfectly consistent with a largely aromatic unit structure.

Further work (Hayatsu *et al.*, 1978a,c; Studier *et al.*, 1978) has compared the structures of lignite, bituminous coal, and anthracite. No polynuclear aromatic compounds were obtained from lignite, providing further evidence against the possibility that $Na_2Cr_2O_7$ oxidation resulted in the pyrolysis formation of polynuclear aromatic compounds. The compounds were again identified using mass spectrometry techniques. Thirty-two aromatic ring systems were identified, 20 of them heteroaromatic. The relative abundances of these species in the three coals studied are shown in Fig. 13. The results show that the ring sizes increase from lignite through bituminous coal to anthracite. Solvent-refined coal (SRC) (Hayatsu *et al.*, 1978c) has been compared with the feed coal from which it was derived—the major finding being that the SRC has a greater degree of aromatic condensation. Gartsman *et al.* (1979) have also examined the benzene polycarboxylic acids in the products of oxidation of Asian coals and have found similar ring structures, some of which are heteroaromatic.

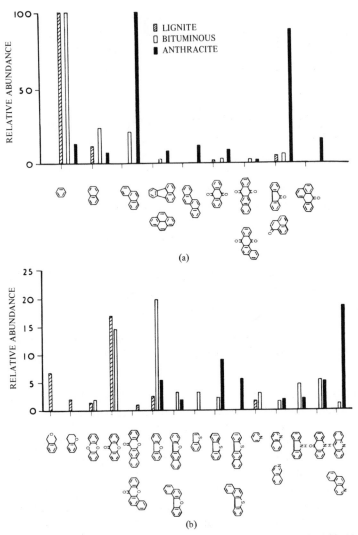

Fig. 13.   Relative abundances of aromatic and heteroaromatic units produced by $Na_2Cr_2O_7$ oxidation of lignite, bituminous coal, and anthracite. Identification based on mass spectrometry of the methyl esters of carboxylic acids and as nonacidic compounds (Hayatsu *et al.*, 1978a,c). (a) Aromatic units (from left to right): benzene, naphthalene, phenanthrene, fluoranthene/pyrene, chrysene, anthraquinone, naphthacenequinone/benzanthraquinone, fluorenone/phenalenone, benzanthrone. (b) Heteroaromatic units (from left to right): phthalan, chroman, dibenzo-*p*-dioxin, xanthone, benzoxanthones, dibenzofuran, benzonaphthofurans, benzothiophene, dibenzothiophene, benzonaphthothiophenes, pyridine, quinoline/isoquinoline, carbazole, acridone, acridine/benzoquinoline.

2. Photochemical Oxidation

Hayatsu *et al.* (1975, 1978a, also Studier *et al.*, 1978) have studied the photochemical oxidative degradation of coals. Almost no photooxidation was observed for anthracite and lignite, and bituminous coals were oxidized in yields of 25 and 30%, respectively. Several aromatic acids were identified and also aliphatic carboxylic acids. The main interest of this work lies in the identification of nitrogen heterocyclic compounds, which were not detected in the $Na_2Cr_2O_7$ studies. These included pyridine, quinoline and/or iso-quinoline, and carbazole (which is stable to photochemical oxidation although sensitive to oxidizing agents).

3. Performic Acid Oxidation

Raj (1976) reported the results of oxidizing coal in aqueous performic acid at 50°C, followed by extraction by sodium hydroxide. The NaOH-soluble fraction (the bulk of the coal that, petrographic examination showed, came mainly from the vitrinite) was reduced to phenols and phenolic acids. Ether extracts of these were examined by gas–liquid chromatography. The ether extract yields were fairly low—representing between 10 and 30% of the original coal, with maximum yield for the youngest coals. Fourier transform IR spectrometry studies of the ether-insoluble product indicated that the ether-soluble and -insoluble products were similar in structure, suggesting that the insoluble material was insufficiently depolymerized. The compounds identified were mainly benzene derivatives, and it was concluded that bicyclic and polycyclic systems were rare. However, it was admitted that this conclusion was open to question. The mechanism of the reaction of coal with performic acid is not clear, and the mechanism of the reduction of the NaOH-soluble material using sodium amalgam is also unknown. Given these doubts it would seem that oxidation with performic acid does not add much to current knowledge of the ring structure of coal.

C. NaOH/Alcohol Hydrolysis

On the basis of studies of NaOH/alcohol hydrolysis of coal (Makabe and Ouchi, 1979; Ouchi *et al.,* 1979) already reported in Section II, the number of rings per unit structure was calculated using conventions developed by Kanda *et al.* (1978). The assumptions behind this method are largely that the aromatic clusters are connected by methylene ($-CH_2-$) groups. The calculation method allows the estimation of (among other parameters)

1. R_t the total ring number,
2. R_a the aromatic ring number,
3. R_n the naphthenic ring number.

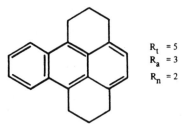

$R_t = 5$
$R_a = 3$
$R_n = 2$

Fig. 14. Ring numbers for a possible unit structure (Kanda *et al.*, 1978).

An example of the values for a possible unit structure is shown in Fig. 14. For the coals studied by Ouchi and others it was found that for each unit structure

1. % C ranged from 71.5 to 87.4,
2. R_t ranged from 1.9 to 5.7,
3. R_a ranged from 1.1 to 4.0,
4. R_n ranged from 0.8 to 1.8.

The results indicate that the lower rank coals have tetralin-like unit structures, and that 80–85% C coals have two to three aromatic rings with 0.5 naphthenic rings. The Shin–Yūbari coal (86.7% C) has a four-ring aromatic structure with two naphthenic rings attached. The calculation method used depended on a knowledge of the carbon aromaticity, which was estimated from ^{1}H NMR data.

D. Solvation Studies

Ruberto *et al.* (1977a,b) have studied coal structure indirectly by "solvating" subbituminous coal in anthracene oils and hydrophenanthrene solvents. The authors define solvation of coal as thermal bond breaking at moderate temperatures in a solvent capable of bond stabilization (formation), usually by the donation of hydrogen. The coal studied was liquefied by solvation under hydrogen at 426°C. The liquids were examined using various techniques to determine their physical and chemical properties. The catalytic solvation of coal with phenanthrene resulted in the production of three-ring aromatic compounds in a very narrow molecular weight range (175–200). These were detected by low-voltage mass spectrometry. When solvation was carried out in the absence of hydrogen (under nitrogen) and catalyst, the yield of one-ring aromatics decreased. The two- and three-ring aromatic compounds obtained could not arise from classical hydrocracking in the absence of hydrogen and catalyst. Therefore, it was suggested that the likely mechanism was hydrogenolysis of aryl oxygen, nitrogen, and short aliphatic linkages giving rise to the simplest "monomers" in the original coal. These monomers are rarely larger than four-ring (pyrene-type) structures in sub-bituminous coal and three condensed aromatic ring structures predominate. Other studies (e.g., Fischer *et al.*, 1978) on extracts of high-volatile bituminous coal prepared by treatment under hydrogen at 450°C show the presence of dihydrophenanthrene units. However, it must be stressed that solvation studies and any degradation procedure at elevated temperatures give only indirect evidence for the existence (or otherwise) of the two- and three-ring aromatic structures in the parent coal. It is likely that the ring structures found by these studies might be too high if it is accepted that the processes used to solvate the coal result in products of higher aromaticity than the parent coal—as indicated by the Studier *et al.* (1978) comparison of SRC and coal and by the Whitehurst *et al.* (1977) finding that concentration of multiple aromatic rings increases with extended reactions of coal products.

E. Comments

The understanding of the aromatic ring structure in coals is still incomplete, but there appears to be a certain amount of progress. Most of the recent studies have indicated that the condensed units in coal are fairly small—often not more than four rings. These values can be obtained by deducing ring structures from elemental composition and aromaticity data or by degrading coal and identifying the products. Indeed, the number of side chains on the products identified from $Na_2Cr_2O_7$ degradation would indicate that the values suggested by the ^{13}C NMR data might err on the high side. These low figures contrast with the values suggested by van Krevelen and also by Hirsch (1954) on the basis of X-ray scattering. These X-ray results have been reevaluated by Scaroni and Essenhigh (1978). They suggest that the number of condensed rings is constant at four or five up to about 90% C, after which it rapidly rises with rank. This result is lower than that originally suggested by Hirsch, who suggested that the condensed ring size varied from four to five in low-rank coals to seven at 89% C. However, some evidence for the presence of larger rings is forthcoming: Drake *et al.* (1978), using Shpol'skii luminescence spectroscopy of coal extracts and coal tar pitch, have identified 12 individual polycyclic hydrocarbons containing from three to ten rings. Rings as large as ovalene and coronene were detected, but only in small quantities. This evidence, although interesting, does not conflict with the evidence for a preponderance of smaller ring structures.

V. ALIPHATIC STRUCTURES

Much of the interpretation of the nature of the aliphatic part of coal seems to have been obtained by simply considering the elemental analysis and the percentage carbon aromaticity (often derived from the percentage hydrogen aromaticity). The percentage of aliphatic carbon and hydrogen is then obtained by difference. There can be a certain amount of circularity involved in such a procedure, especially when carbon aromaticity values are obtained indirectly by assuming that the aromatic groups are connected by $-(CH_2)_n-$ linkages. Chakrabartty and Kretschmer (1972) noted that information about the aliphatic structures in coal was very sketchy and its significance possibly underestimated. Using sodium hypochlorite to oxidatively degrade coal at 60°C, they obtained results which indicated that at least 17% of the total carbon in a low-rank coal is present as activated CH_2 and/or CH_3 groups and that a similar proportion could be activated by nitrating a high-rank coal. The volatile acids obtained from the oxidation were characterized: these were acetic, propionic, and formic acids. This indicated that n-propyl, n-butyl, and ethyl groups exist in coal. It was estimated that there was a minimum of 1–1.5 n-propyl groups per 100 C atoms, representing 3.5% of the total carbon. The n-butyl group was less than 1% of the total carbon in low-rank coals and about 2% in high-rank coals. In 1974, Chakrabartty and Kretschmer provided the surprising conclusion that the sodium hypochlorite oxidation studies showed that between 55 and 60% of the total carbon in coals exists in the sp^3 valence (or fully saturated aliphatic) state. This conclusion was reached by assuming that, in all cases of NaOCl oxidation of coal, carbon dioxide is formed by oxidation of sp^3 carbon only. This coal was described as being essentially nonaromatic in structure. The NaOCl studies were severely criticized on several grounds—including the inability of NaOCl, under the conditions used, to distinguish between sp^2 and sp^3 carbons (Ghosh et al., 1975; Landolt, 1975) such that the carbon dioxide produced could, in part, be derived from aromatic structures. In general, the claimed selectivity for NaOCl was shown to be largely unfounded under the reaction conditions used. However, Mayo and Kirshen (1978a,b; 1979) have treated pyridine-extracted Illinois No. 6 coal at 30°C at pH 11 with repeated treatments with small proportions of hypochlorite. The major product is a mixture of black, bicarbonate-soluble acids of molecular weight above 1000. They note that *excess* hypochlorite leads to the products reported by Chakrabartty. The black acids contain a high proportion of aliphatic carbon, as indicated by CP NMR studies (see Fig. 8) and a low proportion of hydrogen to carbon. It is suggested that the black acids must contain a high proportion of tertiary aliphatic carbon atoms and probably bridged and condensed aliphatic rings. This evidence leads to a representation of the main component of bituminous coal being a three-dimensional, cross-linked structure with aggregates of condensed aromatic, hydroaromatic, and bridged alicyclic rings held together by ether (and some sulfide) links, methylene or polymethylene links, or combinations of these. The acids produced by the NaOCl oxidation appear to retain most of the connecting links in the original coal and indicate that the coal must contain a high proportion of condensed aliphatic rings.

Trifluoroacetic Acid Oxidation

Deno *et al.* (1978a,b) have introduced a method of degrading coal by oxidation with 30% aqueous hydrogen peroxide in trifluoroacetic acid (TFA) with or without the addition of sulfuric acid. This dissolves the coal to form colorless or pale brown solutions in which all the aromatic structures have been destroyed and most (over 70%) of the aliphatic structures preserved. Other oxidizing agents, for example, HNO_3, O_2, $Na_2Cr_2O_7$, attack at the benzylic position and cause oxidative cleavage, but the $H_2O_2/TFA/H_2SO_4$ agent attacks the aromatic rings, leaving the benzylic positions largely untouched. This was demonstrated by studies on model compounds. The products of the $H_2O_2/TFA/H_2SO_4$ oxidation have been examined by proton NMR spectrometry and gas chromatography. Of the original hydrogen 9–55% appears in the products, and about three-quarters of this hydrogen is present in four aliphatic compounds: acetic acid, succinic acid, glutaric acid, and methanol (Table VI). The coals studied ranged from a North Dakota lignite (65.3% C) to a Pittsburgh seam coal (79.6% C). The results of the major components were interpreted as follows:

1. The acetic acid probably arises largely from methyl groups attached to benzene rings, and the estimated percentage of the coal hydrogen in these groups ranges from 1.4% in Pittsburgh seam coal to 9.2% for Illinois No. 6.

2. The absence of isobutyric acid indicates the absence of isopropyl groups.

3. NMR spectra showed no bands in the 0.8–1.2 region, indicating the absence of propyl, butyl, and other higher alkyl groups.

4. The absence of ethanol and higher alcohols is firm evidence that methoxy is the only alkoxy group present since primary and secondary alcohols are inert to the oxidation.

5. The glutaric acid could arise from 1,3-diarylpropane or indan structures or, possibly, 1-substituted tetralins.

6. Succinic acid, which accounts for 4.4–13.4% of the total hydrogen in the coals studied, could arise from $ArCH_2CH_2$ Ar, indans, or possible $—CH_2CH_2CHOH—$ and $—CH_2CH_2CHR—$ systems.

7. The low yield of succinic acid in the models studied indicates that the structures generating succinic acid could account for more hydrogen in the original coals than that calculated from the succinic acid yield.

8. The absence of cyclohexene-1,2-dicarboxylic anhydride suggests that there are no tetralins with unsubstituted aliphatic rings.

TABLE VI

Yield of H Appearing in Products from Coal and H_2O_2/TFA/H_2SO_4 [a]

Substrate	C (%)	H (%)	Moisture (%)	Yield of H (%) appearing in					Total	NMR (1.88– 3.25δ) [c]
				Acetic acid	Succinic acid	Glutaric acid	Methanol			
Pittsburgh seam coal	79.6	5.3	1.5	0.9	4.3	0.5	0		8.6	8.7
Illinois No. 6 coal	70.8	5.2	17.7	6.1	13.4	2.2	0		19.7	23.9
Illinois No. 6 coal (Monterey)	69.7	5.0	12.8	3.2	10.2	[b]	0		20.6	20.6
Lignite, North Dakota	65.3	4.4	35.7	4.2	6.0	[b]	16.2		54.3	54.9

[a] From Deno et al. (1978b).
[b] Not determined.
[c] As there was no H determined outside this range, this represents the percentage of aliphatic hydrogen.

Larsen (1978) notes that the oxidation of the Illinois No. 6 and Pittsburgh coals produced no malonic acid. Malonic acid is produced (9% yield) from the oxidation of diphenylmethane and 9,10-dihydroanthracene (36% yield). This would indicate that there are little or no $Ar—CH_2—Ar$ structures present. The yields obtained with the $H_2O_2/TFA/H_2SO_4$ oxidation were in the range 60–70%. Therefore, the presence or absence of particular compounds does not give the values for the percentage of contribution of the aliphatic structures in coal, but it seems certain that the technique provides a way of characterizing a part of the coal structure which is largely lost in other degradative reactions.

The proton NMR results obtained also indicate the H aromaticity; the hydrogen was observed within the range 1.88–3.52δ, and no bands were observed outside this range. It would appear that all the aliphatic hydrogen relative to the total organic hydrogen in the coal is represented. These values ranged from 8.7% for Pittsburgh seam coal to 54.9% for North Dakota lignite.

The wide range of values obtained, both for the yield of acids and methanol and the percentage of aliphatic hydrogen, emphasizes the varied nature of the molecular structures in the coals studied. It was reported that work is in progress to improve the yields of the reaction and to identify the minor components—this would seem to be important if the lower rank coals are to be properly characterized. Most of the aliphatic hydrogen in the higher rank coals seems to be accounted for in the major products, but there are noticeable discrepancies between the percentage of aliphatic hydrogen observed in the NMR spectra and the yield appearing in the major products for the lower rank coals.

The work in progress is yielding some very interesting results (N. C. Deno, private communication, 1979). First, it is now recognized that no propionic acid was in fact observed; its presence would have indicated arylethyl groups. The most surprising result has been the identification of benzene polycarboxylic acids. The TFA oxidation is supposed to destroy all the aromatic structures. It is significant that completely aromatic molecules such as anthracene, phenanthrene, pyrene, or chrysene do not produce these benzenepolycarboxylic acids on oxidative degradation with TFA. However, 5,12-dihydronaphthacene (Fig. 15) under these conditions produces the 1,2,4,5-benzenetetracarboxylic acid and the 1,2,4-triacid, and also other compounds obtained from the coal oxidation. The results strongly imply that there are extensive regions of hydroaromatic structures present in the original coal that would give rise to these products. Experiments are being carried out on the related dihydrobenzanthracene, and this is expected to yield the 1,2,3,4-tetra acid, and the 1,2,3- and 1,2,4-triacids. If this proves to be the case, then the evidence for the presence of hydroaromatic structures in coal will be substantial.

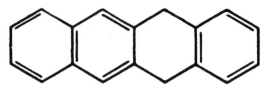

Fig. 15.　5,12-Dihydronaphthacene.

VI. LOW MOLECULAR WEIGHT COMPOUNDS

Coal contains a proportion of volatile matter that is normally trapped within the pores. The most notable of these is methane, principally because of its significance in coal mining safety. However, there are other, not so simple hydrocarbons present in coal, which, if not volatile, can be extracted by various solvents. Vahrman (1972) reported work on the occurrence in coal of material with molecular weights below 1000. Only small amounts of this low MW material can be extracted using low-boiling nonspecific solvents (e.g., benzene, ether, acetone), but Vahrman presented evidence to show that there could be considerable proportions of the low MW material present. These compounds can be detected both in coal extracts and tars, and it was suggested that a substantial part of the primary tars is present, as such, in coal. Vahrman distinguished between two main categories of low MW material: the dark "coal-like" materials, which have a high oxygen content (mostly phenolic —OH), and an aromatic, but partly hydrogenated, skeletal structure; and a range of hydrocarbons, both aliphatic and aromatic. These two classes were called O and H compounds, respectively. Normal Soxhlet extraction was performed in stages up to 250 hr. Samples were also heated stepwise from 200 to 290°C and extracted. The extracts contained alkanes (n-, branched, and cyclo) and traces of monoalkenes; aromatics from one to six rings (mainly alkylated) with benzene and naphthalenes predominating; and "O compounds." The importance of these low MW compounds lies in the fact that they influence structural parameters of the whole coal; the low C/H ratio of H compounds, the high aromaticity of the O compounds, and the low molecular weight must have implications toward the bulk of the coal. Vahrman concluded that molecules with molecular weights below 500 form a larger proportion of the coal structure than had been hitherto believed; the proportion in the coals he studied ranged from 10 to 23% for the total of H and O compounds. The ratio of H and O, however, varied widely.

Raj (1976) reported a study of the alkanes extractable from a restricted selection of bituminous coals. A mixture of benzene and ethanol was used for extracting the coals, and Soxhlet extraction was carried out for a period of 10 days. The extracts were fractionated, and the alkanes studied were contained in the first elute, eluted by pentane. The yields of pentane-soluble material varied with rank. A smooth curve passing through a maximum at 83–84% C is shown in Fig. 16. No obvious reason could be found for this rank dependency. The pentane-soluble extract was further eluted with various solvents, and the fractions were analyzed by IR spectroscopy. The IR spectra showed that the material was highly aliphatic, and other spectral details suggested that certain structural elements or skeletons were of very frequent occurrence in the eluates. This structural similarity is notable since the 14 coals studied included Carboniferous, Cretaceous, and Tertiary coals from various parts of the United States. The alkane fractions were also analyzed by gas chromatography, which revealed that all the coals contained n-alkanes from C_{13} to C_{33} and that the alkanes from the lower rank coals were highly branched to the extent that the branched alkanes predominated over the straight-chain alkanes. Three of the Cretaceous coals showed an interesting degree of odd–even carbon number preference in part of the distribution. The distribution of alkanes from higher rank coals were bimodal around C_{18} : below C_{18} branched and cyclic compounds predominate, while above C_{18} n-alkanes predominate. It is suggested that the branched and/or cyclic alkanes are derived from substances in plants or peat that were alkanes or oxygen substituted, or both, for example, terpenoid substances. Raj reported a very diverse set of alkane distributions and suggested that this diversity is probably due to the diversity of the starting materials rather than differing conditions acting on similar starting materials.

Hayatsu *et al.* (1978a,b; Studier *et al.,* 1978) have studied "trapped" low molecular weight compounds in coal. They believe that the trapped compounds bear some relationship to the macromolecular material that con-

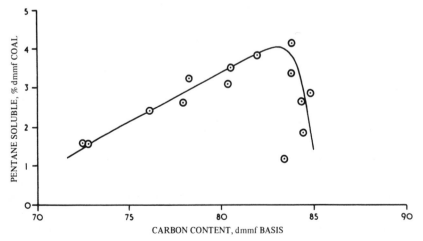

Fig. 16. Pentane-soluble material of coal, as a function of carbon content (Raj, 1976). Dmmf = dry mineral-matter-free.

stitutes the bulk of coal. These lower MW compounds were isolated by vacuum distillation (to 150°C) and solvent extraction. They were then characterized by gas chromatography and by time-of-flight mass spectrometry (GC–TOFMS) and high-resolution mass spectrometry. A lignite (67.3% C), a high-volatile bituminous coal (77.8% C), and an anthracite (91.3% C) were studied, and freshly ground raw coals were used. The trapped volatile compounds isolated from bituminous coal are predominately aliphatic and alicyclic hydrocarbons. The hydrocarbons in the range C_1–C_5 are ten times as abundant as the ones from C_6 to C_{13}. The distribution of compounds from C_6 up was: 37% aliphatic alkanes, alkenes, alkynes, or alkadienes; 29% alicyclic molecules; 29% hydroaromatics and aromatics; and 5% heteroaromatics. No nitrogen-containing compounds or phenols were detected. The lignite and the anthracite contained much less volatile organic materials ($< C_{10}$) than did the bituminous coal—indeed, only traces were detected (except for methane). Alkaline extracts of the lignite and the bituminous coal yielded organic acids which were similarly characterized. These indicated further differences between the coals. The observed differences between the low-molecular portion of the coals have certain implications affecting the understanding of the bulk of the coal. It would appear that, if current views on the coalification process are reasonable, the smaller trapped molecules are produced after the lignite stage and are either lost or incorporated into the bulk of the coal at the anthracite stage. It should be noted that the variation of the quantity of trapped molecules found by Hayatsu and others is similar to the variation in pentane soluble alkanes found by Raj.

Yoshii and Sato (1979) studied the hexane-soluble portion of pyridine extracts obtained at 50°C from Sorachi coal (83.6% C) and isolated five crystalline substances. They identified three of these as alkyl derivatives of picene, one an alkyl derivative of chrysene, and one to be a mixture of C_{28}–C_{30} paraffins. Studies on the methanol-soluble portion (Sato and Yoshii, 1979) were not so conclusive, since the crystalline substances could not be characterized as readily, but it appears that one is an alicyclic- or methyl-substituted unit of 5 catacondensed aromatic rings, one contains an ether linkage and two or three condensed aromatic rings, and a third is a mixture of di- and trimethyl-1,2,5,6-dibenzanthracene.

A. Flash Heating

McIntosh (1976) exposed a vitrain-enriched powdered sample of a high-volatile Utah coal to the intense light flash from a capacitor discharge lamp. Very little of the coal was gasified, but material was discharged from the coal. This was washed out with benzene or acetone, leaving a black, solid residue, which had an IR spectrum very similar to the original coal spectrum. The results were interpreted by suggesting that flash heating results in the evaporation or evolution of the less tightly held molecules in the coal particle. The IR spectra of the extracted material indicated a highly aliphatic substance very different from the coal or the residue. It is suggested that the experimental results are similar to Vahrman's (1972), that the low-energy flash heating has caused the coal to release a disproportionately large concentration of simple aliphatic hydrocarbons, and that the unextracted material is enriched in very high MW substances. These suggestions are based on the assumption that the small molecules are discharged so rapidly that they have undergone very little chemical rearrangement.

B. Supercritical Gas Extracts

Bartle *et al.* (1975) reported studies on the chemical nature of extracts of coal obtained by the use of compressed toluene at 350°C and 10 MPa. The extract represented 17.0% of the coal (NCB* rank code 802: a high-volatile coal, volatile matter over 36%) and was fractionated by solvent extraction, silica gel, and gel permeation chromatography. The soluble fractions amounted to 85% of the extract and were investigated by a variety of methods including high-resolution ^{1}H NMR, ^{13}C NMR, IR spectroscopy, mass spectrometry, and gas chromatography. The presence of isoprenoidal hydrocarbons was confirmed. These included phytane, pristane, norpristane, farnesane, 2,6,10-trimethyltridecane, and 2,6,10-trimethylundecane. Straight-chain alkanes up to C_{31} were found, and there was a predominance of those with odd carbon numbers over those with even carbon numbers. Table VII shows the results of a chromatographic analysis of a petroleum–ether eluate representing 2.6% of the whole extract. This odd/even distribution is similar to that reported by Raj (1976). The presence of isoprenoid hydrocarbons and the relatively low concentration of C_{10}–C_{15} straight-chain alkanes indicate that little thermolysis takes place in the supercritical extraction process. A very low degree of pyrolysis is indicated by the odd over even predominance and the absence of alkenes, but it is suggested that some thermal degradation and rearrangement does occur.

The NMR spectra for the extracts indicate that, for the aromatic compounds extracted, C_1–C_4 alkyl groups and hydroaromatic structures are substituents on the rings. The various structural parameters obtained were used to calculate "average aromatic nuclei" for the extracts. These are shown in Fig. 17. A notable feature of these calculated structures is their "open-chain" and "hydrogenous" nature, which suggests that there is little degradation of the aromatic part of the coal during supercritical extraction. The calculated

*NCB = British National Coal Board.

TABLE VII

GLC Analysis of Petroleum–Ether Eluate of a Supercritical Gas Extract of Coal[a]

Components (straight chain unless noted otherwise)	Eluate (%)
Octane	0.2
Nonane	0.6
Decane	0.2
C_{11}	0.5
Branched paraffins between C_{11} and C_{12}	5.5
C_{12}	0.6
2,6,10-Trimethylundecane	0.1
C_{13}	0.1
Farnesane	0.4
C_{14}	0.8
2,6,10-Trimethyldodecane	1.5
C_{15}	1.4
C_{16}	2.4
Norpristane	2.5
C_{17}	4.1
Pristane	8.4
C_{18}	4.5
Phytane	1.5
C_{19}	4.9
C_{20}	4.2
C_{21}	4.8
C_{22}	4.1
C_{23}	3.7
C_{24}	3.5
C_{25}	4.5
C_{26}	3.4
C_{27}	3.5
C_{28}	2.0
C_{29}	1.1
C_{30}–C_{31}	5.1
Total identified:	80.1

[a] From Bartle *et al.* (1975).

structures have been reexamined by Oka *et al.* (1977), who used a computer-assisted molecular structure construction technique, and Stephens (1979a), who used a C–H–O ternary diagram, but in general similar average structures are obtained. Stephens points out that the calculated structures are remarkably specific and thus should be accepted with caution as allowance should be made for the normal uncertainty of the compositional data and for some of the assumptions made by Bartle *et al.* (1975) when interpreting the [1]H

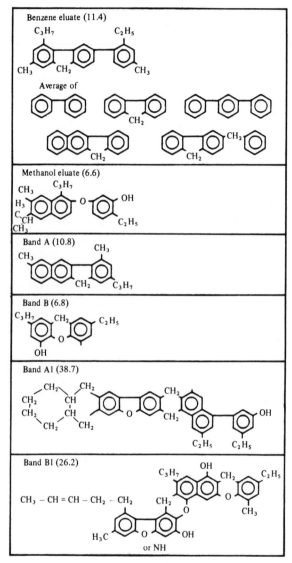

Fig. 17. Average molecules in fractions of supercritical-gas extracts of coal. Figures in parentheses represent the yields as percentages of the whole extract (Bartle *et al.*, 1975).

NMR data. It is suggested (Bartle *et al.*, 1975; Williams, 1977) that the extracts may be representative of the lower molecular weight fractions of the coal as the chemical evidence indicates that the extracted material is the product of the coalification of plant chemicals and not of extensive degradation during extraction. There is a possibility that supercritical gas extraction of coal may simply release the material physically absorbed in the micropore structure but it may also be producing compounds by the degradation of oxygen-containing compounds or long alkyl chains (Bartle *et al.*, 1979a). This latter view is largely confirmed by the work of Blessing and Ross (1978), which indicates that the soluble products are the result of thermolytic cleavage during the supercritical treatment.

According to later studies, supercritical toluene extraction at 400°C in the presence of hydrogen and zinc chloride (Bartle *et al.*, 1979b) yields an extract of which about half is a pentane-soluble oil having an average structure of three aromatic rings with two methyl groups and a naphthenic group as substituents. This extract was obtained in higher yields than in the absence of hydrogen and consisted of smaller molecules on average. It was concluded that the changes result largely from the cleavage of ether and methylene bridges and a certain amount of condensation under the influence of the hydrogen and catalyst. Further studies on Turkish lignite (Bartle *et al.*, 1979c) have suggested the presence of aromatic ether oxygen and single aromatic ring average structures, many of which are substituted by alkyl groups (some at least eight carbon atoms long). The carbon aromaticity for these extracts, at between 34 and 56%, is markedly lower than for bituminous coal extracts (67–73%). These results largely confirm the results of Hayatsu *et al.* (1978a) that the lower MW fraction of lignite is predominantly composed of single aromatic rings but, whereas Hayatsu and his colleagues report an abundance of terpenoid and alicyclic species, Bartle and his associates note the importance of long-chain aliphatic substituents. However, different lignites were examined, and it might be expected that there could be a greater variation in structures present in lower rank coals.

VII. FREE RADICALS

Electron spin resonance (ESR) studies have been carried out on coals—these have indicated the presence of paramagnetic species. The intensity of the signal indicates the number of free radicals in a sample, and the spectral linewidth is due to unresolved hyperfine interactions between unpaired electrons and protons in coal. Earlier work on ESR studies on coal was comprehensively reviewed by Ladner and Wheatley (1965) (their paper also gives an account of ESR theory, which is not attempted here). A more recent account of ESR studies is given by Retcofsky *et al.* (1978), who report an investigation of ESR spectra obtained for vitrains and fusains from a large number of coals. The data show that the concentrations of unpaired electrons in vitrains increase with increasing carbon content up to ~94% after which they decrease rapidly. This is shown in Fig. 18. This is not observed for fusains, where there does not seem to be any correlation with rank. For the vitrains an increase in ESR linewidth with increasing rank is observed for lignite to low-rank bituminous coals; this decreases through the higher rank bituminous coals to the early anthracite stages and then increases markedly (Fig. 19). The ESR linewidths for the fusains are relatively small except for the lowest rank coals. The ESR "*g* values" that arise from spin–orbit coupling are higher than 2.0023, which is the free electron value in the absence of spin–orbit coupling. The *g* values for vitrains and fusains decrease from low to high rank with the exception of the *g* values for vitrains from very high-rank coals (Fig. 20); these are markedly higher. The interpretation of ESR findings in terms of coal structure is fairly tentative; the fact that the concentration of free radicals increases with rank suggests that they are being delocalized over aromatic rings and stabilized by resonance. This stabilization will increase as the ring systems condense, but as electric conductivity increases in coals with more than 94% C, the free radicals will be destabilized. The very broad linewidths of vitrains from high-rank anthracites and metaanthracites are

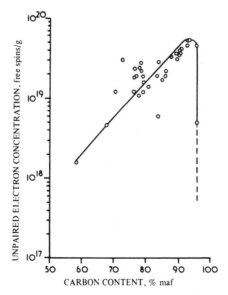

Fig. 18. Concentrations of unpaired electrons as a function of carbon content for vitrains of selected coals (Retcofsky *et al.*, 1978, p. 144).

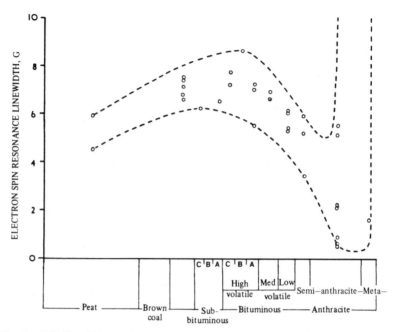

Fig. 19. ESR linewidths as a function of coal rank for vitrains from selected coals (Retcofsky *et al.,* 1978, p. 147).

almost certainly due to the presence of graphite structures. This would also explain the observed high *g* values for high-rank anthracite vitrain. Organic free radicals have higher *g* values if the unpaired electron is stabilized by atoms with high spin–orbit coupling constants. The high *g* values for low-rank coals can be explained by localization of unpaired electrons on heteroatoms. As the carbon content increases and the oxygen content decreases, the *g* values decrease as the radicals become localized on aromatic hydrocarbons until the *g* values increase again with the formation of graphitic structures by condensation of aromatic rings. Figure 21 shows the variation of *g* value with oxygen content. A small anisotropy in the *g* value of some anthracites was observed, and it was suggested that this might be due to some ordering of the condensed aromatic ring structures.

Petrakis and Grandy (1978) studied four coals ranging from sub-bituminous (72.2% C) to bituminous (86.4% C). They are of the opinion that the *g* values are very important structural parameters and that, if used cautiously, the *g* values found in coal could be compared with the *g* values of pure compounds to obtain some indication of the paramagnetic species present in coal and perhaps other structural information. Their comparisons show that for the higher ranking coals the *g* values are typical of those found for typical

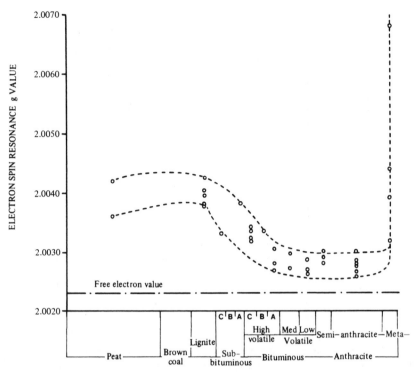

Fig. 20. ESR g values as a function of coal rank for vitrains from selected coals (Retcofsky *et al.*, 1978, p. 150).

π-type aromatic hydrocarbon radicals, but the higher g values for sub-bituminous coal indicates the probability that oxygen atoms are important. These suggestions are in line with those of Retcofsky *et al.* (1978, 1979). Comparisons with pure compounds suggest that these radicals might be of the quinone type in molecules containing four to nine rings or they may be due to methoxy benzene radicals and related compounds. The greater linewidths found for subbituminous coals imply that atoms with attached hydrogens participate more in the molecular orbital of the unpaired electron in sub-bituminous radicals than in the higher ranking coals. It appears very likely, from the four coals studied, that the radicals in subbituminous coal are different from higher rank coal radicals.

Most of the studies on the free radicals in coal have concentrated on interpreting the g value data, although Retcofsky *et al.* (1978) obtained some correlations of linewidths with rank. Further studies and suggested explanations have come from Kwan and Yen (1979), who studied coals of various ranks and derived some parameters of linewidth that are independent of line shape.

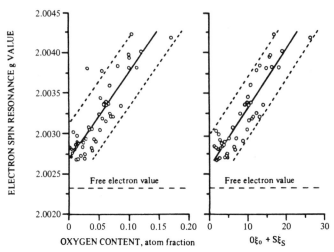

Fig. 21. Functional dependencies of the electron spin resonance g values of vitrains from selected coals (Retcofsky *et al.*, 1979).

They obtained correlations between some of these parameters and the percentage of carbon content. These positive correlations are explained if the paramagnetic aromatic centers in coal increase in size as the rank increases. It was also suggested that the data obtained indicated that the line shape parameters could be used to characterize coals. Electron spin resonance spectra of coals usually consist of a single line with no resolvable fine structure. However, the electron nuclear double resonance (ENDOR) technique can show hyperfine interactions not easily observable in conventional ESR spectra. Recently ENDOR has been applied to coal (Schlick *et al.*, 1978; Miyagawa and Alexander, 1979; Retcofsky *et al.*, 1979). Schlick *et al.* (1978) saw no ENDOR signals at room temperature, but at lower temperatures between 110 and 160 K a signal was obtained in the form of a single peak. The ENDOR line is interpreted as being due to predominately dipolar interaction between an unpaired electron and the surrounding nuclei. It is claimed that the very observation of an ENDOR signal shows interaction between the electron and nearby protons and that the results indicate that the interacting protons are twice removed from the aromatic rings on which, it is assumed, the unpaired electron is stabilized. Miyagawa and Alexander (1979) obtained spectra of vitrain, clarain, and fusain from a bituminous coal. The signals shown in Fig. 22 are quite different for the three ingredients and indicate that the environment of the unpaired electrons is different in each of three, but the nature of this difference is not clear. Retcofsky *et al.* (1979) obtained a detailed spectrum for vitrain-rich Pittsburgh coal (Fig. 23). This contrasts with

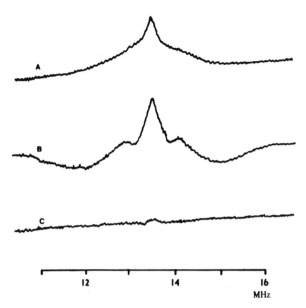

Fig. 22.   ENDOR signals from (A) vitrain, (B) clarain, and (C) fusain from a bituminous coal sample (Miyagawa and Alexander, 1979). Reprinted by permission from *Nature* (*London*) **278,** 40 (1979). Copyright © 1979 Macmillan Journals Ltd.

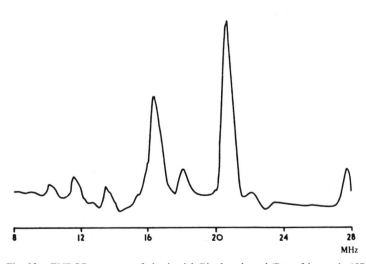

Fig. 23.   ENDOR spectrum of vitrain-rich Pittsburgh coal (Retcofsky *et al.,* 1979).

other, single-line, results, but it was obtained in the absence of air and the hyperfine structure disappeared when air was admitted.

Obviously, any interpretation of ESR spectra is fairly tentative at present, but the interpretations put forward seem to be reasonable. It may also be possible to interpret ENDOR signals in terms of structural parameters, especially if more fine-structure spectra can be obtained.

VIII. FUNCTIONAL GROUPS AND HETEROATOMS

Much of the evidence for the presence or absence of functional groups in coal has been obtained from IR spectra of solid coals, either in thin section or in KBr pellets. Most of the relevant work was carried out in the 1950s and 1960s and is thus outside the scope of this review, being adequately covered in the standard works such as Lowry (1963). A brief résumé of the findings obtained by IR spectrometry is provided by Retcofsky (1978). The spectra show distinct bands, which have been assigned with some confidence to such groups as —OH, —CH_2, —CH_3, and aromatic —CH. The peak obtained at 1600 cm^{-1} has been assigned to aromatic C=C, chelated quinone groups, or heterocyclic nitrogen. The bands at 3030 cm^{-1} and 2920 cm^{-1} have been used to provide a measure of the aromatic hydrogen content, but the calculations involve assumptions including the estimation and choice of specific extinction coefficients for various hydrocarbons. An attempt has been made to use IR data to estimate the carbon aromaticity (Retcofsky, 1977). The best quantitative use of IR spectra appears to be the estimation of the hydroxyl content in coal by measuring the intensity of the band near 1250 cm^{-1} in trimethylsilylated coal. As many of the data and interpretations of the data on the functional groups in coal are accepted, in this section we shall concentrate on aspects of the functionality of coal which have been examined recently.

A. Oxygen

In general the concentration of organic oxygen in coal is assessed by difference: by subtracting the values for C, H, N, S, and ash. Obviously this means that the oxygen figure is subject to the errors incurred in determining the other values. Ruberto and Cronauer (1978) have reported details of the analytical aspects of determining the quantity and type of oxygen in coals and coal liquids. Table VIII shows the results of analyses of the oxygen functionality of a 69.3% C subbituminous coal and a 71.5% C bituminous coal. The results are consistent with the generally agreed upon finding that the oxygen content of coal decreases with rank. The oxygen content of subbituminous coal was referred to in the Ruberto *et al.* (1977b) solvation studies,

TABLE VIII

Analyses of Oxygen Functionality in Coals (wt. % maf Coal Basis)[a]

Coal	Burning Star (bituminous)	Belle Ayr (subbituminous)
Oxygen content as:		
Hydroxylic (—OH)	2.4	5.6
Carboxylic (—COOH)	0.7	4.4
Carbonylic ($=C=O$)	0.4	1.0
Etheric (—O—)	2.8	0.9
Total:	6.3	11.9
Oxygen by difference:		
Ash basis	5.9	16.2
Mineral matter basis	—	16.0

[a] From Ruberto and Cronauer (1978, p. 61). Maf = moisture ash-free.

where it was concluded that a significant portion of the oxygen occurs in saturated ether functional groups α or β to the aromatic moieties or as furan systems. Hayatsu *et al.* (1978a,c) found significant quantities of phthalan and xanthone in lignite; xanthone and dibenzofuran in bituminous coal, and dibenzofuran in anthracite (see Fig. 13).

1. Hydroxyl Groups

The importance of phenolic hydroxyl and its characteristics was pointed out by Given in 1975 and recently results of studies on the hydroxyl groups in coals have been reported (Abdel-Baset *et al.,* 1978; Yarzab *et al.,* 1979). Sophisticated statistical analyses on data from 52 coals were carried out. The hydroxyl contents of the new coals were determined by acetylation with ^{14}C-labeled acetic anhydride. Only coals with a significant hydroxyl content were studied; anthracite and medium- and low-volatile coals were rejected and lignites were considered to be too different chemically. The range of coals examined included subbituminous and high-volatile bituminous coals from three regions of the United States (eastern, interior, and western). Two parameters were examined against a range of rank parameters; these were (% O as OH)/(coal organic matter) and (% O as OH)/(total organic oxygen content). It was found that OH/coal is a rank parameter in its own right—it decreases with increasing degree of coalification—but that OH/O is a rather poor rank parameter. These results implied that the sum of the other oxygen functional groups present (carbonyl, ether, heterocyclic oxygen) must change in a rather random manner with rank. Regression analysis of the data revealed that the coals of the three regions constitute at least two distinct populations of materials, and it is believed that it is the coals of the interior province that are distinctly different. It was concluded that the gross differences in paleoenvironments and geological differences had significantly influenced the geochemistry of the coals, leading to the formation of sets of materials having systematic differences in chemical structure.

2. Distribution of Oxygen Groups

Sternberg (1975) noted that one of the bad features of the Wiser model of coal (see Fig. 32, Section X) was that it gives the impression that the coal molecule is amphoteric with acidic and basic groups randomly distributed. Sternberg did not believe that such an amphoteric structure was correct. Work by Alemany *et al.* (1978) has suggested that the distribution of functional groups in coal is not uniform among the molecular weight fractions obtained from the butylated coal examined by Sun and Burk (1975). The fractions were examined by ^{1}H and ^{13}C NMR. The spectra of the low MW fractions indicated that there were few, if any, oxygen atoms present, whereas the spectra of the high MW fractions indicated that there were no oxygen atoms present. In the medium MW fractions, most of the oxygen atoms appeared to originate from phenolic or diaryl ether fragments, and as the MW increased a significant proportion appeared to arise from aralkyl and dialkyl ethers. No oxygen atoms appeared to have originated from alkyl benzene ethers or dibenzyl ethers. These results clearly indicate that the structures of the butylated reaction products depend on the molecular weights of the molecules, and consequently the distribution of oxygen in the original coal is not uniform.

B. Sulfur

Sulfur has a similar chemistry to oxygen, but much less is known about the organic sulfur in coal compared with the organic oxygen. Heterocyclic sulfur compounds have been identified in coals (see Fig. 13; Hayatsu *et al.*, 1978a,c) but there have been very few studies which have concentrated on the organic sulfur functional groups in coal. However, Attar and Dupuis (1979) have reported data on these functionalities in different coals. Table IX shows the distribution of organic sulfur groups in five coals and shows that the content of thiols (—SH) is substantially larger in lignites and high-volatile bituminous coals than in low-volatile bituminous coals. The fraction of aliphatic sulfides (R—S—R) remains approximately constant at 18–25%. The data indicate that larger fractions of the organic sulfur are present as thiophenic sulfur in higher ranked coals than in lower ones. The data on —SH, R—S—R, and thiophenic sulfur are explained by suggesting that the coalification process

TABLE IX

Distribution of Organic Sulfur Groups in Five Coals[a]

Coal	Organic S (wt. %)	Organic S accounted (%)	Thiolic	Thiophenolic	Aliphatic sulfide	Aryl sulfide	Thiophenes[b]
Illinois	3.2	44	7	15	18	2	58
Kentucky	1.43	46.5	18	6	17	4	55
Martinka	0.60	81	10	25	25	8.5	21.5
Westland	1.48	97.5	30	30	25.5	—	14.5
Texas lignite	0.80	99.7	6.5	21	17	24	31.5

[a] From Attar and Dupuis (1979).
[b] Corrected for "uncounted for" sulfur.

causes the organic sulfur to change from —SH through R—S—R to thio-
phenes in condensation reactions.

C. Nitrogen

Very little appears to be reported on the forms of organic nitrogen in coal.
Hayatsu *et al.* (1975) reported experimental difficulties in isolation of
nitrogen-containing units from the $Na_2Cr_2O_7$ oxidation products, but some
were identified in the photochemical oxidation products (see Fig. 12). Later
work by this group (Hayatsu *et al.,* 1978a,c) produced better results—notably
in showing the relative abundance of acridine/benzoquinoline units in
anthracite.

IX. STRUCTURAL CHANGES WITH RANK

Coal rank has been briefly alluded to in the introduction, but it is probably
worth pointing out again that rank, although usually considered to be a continuous parameter, cannot be satisfactorily represented by any single
numerical parameter (Given, 1975; Neavel, 1979). These parameters include
carbon content, oxygen content, vitrinite reflectance, volatile matter content,
density, surface area, and moisture content. Rank, even if it is a continuous
parameter, is certainly not a simple progression. Many of the data discussed in
this report have been examined as a function of rank, or at least, one of the
parameters of rank. A common rank parameter, useful from a chemical point
of view, is the carbon content. However, it should be noted that however
straight a line might be when a set of values is plotted against a single rank
parameter it need not, and probably will not, be straight when plotted against
another. An interesting approach to the problem of plotting data against one
parameter has been revived by Stephens (1979a,b), who suggests using a
C–H–O ternary diagram of percentage of atomic composition, and Battaerd
and Evans (1979), who suggest using the bond-equivalent percentages. Many
of the changes in structural parameters have already been mentioned or are
apparent in the preceding sections; therefore, only a brief overview will be
attempted here.

A. Molecular Weight

Wachowska's (1979) results (see Table II) for the molecular weights of
alkylated coals show an increase from 500–800 for the lowest rank coals to
1300–2000 for the higher ranks. The result for anthracite is around 450, but
this is based on a very low percentage of the alkylated anthracite soluble in

benzene. The results of Larsen and Kovac (1978) based on solvent swelling studies show a sudden drop in MW values with high-rank coals. This sudden drop, according to the authors, is to be expected due to a more graphite-like, highly cross-linked structure in the highest rank coals. The results predict a minimum in the cross-link density at ~86% C. Belyi and Voitkovskii (1979) have demonstrated that the relationship between the molecular weights of the fundamental mean structural units in coals to the rank is valid only for a minimum of 89% carbon.

B. Carbon Aromaticity

A consensus on the aromaticity values reported would be that there appears to be a definite increase with rank. The values reported range from 40–50% for low-rank lignitic and subbituminous coals to nearly 100% for anthracites. Interestingly, Miknis et al. (1979) obtain a correlation between f_a and the atomic H/C ratio, but Whitehurst (1978) reports that there is relatively little correlation in coal and coal liquids. However, there is some correlation between aromaticity and the weight percentage of carbon. As ^{13}C NMR is now apparently the standard means of measuring aromaticity, an interesting correlation between the ^{13}C linewidth and the carbon content could be noted here—there is a pronounced inflection in the plot of this value against carbon content at 93% C (Fig. 24). The data are only from four coals, however (Retcofsky and Friedel, 1973).

Heating Value and Aromaticity

Maciel et al. (1979; also Miknis et al., 1979) have attempted to relate the heating values of coals to structural features. Heating values tend to increase with rank, as do aromaticity values, but this is not an expected relation. An in-

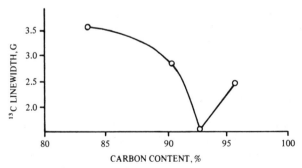

Fig. 24. Variation of ^{13}C linewidth with carbon content (Retcofsky and Friedel, 1973). [Reprinted with permission from *J. Phys. Chem.* **77**, 71 (1973). Copyright 1973 American Chemical Society.]

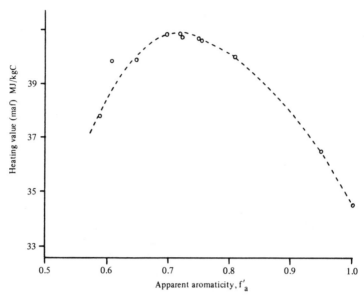

Fig. 25.   Plot of heating value as a function of apparent carbon aromaticity derived from ^{13}C NMR spectra (Maciel *et al.,* 1979).

crease in f_a would be expected to accompany a decrease in heating value (if the heats of combustion of aromatic and alicyclic compounds are compared, the alicyclic compound usually has the higher heat of combustion). No simple correlation between heating value and aromaticity was found. However, if the heating value *per kilogram* of carbon is plotted against the aromaticity, a smooth curve with a maximum at 40.70 MJ/kg carbon for f_a 70–75% is obtained (Fig. 25). These results are interpreted by assigning the falloff at higher f_a to the expected decrease in heating value with increase in aromaticity and the falloff at lower f_a to the increase in oxygen content in coals of lower aromaticity. There are thus two competing influences on the heat of combustion parameter: the increase in aromaticity with rank and the decrease in oxygen with rank.

C.   Aromatic Ring Structures

It is generally agreed that the average size of ring clusters in higher rank coals is greater than that in lower rank coals (Gerstein *et al.,* 1979). This is largely confirmed by, for example, the results obtained by Hayatsu and others (1978a,c), which show this increase in ring size from lignite through bituminous to anthracite coals. Benzene derivatives predominate in lignite and bituminous coals and phenanthrenes predominate in anthracite (see Fig. 13).

Whether this increase can be considered linear or regular is open to question; the NaOH/alcohol hydrolysis studies (Makabe and Ouchi, 1979) might indicate that the ring structures do not increase markedly until about 82% carbon is reached.

D. Low Molecular Weight Compounds

The results of Hayatsu *et al.* (1978a,b) showed that lignite and anthracite contain much less trapped organic material ($< C_{10}$) than the bituminous coal studied. The result for lignite is at first sight surprising, given that a general decrease of volatile matter is among the parameters for increasing rank. However, the "volatile matter" used as a rank parameter is largely a misnomer; it is a measure of weight loss by decomposition at 900°C, not of the composition of the original coal. The results obtained by Raj (1976) are largely in agreement with these observations (see Fig. 16). Hayatsu and his colleagues suggest that this variation indicates that the smaller trapped molecules are produced after the lignite stage and either lost or incorporated into the bulk of the coal at the anthracite stage. Their study of organic acids trapped in coals (Hayatsu *et al.*, 1978b) led to the conclusion that in the coalification reaction there was (1) extensive degradation of phenols, (2) loss of carboxyl groups, and (3) dehydrogenation and partial dealkylation of side chains.

E. Free Radicals

Retcofsky *et al.* (1978) in the ESR studies of coals noted that the concentrations of unpaired electrons in vitrains increase with increasing carbon content up to about 94% and then decrease rapidly. The behavior of ESR spectra with coal rank is very interesting and is covered in more detail in Section VII (see Figs. 18–21). An interesting piece of additional information comes from Yokono and Sanada (1978), who examined coals using pulsed NMR and ESR and measured the nuclear relaxation times. The proton spin–lattice relaxation time varies with the rank of the coal and shows a maximum value at about 86% carbon (Fig. 26). It is suspected that this change may stem from some chemical structural change owing to the coalification reaction. The decrease at ranks higher than 86% carbon is probably closely connected with the rapid increase of free radicals in high-rank coals.

F. Anthracites

A notable aspect of the changes in structural parameters with rank is the often pronounced changes in properties as the very high-rank coals are approached. Retcofsky and Friedel (1973) note that plots of resistivity, free

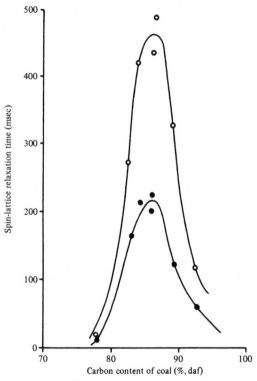

Fig. 26. Relationship between ^{1}H spin–lattice relaxation time and rank of coal. The lower curve was obtained in the presence of air, the upper *in vacuo* (Yokono and Sanada, 1978).

radical concentration, and ^{13}C linewidths in NMR spectra against percentage of carbon content all show a pronounced inflection point near a carbon content value of 93%. The pronounced changes are also evident in some of the other parameters discussed in this report. The physicochemical properties of anthracites have been examined by Artemov and Peresun'ko (1975) in an attempt to estimate the methane outburst hazard in underground coal mining. They observed two distinct phases in the formation of anthracite, noting that various indices of rank had a point of inflection at 95.5% C. X-ray studies of anthracite with a carbon content above this value reveal a markedly more graphitic structure than in ones below. Infrared studies show the absence of aliphatic CH_x groups at 95.5% C and above. These points of inflection were explained by suggesting that anthracite is formed in two distinct coalification stages. The first stage is relatively smooth and can be considered as the logical extension of the "normal" coalification process; this involves the loss of hydrogen through the loss of CH_x groups and the increase in the number of aromatic systems. The second, faster stage, which takes place in a jumplike fashion, involves the graphitization of anthracite by the loss of aromatic hydrogen and the accelerated growth of the aromatic systems. These chemical changes should be considered in the light of the generally accepted geological evidence that anthracites only appear in areas subject to tectonic influences and that this additional temperature and pressure is needed for coals to meta-morphose to anthracite (see Parks, in Lowry, 1963, p. 8). If this two-stage process is acceptable, then many of the rapid changes in structural parameters at very high rank may be explained. This may also mean that comparisons of structural parameters present in anthracites and those in other coals may not be strictly valid if "extraordinary" forces have contributed to the coalifica-tion such that the rank of an anthracite is abnormally high.

G. Brown Coal/Lignite

There is much less reported work on brown coals (or *lignites* according to usage in the United States) than there is on bituminous coals, but recently these coals have attracted some interest, for example, in Australia, where there are large deposits of these low-rank coals. Camier and Siemon (1978, 1979) have provided succinct accounts of the application of X-ray crystal-lography to brown coal structure and on the literature relating to the deter-mination of the molecular weight of brown coals. They note (Camier and Siemon, 1978) that the currently accepted concept of the submicroscopic structure of Victorian brown coal is as a micellar colloid; however, they point out that this theory is questionable. They found that the structure of brown coals needs to be studied in the light of modern techniques. More recently Camier *et al.* (1979) have studied the physical structure of a brown coal (65.6% C) by alkali digestion, followed by particle size analysis using gravita-tional sedimentation and ultracentrifugation. The presence of geometrically uniform "rods" and the persistent appearance of particles in the same size range were noteworthy, and a speculative hypothesis was put forward to ex-plain the findings. It was suggested that the first step of the coalification reac-tion is the destruction of cellulose and the degradation of lignin to humic acids. These acids can polymerize by condensation reactions. Woody residues can intermingle with condensation polymers, which are swollen by the water-soluble products from the degradation of cellulose and lignin. Local changes in carbon content brought about by condensation polymerization may be suf-ficiently large for phase separation to occur, and the rods may be formed from local high-carbon regions. Thus, the resulting brown coal can be considered as a gel of humic acid molecules swollen by water and incorporating the rods and detrital matter such as pollen and plant remains. These particles are bound together by weak nonregenerable bonds such as van der Waals forces.

Hooper and Evans (1979) depolymerized Victorian brown coals (65–

70% C) using phenol, which, although having a relatively weak action on very high rank coals, is very effective for brown coal. The depolymerized coal was solvent fractionated and the fractions characterized by IR spectrophotometry and ^{1}H NMR. Twenty-five percent of the coal had been calculated to be oxygen, and the analysis revealed that 5% was phenolic, 5% carboxylic, and 3% carbonyl, leaving the remaining 12% unaccounted for. The remainder would probably include some ethers and heterooxygen groups. Of the carbon, 17 g/100 g coal was in monoaromatic, 28 g in two-ring aromatic, and 10 g in three-ring aromatic structures. This result, especially the predominance of two-ring aromatics, is noteworthy especially when compared to the relative abundances found by Hayatsu *et al.* (1978a,c; see Fig. 13). A very high methylene bridge content was found accounting for more than 25% of the total hydrogen. This strongly indicates that there is more than one bridge per aromatic group and probably that there is a high degree of cross-linking in a three-dimensional network.

It is in the low-rank coals that one should expect to see chemical evidence of the first stages of coalification. Some of this evidence has been obtained by Hayatsu *et al.* (1979a,b). Using a CuO–NaOH oxidation procedure they were able to isolate phenolic acids which show a relationship between lignins and coals. Large amounts of *p*-hydroxy- and 3,4-dihydroxybenzoic acids (regarded as characteristic lignin oxidation products) were identified. This work shows that lignin-like polymers are incorporated into the macromolecules of coals and are identifiable in lower rank coals.

X. MACROMOLECULAR SKELETAL STRUCTURES

In preceding sections we have examined individual aspects or parameters of the molecular structure of the organic matter in coal, but no attempt has been made to coordinate those parameters into a "coal structure" as such. The concept of a unique structure or a simple repetitive structure cannot really be justified given the extensive heterogeneity of coals. However, the fact that coals are heterogeneous as a group and, indeed, heterogeneous mixtures individually, does not mean that there cannot be a concept of an outline macromolecular structure. This basic skeleton of the coal structure should really be limited to the vitrinite in coal or to vitrinite-rich coals. As pointed out by Aczel *et al.* (1976), there are currently three models which claim to represent the organic structure of coal. They are

1. the aliphatic/polyamantane model,
2. the aromatic/hydroaromatic model, and
3. the molecular sieve model.

A requirement which must be made of each model is that it shall explain as many of the known structural parameters as possible and, preferably, predict the chemical behavior of coal under certain reaction conditions.

A. Polyamantane Structures

The polyamantane structure resulted from the sodium hypochlorite oxidation studies of Chakrabartty and Kretschmer (1972, 1974) and was proposed by Chakrabartty and Berkowitz (1974). The low aromaticity values obtained from the NaOCl studies led these authors to suggest that the predominant skeletal structures of coal are composed of modified bridged tricycloalkane configurations. They noted that the interpretation of an aromatic/hydroaromatic structure often stemmed from an interpretation of, for example, X-ray diffraction data on the *presumption* that the coal was basically aromatic. These presumptions meant that the alternative aliphatic structures were generally ignored. Aczel *et al.* (1976) and Speight (1978) have pointed out that nonaromatic structures had been postulated in the 1950s (e.g., see Francis, 1961, p. 749), but generally these structures received little support, and it is true that the early studies largely ignored any nonaromatic or nonhydroaromatic interpretation of coal structure. Chakrabartty and Berkowitz suggested a structural type which claimed to encompass all ranks of coal up to about 90% C. The low aromaticity values used and the known H/C ratio in coals ruled out predominately aromatic or straight- or branched-chain hydrocarbons but were reasonable for a diamond-type skeletal carbon arrangement. The structure suggested for coals had at least 50% of the total carbon in complex bridged cycloalkanes, about 10% in alkane chains, and about 26–30% in methine-substituted benzene rings. Hypothetical polyamantane configurations simulating coals from 76 to 90% C were presented (Fig. 27). The model requires that the polyamantane units increase in size with increase in rank and that benzenoid carbons occupy the periphery of the units. The units would have to aromatize as bituminous coals pass to the anthracite stage. The coalification process was interpreted as first producing a stable aliphatic carbon skeleton from the original plant matter and then producing individual benzene rings.

The preceding sections on aromaticity and aliphatic structures have stressed the criticisms of NaOCl studies. Further criticism came from Aczel *et al.* (1975, 1976, 1979), who examined the stability of adamantane and its derivatives to coal liquefaction conditions. On the basis that adamantane-type structures are only found in trace amounts in coal liquids it was suggested that the polyamantane structures must be unstable under liquefaction conditions. Adamantane, 1-adamantanol, 1-adamantane carboxylic acid, adamantanone, 2-phenyladamantane, and diamantane were found to be stable

SKELETAL ARRANGEMENT

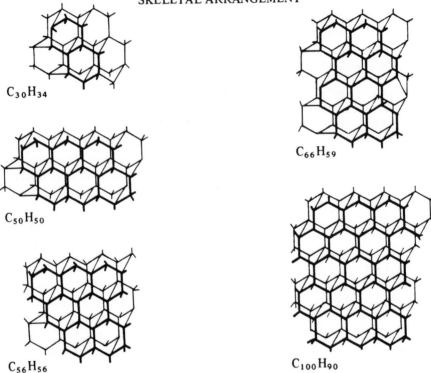

Fig. 27. Hypothetical polyamantane configurations simulating coals from 76 to 90% C (Chakrabartty and Berkowitz, 1974).

when "liquefied" at 425°C for 130 min in tetralin. Adamantane was 96% recovered, although the recovery from an adamantane/Illinois No. 6 coal mixture was only 76%. This was ascribed to losses in handling—although the adamantane could possibly have become attached to the cyclohexane-insoluble portion of the coal. Diamantane was 96% recovered, but the oxygen-substituted adamantanes were recovered as adamantane itself. The results would suggest that a predominantly polyamantane structure is not a good model of coal structure, although Chakrabartty and Berkowitz (1976b) point out that these results show the stability of the adamantane nucleus *per se* and do not rule out the possibility that *extended* polyamantane structures exist. They added that the model was of secondary importance compared with a knowledge of the carbon valency states in coal.

Recently, there has been a certain amount of support for the concept of extended cycloalkane structures in coal. Mayo and Kirshen (1978a,b, 1979), on the basis of NaOCl oxidation under more carefully controlled conditions, have noted that the "black acids" obtained must contain a high proportion of tertiary aliphatic carbon atoms, and probably of bridged and condensed aliphatic rings. However, it is pointed out that the probability of formation of regular polyamantane ring structures is unlikely and that the bridged tricyclo-alkane structures in bituminous coal are highly irregular. Whitehurst *et al.* (1977; also Farcasiu, 1979) have suggested that subbituminous Wyodak coal has structural parameters consistent with the presence of some polycon-densed aliphatic structures. An "average structure" for this coal is shown in Fig. 28. Farcasiu (1979) also noted that the results of Deno *et al.* (1978b) for the same Wyodak coal are not reproducible because the $H_2O_2/BF_3/H_2SO_4$ oxidation is only selective if polycondensed aliphatic rings are not present. It must be pointed out, however, that the results of Mayo and his colleagues and Whitehurst and his associates are not truly independent since the f_a values were obtained on the same CP NMR spectrometer at the University of California at Berkeley and the results were interpreted in the same way (as described by Pines and Wemmer, 1978). Work carried out by Doğru *et al.* (1978) on reductively ethylated coals (a 67.2% C lignite and 89% C bituminous coal) has yielded results from ^{13}C NMR, which suggests that the coals have structures consistent with mobile lightly substituted aromatic rings and paraffin chains attached to a comparatively rigid structure which could be an alicyclic network.

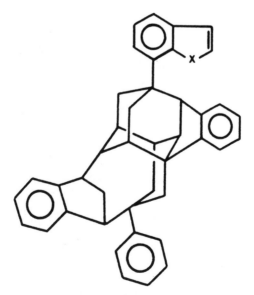

Fig. 28. Average carbon skeleton structure for Wyodak solvent-refined coal (Whitehurst *et al.*, 1977).

B. Aromatic/Hydroaromatic Structures

The aromatic/hydroaromatic model of coal is the "generally accepted" model, and it presents a picture of a skeletal structure which consists of clusters of condensed aromatic nuclei. The size of the clusters ranges from one ring to several but on average is about three rings for coals between 80 and 90% C. Above 90% the cluster size increases rapidly as the anthracites are approached. The clusters are connected by hydroaromatic linkages with a small number of heteroatom (mainly oxygen) linkages (Wyss, 1977). One of the well-known models was proposed by Given (1960) (Fig. 29) but, as he pointed out (Given, 1975), it relies mainly on the X-ray studies of Hirsch (1954) and on the physical constitutional analysis of van Krevelen. Given points out that the X-ray diffraction work was interpreted by assuming that 75% of the carbon was in aromatic rings. In Section IV in this article we have shown that modern methods have largely justified that assumption, at least for the higher rank coals. The oxidative degradation studies of Hayatsu *et al.* (1975, 1978a,c) also support a largely aromatic structure for the skeletal structures of hard coals. The models of coal are constructed by considering the elemental composition and the proportions of aromatic carbon and hydrogen. Many of the model structures proposed, for example Ladner's (in Gibson, 1978), are "two dimensional" although the presence of folded rings such as 9,10-dihydrophenanthrene means that the structures are not flat (Fig. 30). Interestingly,

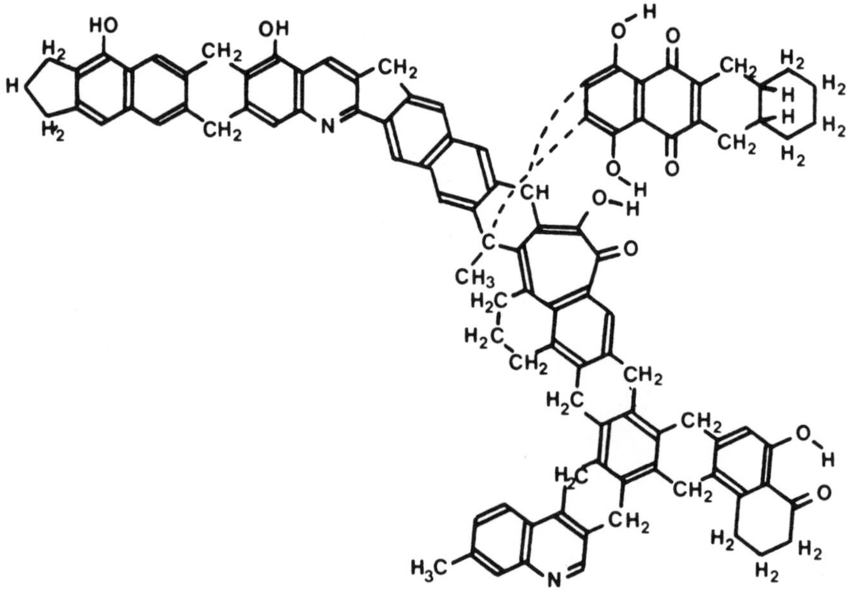

Fig. 29. Proposed structural elements of coal (Given, 1960).

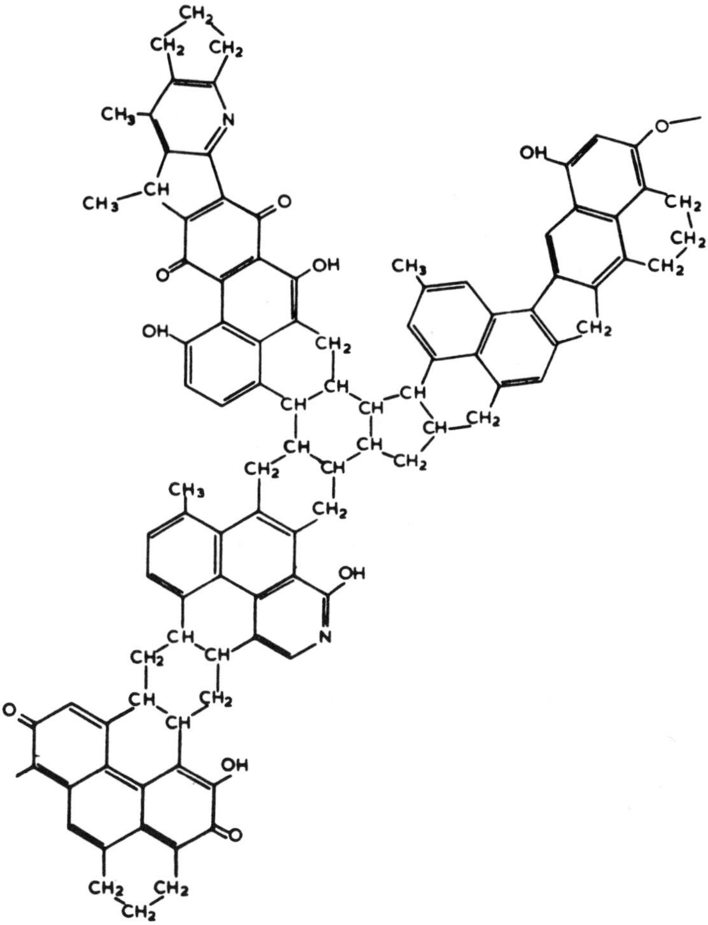

Fig. 30. Proposed molecular model of an 82% C vitrinite (by W. R. Ladner, 1963, in Gibson, 1978).

the Given model is three-dimensional. Pitt (1979) has proposed models of coal structure based on 9,10-dihydrophenanthrene structures (Fig. 31) and that these molecules exist in a tangled state in coal. The suggestion of a tangled network contrasts with the linked network proposed by Larsen and Kovac (1978), discussed in Section II. Pitt notes that his models contain slightly more hydrogen than the corresponding coals and suggests that the difference would become smaller if the models were dimerized. It is also the case that the introduction of some bridging structures would reduce the hydrogen content. Pitt reports the "testing" of some model structures, prepared by condensation of various aromatic molecules with ^{14}C-labeled formaldehyde, by con-

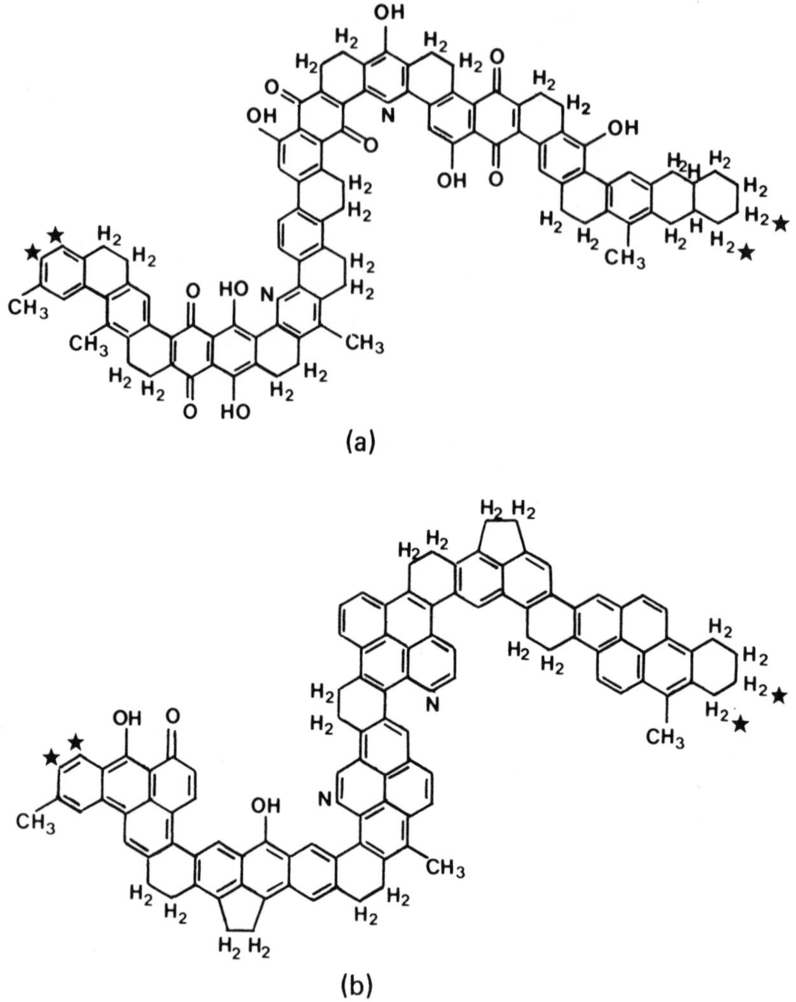

Fig. 31. Model structures for 80% C vitrain (a) and 90% C vitrain (b). Asterisks indicate where dimerization could occur. Reprinted with permission from Pitt (1979). Copyright: Academic Press, Inc. (London) Ltd.

sidering their behavior when pyrolyzed, and it appears that hydroaromatic structures would probably behave in the right manner during pyrolysis. This is an important step; it ought to be a requirement of any model that it should not only explain the physical structural parameters such as elemental composition and aromaticity values but should predict chemical properties (or "behave" chemically in a similar manner to the substance being modeled). This is one of the strengths of the so-called "Wiser model" (Fig. 32), which

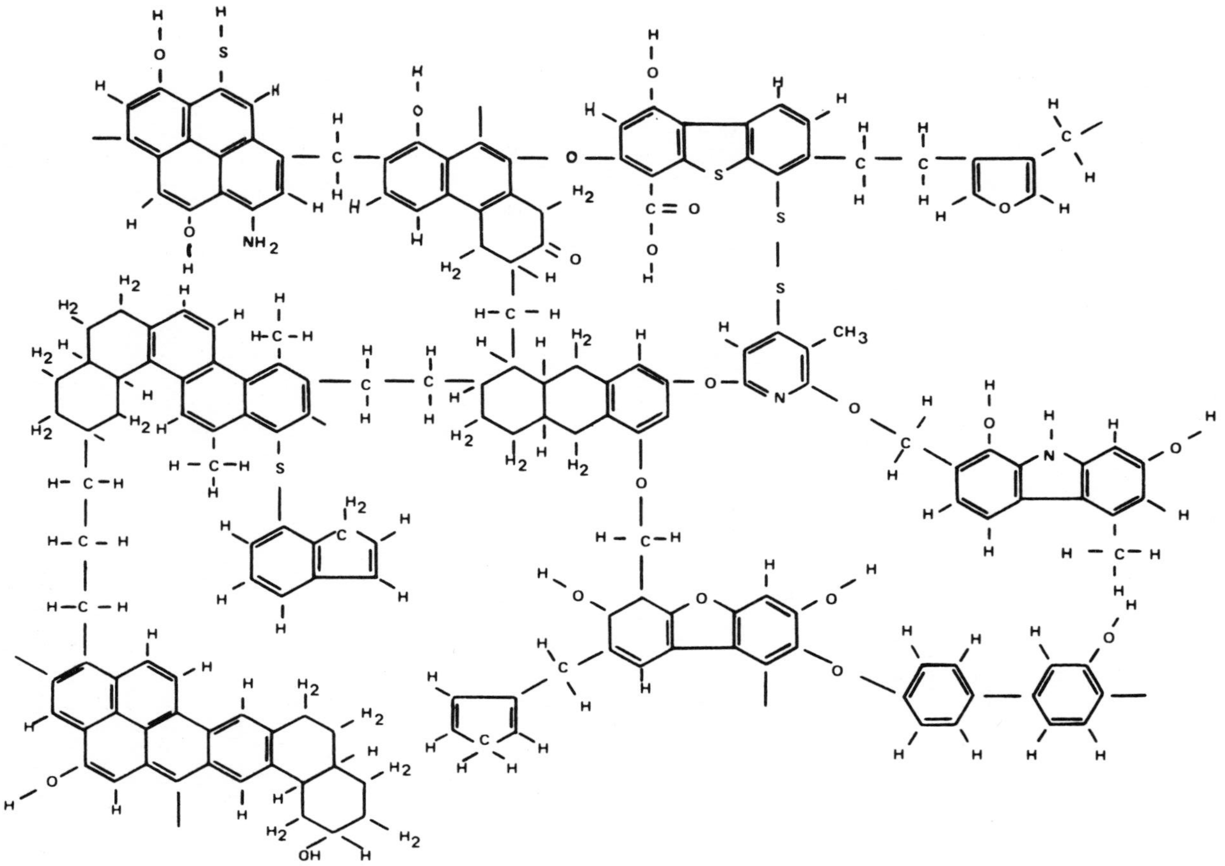

Fig. 32. Schematic representation of structural groups and connecting bridges in bituminous coal (by W. H. Wiser, in Sternberg, 1975).

modifies the extensive hydroaromatic structures by introducing relatively weak bonds between the aromatic units. These bonds account for the relatively rapid degradation of coal into smaller soluble fragments. Surprisingly, the hydroaromatic model has not been *directly* tested or confirmed and has arisen largely because it appears to be a "reasonable" interpretation of the parameters of coal structure. However, there appears to be new evidence which strongly suggests that the hydroaromatic model is largely correct for the majority of coals; recent work by N. C. Deno (private communication, 1979) has shown that hydroaromatic molecules are excellent models for coal under the conditions of oxidative degradation by trifluoroacetic acid. This work, which is still continuing, suggests that hydroaromatic structures are the necessary precursors of the reaction products obtained. The similarity between 5,12-dihydronaphthacene and the "monomer" proposed by Given (1960) is very striking indeed.

C. Molecular Sieve Structures

The molecular sieve model of coal structure considers coal as consisting of relatively light, hydrogen-rich molecules trapped in a hydrogen-poor carbon matrix (Aczel *et al.,* 1976). The basis for this model is the work reported in Section VI on the low molecular weight compounds of coal. A corollary of the presence of large numbers of these low MW "trapped" compounds is that since they are relatively hydrogen rich and "light," the macromolecular skeleton must consequently be hydrogen poor and of high molecular weight. This could imply that, in the matrix, the ratio of naphthenic to aromatic rings is higher or that there are bridged structures present. Marzec *et al.* (1979a,b) have reported studies on the solvent extraction of high-volatile bituminous coal at ambient temperature with various solvents. The yields varied between 0 and 35% and an attempt was made to correlate the yields with the electron-donor and -acceptor properties of the solvents. The results are shown in Table X. The results indicate that the trend of extract yield increases with the electron-donor properties of the solvent and the electron-acceptor properties affect the yield only to the extent that there is a correlation between the DN − AN values (these values are quantitative measures of the electron-donor and -acceptor properties of the solvents). The results can be interpreted in the light of a macromolecular network containing an amount of smaller volatile or extractable molecules. The macromolecular network is considered to be a linked, three-dimensional molecule as proposed by Larsen and Kovac (1978). This macromolecular network binds the trapped molecules by "donor–acceptor" bonds as shown in Fig. 33. It is suggested that the extractable molecules fill the pores in the network and are sited next to electron-donor and -acceptor centers in the network. Solvents "dislodge" these trapped mole-

TABLE X

Yields of Coal Extracts and Electron-Donor/Acceptor Properties of Solvents[a]

Solvent	Yield of coal extract (wt. % daf coal)	DN	AN	DN − AN
n-Hexane	0.0	0	0	0
Water	0.0	33.0	54.8	− 21.8
Formamide	0.0	24.0	39.8	− 15.8
Acetonitrile	0.0	14.1	19.3	− 5.2
Nitromethane	0.0	2.7	20.5	− 17.8
Isopropanol	0.0	20.0	33.5	− 13.5
Acetic acid	0.9	—	52.9	—
Methanol	0.1	19.0	41.3	− 22.3
Benzene	0.1	0.1	8.2	− 8.1
Ethanol	0.2	20.5	37.1	− 16.6
Chloroform	0.35	—	23.1	—
Dioxane	1.3	14.8	10.8	+ 4.0
Acetone	1.7	17.0	12.5	+ 4.5
Tetrahydrofuran	8.0	20.0	8.0	+ 12.0
Diethyl ether	11.4	19.2	3.9	+ 15.3
Pyridine	12.5	33.1	14.2	+ 18.9
Dimethylsulfoxide	12.8	29.8	19.3	+ 10.5
Dimethylformamide	15.2	26.6	16.0	+ 10.6
Ethylenediamine	22.4	55.0	20.9	+ 34.1
1-Methyl-2-pyrrolidinone	35.0	27.3	13.3	+ 14.0

[a] From Marzec et al. (1979a).

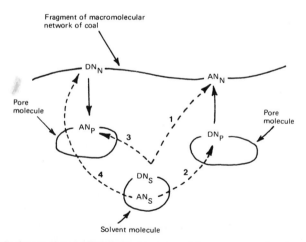

Fig. 33. Coal extraction model showing two types of donor–acceptor bonds present in coal. 1, 2, 3, and 4 represent possible routes of donor–acceptor bond formation between solvent molecule and electron-donor or -acceptor centers in coal (Marzec et al., 1979a).

cules by forming donor-acceptor bonds of higher energy. The aromatic/ hydroaromatic model of coal structure does contain possible centers of this type, for condensed aromatic systems have electron-acceptor properties.

Ebert *et al.* (1979) have used magnetic resonance techniques to study labeled "guest" molecules in coal. Dry coal samples were exposed to vapors of labeled solvents such as C_6D_6 and C_5D_5N. It was noted that more C_5D_5N was taken up by the coals than C_6D_6. ^{2}H NMR studies of the imbibed solvents indicated that the C_6D_6 was relatively free to spin (but was probably sterically confined) but that the C_5D_5N molecules showed an absence of rapid motion, indicating some chemical interaction between the molecules and the coal matrix. These results seem to be consistent with Marzec's findings: pyridine with a DN value of 33.1 yielded 12.5% of the coal as extract, whereas benzene with DN of 0.1 yielded only 0.1%.

Yokono and Sanada (1978) reported nuclear magnetic relaxation studies of coal. They noted that the spin–spin relaxation times indicated that there were two relaxation regions—an immobile region and a relatively free one at room temperature. The results for three Japanese coals are shown in Table XI. It is noteworthy that the mobile region is greater for the 86% C coal than for the 77% or 93% C coals; this is in line with the results reported by Hayatsu *et al.* (1978a,b).

Further indications that there might be two distinct sets of molecules in coal are revealed by the MW distribution obtained by Hombach *et al.* (1979; see Fig. 4), which shows two sets of MW distributions.

D. Structural Models

Any discussion of the macromolecular skeleton of coal tends to be accompanied by illustrations of "model structures." It is worth pointing out that several authors have stressed that these models do not represent the real structure but merely the kinds of structures to be expected in coal (e.g., Whitehurst, 1978). Coal structure, as understood at present, is known only as a series of parameters and a few identified "building blocks," some of which have been discussed in this report. The model structures are interpretations of possible combinations of these factors. Some authors (e.g., Larsen and Kovac, 1978) are not in favor of the use of these "average structures" at all, pointing out that they are easy to derive but that a knowledge of the extremes of the range of structures present is much more useful in predicting the chemical behaviour of coal. The very pronounced heterogeneity of coals would tend to support this view. The dangers of average structures might be illustrated by considering a mixture of naphthalene ($C_{10}H_8$) and decalin ($C_{10}H_{18}$) in the ratio of 3:2—the "average molecule" of this mixture would be $C_{10}H_{12}$ and the carbon aromaticity is 60%. In the absence of other informa-

TABLE XI

Volume Fraction and Spin–Spin Relaxation Time (T_{2n}) of the Two
Phases of Coals at Room Temperature[a]

Coal	C (%)	T_{2n} (μsec)	Fraction (%)
Taiheiyo	77	10.0	70
		24.0	30
Yubarishinko	86	10.8	59
		21.3	41
Hongei	93	11.0	76
		17.5	24

[a] From Yokono and Sanada (1978).

tion this would suggest that the mixture is tetralin (a pure compound). The average structures may only represent some aspects of the true structure and may in fact distort the current knowledge of coal structure by ignoring the diversity of coals. The methods of averaging may also be distorted by using averaged results, for example, MW values, which are themselves distortions of the "true" values. The other danger of coal structural models is that, as Pitt (1979) points out, the construction of models may lead to undue confidence that the chemical constitution of coal is understood. However, if these factors can be borne in mind, it is possible that the coal models could provide a valuable insight into the structures prevalent in coals. The models also have a potentially useful *negative* side; as an example Sternberg's (1975) comments on the Wiser model could be noted. The Wiser model gives the impression that the coal molecules are amphoteric with acidic and basic groups randomly distributed—a view with which Sternberg disagrees. Recent work by Alemany *et al.* (1978) would also side with Sternberg—it strongly suggests that the distribution of the functional groups in coal is not uniform.

E. Comments

It would be rash to point to any of the models presented and choose one as the representation of the molecular structure of coal. In fairness, the idea of *regular* polyamantane structures can be discounted. The molecular sieve model does not really explain the structure of the coal matrix and for that modified hydroaromatic structure or bridged tricycloalkanes need to be considered. It is also unfortunate that the models tend to be seen as competing interpretations. The "classical" hydroaromatic view is largely the result of earlier studies on the relatively high-rank coals of Britain and Germany, and it should be no reflection on its success as a model to suggest that, at least in the lower rank coals, other structures such as bridged tricycloalkanes ought to be considered. Certainly the molecular sieve model has to be modified for different ranks of coal and can be reasonably "superimposed" on other models of coal structure. Indeed, given the diversity of coals, the variety of their starting materials, and their differing geological conditions, it might be more surprising to discover a model which accounted for the whole range of coals than it would be to discover a need to suggest different structures for different coals.

XI. CONCLUSIONS

The work reported in this review has covered a wide variety of investigations into many and various aspects of coal science but, regrettably, it is not possible to say that a coherent picture has emerged. Even if all the research reported was of a consistently high quality, there would remain the virtually insurmountable problem of comparing the results of experiments performed on different coals. The complexity and heterogeneity of coals is the most persistent "problem" in discussing any topic in coal science. The resurgence of interest in coal science, largely within the time scale of the work quoted in this review, has, in general, not been accompanied by any increase in our understanding of the molecular structure of coal. More is known today than yesterday, but as yet the facts (or at least results) have accumulated and have illustrated rather than clarified the lack of understanding of the structure of coal. It would be quite wrong to say that if this is the case then this latest research should not have been undertaken; if recent research has shown anything, it is that earlier theories were based more on supposition than evidence and if the state of knowledge today is little better than that of 30 years ago, then at least recent research has revealed the depths of misunderstanding.

It is fairly obvious that the work reported does not always reach that standard of consistent high quality assumed earlier, but it is not always obvious which research deserves to be considered less seriously. The main problem again is that the reported work is, too often, not comparable with other studies. The (U.S.) National Research Council Committee on Chemical Sciences (1979, p. 36) also noted this problem:

> It is fair to comment that the literature on coal contains too many papers that report results of one particular detailed study of just one coal, or of only a few coals, without presenting data from parallel experiments performed on the same coals. Coal literature describes too many studies that stand in isolation, with no follow-up research in which valuable procedures are applied to additional coals.

Neavel (1979, p. 76) is even more scathing, pointing out that coal science is still mired in the 1950s and that: "The scattered probes of coal structural properties on a bewildering range of poorly selected, poorly collected, poorly prepared, poorly preserved and poorly characterized coal samples will lead us only into further confusion." To this one may add Larsen's comment made in 1975 and still valid today: "We know enough about coal structure to be of help in some respects, but in many of the important areas our knowledge is totally inadequate" (Larsen, 1975, p. 9).

Coal science is not alone in its state of disarray. Speight and Moschopedis (1979), in making some observations on the state of research on the molecular structure of petroleum asphaltenes, report a body of knowledge that suffers from similar defects. They note that attempts made to describe the total structures of petroleum asphaltenes in accordance with NMR data and the results of chemical analyses probably represent extreme uses of spectroscopic and analytical techniques. This point, while adversely criticizing current hypotheses, has a secondary but equally important validity. It is simply that coal science, or rather the coal scientist, is not yet well catered to by available chemical techniques. Chemistry as a science has largely responded to problems posed by single molecules of whatever size or complexity as long as there is a semblance of regularity. It is this lack of regularity in coal itself which means that the coal chemist is forced to extend the boundaries of chemical techniques and knowledge in the examination of coal. It is to be expected, therefore, that a subject which has not yet been incorporated into the corpus of an established discipline should be in a state of confusion, disarray, and promise. For certainly there has been worthwhile research and valuable results, some of which are reported here. It is also true to say, given the growing realization among coal scientists themselves of the state of confusion, that most current and planned research programs are trying to avoid the many pitfalls. This review has attempted to bring together some of the many results obtained in the recent past. If it has failed to weld these results into a coherent picture, that failure may be in part a reflection of the current state of the art in coal science.

REFERENCES

This list of references to the literature reviewed in this report has been extended slightly to include some articles not referred to in the text. Those who are interested in keeping this list up to date may find it useful to scan the section "Chemical properties and analysis" in *Coal Abstracts*.

Abdel-Baset, Z., Given, P. H., and Yarzab, R. F. (1978). *Fuel* 57(2), 95–99.
Aczel, T., Gorbaty, M. L., Maa, P. S., and Schlosberg, R. H. (1975). *Fuel* 54(4), 295.
Aczel, T., Gorbaty, M. L., Maa, P. S., and Schlosberg, R. H. (1976). *Prepr. Coal Chem. Workshop, 1976* CONF-760885, Paper 12, pp. 165–173.
Aczel, T., Gorbaty, M. L., Maa, P. S., and Schlosberg, R. H. (1979). *Fuel* 58(3), 228–230.
Alemany, L. B., King, S. R., and Stock, L. M. (1978). *Fuel* 57(12), 738–748.

Alemany, L. B., Handy, C. I., and Stock, L. M. (1979). *Prepr. Pap.—Am. Chem. Soc., Div. Fuel Chem.* **24**(1), 156–165.

Artemov, A. V., and Peresun'ko, T. F. (1975). *Solid Fuel Chem.* **9**(3), 28–33.

Attar, A., and Dupuis, F. (1979). *Prepr. Pap.—Am. Chem. Soc., Div. Fuel Chem.* **24**(1), 166–177.

Bartle, K. D., and Jones, D. W. (1978). *In* "Analytical Methods for Coal and Coal Products" (C. Karr, Jr., ed.), Vol. 2, pp. 104–160. Academic Press, New York.

Bartle, K. D., Martin, T. G., and Williams, D. F. (1975). *Fuel* **54**(4), 226–235.

Bartle, K. D., Jones, D. W., Pakdel, H., Snape, C. E., Calimli, A., Olcay, A., and Tugrul, T. (1979a). *Nature (London)* **277**, 284–287.

Bartle, K. D., Ladner, W. R., Martin, T. G., Snape, C. E., and Williams, D. F. (1979b). *Fuel* **58**(6), 413–422.

Bartle, K. D., Calimli, A., Jones, D. W., Matthews, R. S., Olcay, A., Pakdel, H., and Tugrul, T. (1979c). *Fuel* **58**(6), 423–428.

Bartle, K. D., Collin, G., Stadelhofer, J. W., and Zander, M. (1979d). *J. Chem. Technol. Biotechnol.* **29**(9), 531–551.

Bartuska, V. J., Maciel, G. E., Schaefer, J., and Stejskal, E. O. (1976). *Prepr. Coal Chem. Workshop, 1976* CONF-760885, Paper 17, pp. 220–228.

Bartuska, V. J., Maciel, G. E., Schaefer, J., and Stejskal, E. O. (1977). *Fuel* **56**(4), 354–358.

Bartuska, V. J., Maciel, G. E., and Miknis, F. P. (1978a). *Prepr. Pap.—Am. Chem. Soc., Div. Fuel Chem.* **23**(2), 19–23.

Bartuska, V. J., Maciel, G. E., Miknis, F. P., and Netzel, D. A. (1978b). *Prepr. Pap.—Am. Chem. Soc., Div. Fuel Chem.* **23**(4), 132–137 (CONF-780902-P2).

Battaerd, H. A. J., and Evans, D. G. (1979). *Fuel* **58**(2), 105–108.

Belyi, A. A., and Voitkovskii, Yu. B. (1979). *Khim. Tverd. Topl. (Moscow)* No. 1, pp. 21–25.

Bimer, J., Given, P. H., and Raj, S. (1978). *ACS Symp. Ser.* **71**, 86–99.

Blessing, J. E., and Ross, D. S. (1978). *ACS Symp. Ser.* **71**, 171–185.

Brown, J. K., and Ladner, W. R. (1960). *Fuel* **39**, 87–96.

Camier, R. J., and Siemon, S. R. (1978). *Fuel* **57**(8), 508–510.

Camier, R. J., and Siemon, S. R. (1979). *Fuel* **58**(1), 67–69.

Camier, R. J., Siemon, S. R., Battaerd, H. A. J., and Stanmore, B. R. (1979). *Prepr. Pap.—Am. Chem. Soc., Div. Fuel Chem.* **24**(1), 203–209.

Chakrabartty, S. K. (1975). *In* "The Fundamental Organic Chemistry of Coal" (J. W. Larsen, ed.), PB-264119, pp. 89–104. University of Tennessee, Knoxville.

Chakrabartty, S. K. (1978). *ACS Symp. Ser.* **71**, 100–107.

Chakrabartty, S. K., and Berkowitz, N. (1974). *Fuel* **53**(4), 240–245.

Chakrabartty, S. K., and Berkowitz, N. (1976a). *Nature (London)* **261**, 76–77.

Chakrabartty, S. K., and Berkowitz, N. (1976b). *Fuel* **55**(4), 362–363.

Chakrabartty, S. K., and Kretschmer, H. O. (1972). *Fuel* **51**(2), 160–163.

Chakrabartty, S. K., and Kretschmer, H. O. (1974). *Fuel* **53**(2), 132–135.

Chung, K. E., Anderson, L. L., and Wiser, W. H. (1979). *Fuel* **58**(12), 847–852.

Collins, C. J., Hombach, H. P., Benjamin, B. M., Roark, W. H., Maxwell, B., and Raaen, V. F. (1979). *Prepr. Pap.—Am. Chem. Soc., Div. Fuel Chem.* **24**(1), 191–196.

Deno, N. C., Greigger, B. A., and Stroud, S. G. (1978a). *Prepr. Pap.—Am. Chem. Soc., Div. Fuel Chem.* **23**(2), 54–57.

Deno, N. C., Greigger, B. A., and Stroud, S. G. (1978b). *Fuel* **57**(8), 455–459.

Deno, N. C., Greigger, B. A., Jones, A. D., Rakitsky, W. G., and Stroud, S. G. (1979). *In* "Coal Structure and Coal Liquefaction," Final report, EPRI-AF-960. Electric Power Res. Inst., Palo Alto, California.

Doğru, R., Erbatur, G., Gaines, A. F., Yurum, Y., Icli, S., and Wirthlin, T. (1978). *Fuel* **57**(7), 399–404.

Drake, J. A. G., Jones, D. W., Causey, B. S., and Kirkbright, G. F. (1978). *Fuel* 57(11), 663–666.

Dryden, I. G. C. (1979). *Fuel* 58(12), 902.

Ebert, L. B., Long, R. B., Schlosberg, R. H., and Silbernagel, B. G. (1979). *Prepr. Pap.—Am. Chem. Soc., Div. Fuel Chem.* 23(4), 104–108.

Elliott, M. A. (1981). "Chemistry of Coal Utilization," 2nd Suppl. Vol. Wiley, New York.

Ensminger, J. T. (1977). *In* "Environmental, Health and Control Aspects of Coal Conversion" (H. M. Braunstein *et al.*, eds.), ORNL/EIS-94, Vol. 1, pp. 2-1—2-104. Oak Ridge Natl. Lab., Oak Ridge, Tennessee.

Farcasiu, M. (1979). *Prepr. Pap.—Am. Chem. Soc., Div. Fuel Chem.* 24(1), 121–130.

Fischer, P., Stadelhofer, J. W., and Zander, M. (1978). *Fuel* 57(6), 345–352.

Francis, W. (1961). "Coal: Its Formation and Composition." Arnold, London.

Franz, J. A., Morrey, J. R., Campbell, J. A., Tingey, G. L., Pugmire, R. J., and Grant, D. M. (1973). "Inferences on the Structure of Coal: C-13 nmr and IR Spectroscopy," BNWL-SA-5457. Battelle Pac. Northwest Labs., Richland, Washington.

Fuller, E. L. (1979). *Prepr. Pap.—Am. Chem. Soc., Div. Fuel Chem.* 24(2), 358–363.

Gartsman, B. B., Rumyantseva, Z. A., Grishin, N. N., and Zaikin, V. G. (1979). *Khim. Tverd. Topl. (Moscow)* No. 2, pp. 105–111.

Gerstein, B. C., Chow, C., Pembleton, R. G., and Wilson, R. C. (1977). *J. Phys. Chem.* 81(6), 565–570.

Gerstein, B. C., Ryan, L. N., and Murphy, P. D. (1979). *Prepr. Pap.—Am. Chem. Soc., Div. Fuel Chem.* 24(1), 90–95.

Ghosh, G., Banerjee, A., and Mazumdar, B. K. (1975). *Fuel* 54(4), 294–295.

Gibson, J. (1978). *J. Inst. Fuel* 51, 67–81.

Given, P. H. (1960). *Fuel* 39, 147–153.

Given, P. H. (1975). *In* "The Fundamental Chemistry of Coal" (J. M. Larsen, ed.), PB-264119, pp. 42–57. University of Tennessee, Knoxville.

Given, P. H. (1978). *Proc. Coal Chem. Workshop, 1978* CONF-780372, pp. 53–67.

Hayatsu, R., Scott, R. G., Moore, L. P., and Studier, M. H. (1975). *Nature (London)* 257, 378–380.

Hayatsu, R., Scott, R. G., Moore, L. P., and Studier, M. H. (1976). *Nature (London)* 261, 77.

Hayatsu, R., Winans, R. E., Scott, R. G., Moore, L. P., and Studier, M. H. (1978a). *Fuel* 57(9), 541–548.

Hayatsu, R., Winans, R. E., Scott, R. G., Moore, L. P., and Studier, M. H. (1978b). *Nature (London),* 275, 116–118.

Hayatsu, R., Winans, R. E., Scott, R. G., Moore, L. P., and Studier, M. H. (1978c). *ACS Symp. Ser.* 71, 108–125.

Hayatsu, R., Winans, R. E., McBeth, R. L., Scott, R. G., Moore, L. P., and Studier, M. H. (1979a). *Prepr. Pap.—Am. Chem. Soc., Div. Fuel Chem.* 24(1), 110–120.

Hayatsu, R., Winans, R. E., McBeth, R. L., Scott, R. G., Moore, L. P., and Studier, M. H. (1979b). *Nature (London)* 278, 41–43.

Heredy, L. A. (1979). *Prepr. Pap.—Am. Chem. Soc., Div. Fuel Chem.* 24(1), 142–155.

Hirsch, P. B. (1954). *Proc. R. Soc., London, Ser. A* 226, 143–169.

Hodek, W., and Koelling, G. (1973). *Fuel* 52(3), 220–225.

Hodek, W., Meyer, F., and Koelling, G. (1977). *Prepr., Div. Pet. Chem., Am. Chem. Soc.* 22(2), 462–470.

Hombach, H. P., Hodek, W., and Koelling, G. (1979). *Erdoel Kohle, Erdgas, Petrochem.* 32(3), 134.

Hooper, R. J., and Evans, D. G. (1979). *Prepr. Pap.—Am. Chem. Soc., Div. Fuel Chem.* 24(1), 131–141.

Huston, J. L., Scott, R. G., and Studier, M. H. (1976). *Fuel* 55(4), 281–286.

Ignasiak, B. S., and Gawlak, M. (1977). *Fuel* 56(2), 216–222.

Ignasiak, B. S., Ignasiak, T. M., and Berkowitz, N. (1975). *Rev. Anal. Chem.* **2**(3), 278–298.

Ignasiak, B. S., Chakrabartty, S. K., and Berkowitz, N. (1978a). *Fuel* **57**(8), 507–511.

Ignasiak, B. S., Fryer, J. F., and Jadernik, P. (1978b). *Fuel* **57**(10), 578–584.

Ignasiak, B. S., Carson, D., and Gawlak, M. (1979). *Fuel* **58**(11), 833–834.

Ivashchenko, L. I., Pershko, A. A., Radaev, E. F., and Khabalov, V. V. (1974). *Solid Fuel Chem.* **8**(1), 13–16.

Kanda, N., Itoh, H., Yokoyama, S., and Ouchi, K. (1978). *Fuel* **57**(11), 676–680.

Kukharenko, T. A. (1977). *Solid Fuel Chem.* **11**(3), 57–62.

Kwan, C. L., and Yen, T. F. (1979). *Anal. Chem.* **51**(8), 1225–1229.

Ladner, W. R., and Snape, C. E. (1978). *Fuel* **57**(11), 658–662.

Ladner, W. R., and Wheatley, R. (1965). *Br. Coal Util. Res. Assoc. Mon. Bull.* **29** (7), 201–231.

Landolt, R. G. (1975). *Fuel* **54**(4), 299.

Larsen, J. W. (1975). *In* "The Fundamental Organic Chemistry of Coal" (J. W. Larsen, ed.), PB-264119, pp. 1–9. University of Tennessee, Knoxville.

Larsen, J. W. (1978). *Proc. Coal Chem. Workshop, 1978* CONF-780372, pp. 39–51.

Larsen, J. W., and Choudhury, P. (1979). *J. Org. Chem.* **44**(16), 2856–2859.

Larsen, J. W., and Given, P. H. (1978). *Proc. Coal Chem. Workshop, 1978* CONF-780372, pp. 69–72.

Larsen, J. W., and Kovac, J. (1978). *ACS Symp. Ser.* **71**, 36–49.

Larsen, J. W., and Kuemmerle, E. W. (1976). *Fuel* **55**(3), 162–169.

Larsen, J. W., and Urban, L. O. (1979). *J. Org. Chem.* **44**(18), 3219–3222.

Larsen, J. W., Choudhury, P., and Urban, L. O. (1978). *Prepr. Pap.—Am. Chem. Soc., Div. Fuel Chem.* **23**(4), 181–184.

Lazarov. L., and Angelova, G. (1976). *Solid Fuel Chem.* **10**(3), 12–18.

Lowry, H. H., ed. (1963). "Chemistry of Coal Utilization," Suppl. vol. Wiley, New York.

Maciel, G. E., Bartuska, V. J., and Miknis, F. P. (1978). *Prepr. Pap.—Am. Chem. Soc., Div. Fuel Chem.* **23**(4), 120–123.

Maciel, G. E., Bartuska, V. J., and Miknis, F. P. (1979). *Fuel* **58**(5), 391–394.

McIntosh, M. J. (1976). *Fuel* **55**(1), 59–62.

Makabe, M., and Ouchi, K. (1979). *Fuel* **58**(1), 43–47.

Makabe, M., Hirano, Y., and Ouchi, K. (1978). *Fuel* **57**(5), 289–292.

Marzec, A. (1979). *Koks, Smola, Gaz* **24**(2/3), 42–47.

Marzec, A., Juzwa, M., Betlej, K., and Sobkowiak, M. (1979a). *Fuel Process. Technol.* **2**(1), 35–44.

Marzec, A., Juzwa, M., and Sobkowiak, M. (1979b). *In* "Gasification and Liquefaction of Coal," COAL/SEM. 6/R.62. Economic Commission for Europe, United Nations.

Mayo, F. R. (1975). *Fuel* **54**(4), 273–275.

Mayo, F. R., and Kirshen, N. A. (1978a). *Fuel* **57**(7), 405–408.

Mayo, F. R., and Kirshen, N. A. (1978b). *In* "Homogeneous Catalytic Hydrocracking Processes for Conversion of Coal to Liquid Fuels: Basic and Exploratory Research," Bienn. Rep., FE-2202-24, pp. 97–119. U.S. Department of Energy, Washington, D.C.

Mayo, F. R., and Kirshen, N. A. (1979). *Fuel* **58**(10), 698–704.

Mayo, F. R., Huntington, J. G., and Kirshen, N. A. (1976). *Prepr. Coal Chem. Workshop, 1976* CONF-760885, Paper 14, pp. 189–201.

Mayo, F. R., Huntington, J. G., and Kirshen, N. A. (1978). *ACS Symp. Ser.* **71**, 126–130.

Miknis, F. P., Maciel, G. E., and Bartuska, V. J. (1979). *Prepr. Pap.—Am. Chem. Soc., Div. Fuel Chem.* **24**(2), 327–333.

Miyagawa, I., and Alexander, C. (1979). *Nature (London)* **278**, 40–41.

National Research Council Committee on Chemical Sciences (1979). "The Department of Energy: Some Aspects of Basic Research in the Chemical Sciences." Natl. Acad. Sci., Washington, D.C.

Neavel, R. C. (1979). *Prepr. Pap.—Am. Chem. Soc., Div. Fuel Chem.* **24**(1), 73–82.

Niemann, K., and Hombach, H. P. (1979). *Fuel* **58**(12), 853–856.

Oka, M., Chang, H. C., and Gavalas, G. R. (1977). *Fuel* **56**(1), 3–8.

Ouchi, K., Iwata, K., Makabe, M., and Itoh, H. (1979). *Prepr. Pap.—Am. Chem. Soc., Div. Fuel Chem.* **24**(1), 185–190.

Petrakis, L., and Grandy, D. W. (1978). *Anal. Chem.* **50**(2), 303–308.

Pines, A., and Wemmer, D. E. (1976). *Prepr. Coal Chem. Workshop, 1976* CONF-760885 Paper 18, pp. 229–232.

Pines, A., and Wemmer, D. E. (1978). *Prepr. Pap.—Am. Chem. Soc., Div. Fuel Chem.* **23**(2), 15–18.

Pines, A., Gibby, M. G., and Waugh, J. S. (1972). *Chem. Phys. Lett.* **15**(3), 373–376.

Pines, A., Gibby, M. G., and Waugh, J. S. (1973). *J. Chem. Phys.* **59**(2), 569–590.

Pitt, G. J. (1979). *In* "Coal and Modern Coal Processing: An Introduction" (G. J. Pitt and G. R. Millward, eds.), pp. 27–50. Academic Press, New York.

Raj, S. (1976). Ph.D. Thesis, Pennsylvania State University, University Park.

Retcofsky, H. L. (1975). *In* "The Fundamental Organic Chemistry of Coal" (J. W. Larsen, ed.), PB-264119, pp. 59–79. University of Tennessee, Knoxville.

Retcofsky, H. L. (1977). *Appl. Spectrosc.* **31**(2), 116–121.

Retcofsky, H. L. (1978). *DOE Symp. Ser.* **46**, 79–97.

Retcofsky, H. L., and Friedel, R. A. (1973). *J. Phys. Chem.* **77**(1), 68–71.

Retcofsky, H. L., and Friedel, R. A. (1976). *Fuel* **55**(4), 363–364.

Retcofsky, H. L., and Link, T. A. (1978). *In* "Analytical Methods for Coal and Coal Products" (C. Karr, Jr., ed.), Vol. 2, pp. 161–207. Academic Press, New York.

Retcofsky, H. L., and VanderHart, D. L. (1978). *Fuel* **57**(7), 421–423.

Retcofsky, H. L., Schweighardt, F. K., and Hough, M. (1977). *Anal. Chem.* **49**(4), 585–588.

Retcofsky, H. L., Thompson, G. P., Hough, M., and Friedel, R. A. (1978). *ACS Symp. Ser.* **71**, 142–155.

Retcofsky, H. L., Hough, M. R., and Clarkson, R. B. (1979). *Prepr. Pap.—Am. Chem. Soc., Div. Fuel Chem.* **24**(1), 83–89.

Ross, D. S., and Blessing, J. E. (1979a). *Fuel* **58**(6), 433–437.

Ross, D. S., and Blessing, J. E. (1979b). *Fuel* **58**(6), 438–442.

Ruberto, R. G., and Cronauer, D. C. (1978). *ACS Symp. Ser.* **71**, 50–70.

Ruberto, R. G., Cronauer, D. C., Jewell, D. M., and Kalkunte, S. S. (1977a). *Fuel* **56**(1), 17–24.

Ruberto, R. G., Cronauer, D. C., Jewell, D. M., and Kalkunte, S. S. (1977b). *Fuel* **56**(1), 25–32.

Rus'yanova, N. D., and Erkin, L. I. (1978). *Khim. Tverd. Topl. (Moscow)* No. 6, pp. 3–15.

Sato, Y., and Yoshii, T. (1979). *Fuel* **58**(8), 619–621.

Scaroni, A. W., and Essenhigh, R. H. (1978). *Prepr. Pap.—Am. Chem. Soc., Div. Fuel Chem.* **23**(4), 124–131.

Schlick, S., Narayana, P. A., and Kevan, L. (1978). *J. Am. Chem. Soc.* **100**(11), 3322–3326.

Shapiro, M. D., and Al'terman, L. S. (1977). *Solid Fuel Chem.* **11**(3), 13–17.

Speight, J. G. (1978). *In* "Analytical Methods for Coal and Coal Products" (C. Karr, Jr., ed.), Vol. 2, pp. 75–101. Academic Press, New York.

Speight, J. G., and Moschopedis, S. E. (1979). *Prepr., Div. Pet. Chem., Am. Chem. Soc.* **24**(4), 910–923.

Stephens, J. F. (1979a). *Prepr. Pap.—Am. Chem. Soc., Div. Fuel Chem.* **24**(2), 347–357.

Stephens, J. F. (1979b). *Fuel* **58**(7), 489–494.

Sternberg, H. W. (1975). *In* "Research in Coal Technology: The University's Role," CONF-741091, pp. 11–57—11–72. State University of New York, Buffalo.

Sternberg, H. W., and Delle Donne, C. L. (1974). *Fuel* **53**(3), 172–175.

Sternberg, H. W., Delle Donne, C. L., Pantages, P., Moroni, E. C., and Markby, R. E. (1971). *Fuel* **50**(4), 432–442.

Stock, L. M. (1978). "Coal Anion Structure and Chemistry of Coal Alkylation," Quarterly Reports, COO-4227. University of Chicago, Chicago, Illinois.

Studier, M. H., Hayatsu, R., and Winans, R. E. (1978). *In* "Analytical Methods for Coal and Coal Products" (C. Karr, Jr., ed.), Vol. 2, pp. 43–74. Academic Press, New York.

Sun, J. Y., and Burk, E. H. (1975). *In* "The Fundamental Organic Chemistry of Coal" (J. W. Larsen, ed.), PB-264119, pp. 80–89. University of Tennessee, Knoxville.

Tingey, G. L., and Morrey, J. R. (1973). "Coal Structure and Reactivity," TID-26637. Battelle Pac. Northwest Labs., Richland, Washington.

Tingey, G. L., Morrey, J. R., Campbell, J. A., and Franz, J. A. (1975). *Prepr. Pap.—Am. Chem. Soc., Div. Fuel Chem.* **20**(3), 11 (abstr. only).

U.S. Department of Energy (1979). "Coal Chemistry Research: Summary of DOE-BES/FE Research Projects, May 30–31, 1979." U.S. Department of Energy, Washington, D.C.

Vahrman, M. (1970). *Fuel* **49**(1), 5–16.

Vahrman, M. (1972). *Chem. Br.* **8**, 16–24.

VanderHart, D. L., and Retcofsky, H. L. (1976a). *Fuel* **55**(3), 202–204.

VanderHart, D. L., and Retcofsky, H. L. (1976b). *Prepr. Coal Chem. Workshop, 1976.* CONF-760885, Paper 15, pp. 202–218.

Van Krevelen, D. W. (1961). "Coal: Typology—Chemistry—Physics–Constitution." Elsevier, Amsterdam.

Wachowska, H. (1979). *Fuel* **58**(2), 99–104.

Wachowska, H. M., Nandi, B. N., and Montgomery, D. S. (1979). *Fuel* **58**(4), 257–263.

Whitehurst, D. D. (1978). *ACS Symp. Ser.* **71**, 1–35.

Whitehurst, D. D., Farcasiu, M., and Mitchell, T. O. (1976). *In* "The Nature and Origin of Asphaltenes in Processed Coals," EPRI-AF-252. Electric Power Res. Inst., Palo Alto, California.

Whitehurst, D. D., Farcasiu, M., and Mitchell, T. O. (1977). *In* "The Nature and Orgin of Asphaltenes in Processed Coals." EPRI-AF-480. Electric Power Res. Inst., Palo Alto, California.

Williams, D. F. (1977). *In* "Meeting on Basic Coal Science," ICTIS/M0015, IEA Working Party on Coal Technology, Paper 4 (available from IEA Coal Research).

Winans, R. E., Hayatsu, R., Scott, R. G., Moore, L. P., and Studier, M. H. (1976). "Examination and Comparison of Structure: Lignite, Bituminous, and Anthracite Coal," CONF-760885-1. Argonne Natl. Lab., Argonne, Illinois.

Winans, R. E., Hayatsu, R., McBeth, R. L., Scott, R. G., Moore, L. P., and Studier, M. H. (1979). *Prepr. Pap.—Am. Chem. Soc., Div. Fuel Chem.* **24**(2), 196–207.

Wyss, W. F. (1977). *In* "Meeting on Basic Coal Science," ICTIS/M0019, IEA Working Party on Coal Technology, Paper 8 (available from IEA Coal Research).

Yarzab, R. F., Abdel-Baset, Z., and Given, P. H. (1979). *Geochim. Cosmochim. Acta* **43**, 281–287.

Yokono, T., and Sanada, Y. (1978). *Fuel* **57**(6), 334–336.

Yoshii, T., and Sato, Y. (1979). *Fuel* **58**(7), 534–538.

Zilm, K. W., Pugmire, R. J., Grant, D. M., Wood, R. E., and Wiser, W. H. (1979). *Fuel* **58**(1), 11–16.

The Reductive Alkylation Reaction

LEON M. STOCK

Department of Chemistry
University of Chicago
Chicago, Illinois

I.	Introduction	161
II.	The Chemistry of the Reaction	162
	A. The Reduction Process	162
	B. The Cleavage Reaction	169
	C. The Alkylation Reaction	181
III.	Methods and Procedures	187
	A. The Reduction Process	187
	B. The Alkylation Reaction	197
	C. The Isolation of the Products	202
	D. Comparison of the Methods	210
IV.	Applications of Reduction, Alkylation, and Reductive Alkylation	218
	A. The Reduction of Coal	218
	B. The Alkylation of Coal	242
	C. The Reductive Alkylation of Coal	248
	References	279

I. INTRODUCTION

In 1968, Sternberg and his associates reported that the treatment of Pocahontas lvB coal with alkali metals in the presence of naphthalene in tetrahydrofuran generated anionic coal fragments with between 10 and 15 negative charges per 100 carbon atoms (Sternberg and Delle Donne, 1968; Sternberg *et al.*, 1970, 1971). Alkylation of the anionic material with methyl, ethyl, or *n*-butyl iodide gave polyalkylated products that had 7–9 additional alkyl groups per 100 carbon atoms. About 90% of the product had molecular

COAL SCIENCE
Volume 1

weights less than 20,000. More important, the alkylated products were much more soluble than the original coal in hexane and benzene. Sternberg proposed that the reaction occurred via an electron transfer process in which naphthalene served as a vehicle for the transfer of electrons from the insoluble metal to the insoluble coal. The anionic coal fragments that are formed in this reaction have a rich chemistry. Labile carbon–oxygen and carbon–carbon bonds cleave under the experimental conditions. However, the conditions are mild enough to ensure that extensive structural rearrangement reactions do not occur. Nevertheless, the cleavage reactions and the reduction reactions of the aromatic compounds in the coal disrupt the coal structure sufficiently that the products formed in the subsequent alkylation reactions are significantly soluble in organic solvents.

Sternberg's discovery of a mild procedure for the conversion of coal to lower molecular weight materials that are soluble in common solvents represented an important advance because many previously obscured issues could readily be addressed through the study of the more soluble alkylated coals. In addition, the study of the factors governing the course of the reaction also provide basic information concerning the structure of the coal. In the past decade, many coal scientists have examined the reaction to gain a more fundamental understanding of the chemistry of coal.

This review will be primarily concerned with the reductive alkylation reaction and with other pertinent procedures such as reduction without alkylation and alkylation without reduction because, although many electrophilic reagents react rapidly with the coal polyanion, almost all of the research undertaken in this area has focused on the reactions of these anions with proton donors and with alkylating agents. Reactions such as acylation, carboxylation, and hydroxyalkylation have not as yet been studied as thoroughly. However, this situation will change rapidly if the successes observed by Niemann and Richter (1979) in their preliminary study of the hydroxymethylation and carboxylation of a bituminous coal [89% C daf (dry ash-free)] can be extended. All these reactions can provide a broad spectrum of information concerning the structural characteristics and reactivity of coal.

II. THE CHEMISTRY OF THE REACTION

A. The Reduction Process

The chemical reduction of coal can be accomplished in several different ways. The three methods that have been used most frequently to initiate reductive alkylation reactions involve an alkali metal and an electron transfer agent in an ether such as tetrahydrofuran, a blue solution of an alkali metal in an ether such as triglyme,[1] or a blue solution of an alkali metal in liquid ammonia. Most investigators have followed the suggestion of Sternberg and his co-workers that an electron transfer agent be employed in the reactions in etheral solvents to increase the rate of the transfer of the electron from the in-

$$\text{Coal} + n\text{K} \xrightarrow{\text{C}_{10}\text{H}_8} n\text{K}^+, \text{Coal}^{n-}$$

soluble potassium to the insoluble coal. Naphthalene has been frequently employed as the electron transfer agent. Both the anion radical and the dianion are formed under the experimental conditions.[2] Although diglyme has been used as a solvent for this reaction, most investigators have followed

$$\text{C}_{10}\text{H}_8 + \text{K} \rightleftharpoons \text{C}_{10}\text{H}_8^{\cdot -}, \text{K}^+$$

$$\text{C}_{10}\text{H}_8^{2-}, \text{K}^+ + \text{K} \rightleftharpoons \text{C}_{10}\text{H}_8^{2-}, 2\text{K}^+$$

Sternberg's procedure and used tetrahydrofuran.

The reduction reaction can also be accomplished by blue solutions of the alkali metals in ether and in ammonia. The reactivity of these solutions can be

$$\text{M} \rightleftharpoons \text{M}^+ + \text{e}^-$$

gauged by the potentials for lithium, sodium, and potassium in liquid ammonia, which are 2.34, 1.89, and 2.04 V, respectively, at 25°C (Jolly, 1956). The elimination of naphthalene from the reaction mixture considerably simplifies the methods used for the isolation of the alkylated coal product. Consequently, homogeneous solutions of alkali metal cations and solvated electrons have been used more frequently in the more recent work.

The chemistry of lithium, sodium, and potassium in liquid ammonia and in liquid solutions of low molecular weight amines such as methyl- and ethylamine and the reactions of organic compounds in these solutions have been discussed quite thoroughly in several critical reviews and monographs (Smith, 1963; House, 1972). On the other hand, the blue solutions of alkali metals in ethers have not received as much attention. Potassium and potassium from sodium–potassium alloy dissolve in ethers to a limited extent to form relatively stable blue solutions of potassium ions and solvated electrons below 0°C (Down *et al.,* 1957, 1959; Cafasso and Sundheim, 1959). The solutions become unstable near ambient temperatures as the blue color fades and reactions with the solvent occur. The concentration of potassium in

---

[1] The glyme notation will be used in this review. The prefix (tri) designates the number of 1,2-dihydroxyethane (gly) units in the structure and the suffix (me) designates the end group as methyl. To illustrate, $CH_3OCH_2CH_2OCH_3$ is monoglyme and $CH_3CH_2OCH_2CH_2OCH_2CH_2$-$OCH_2CH_3$ is diglyet.

[2] The nomenclature used by *Chemical Abstracts* will be used for the anions and dianions of the aromatic hydrocarbons. For example, the anion radical of naphthalene is naphthalene(-1) and the sodium salt of the dianion of anthracene is sodium anthracene(-2).

TABLE I

**Association Constants for Sodium and Potassium Ions
with Ethers in Methanol at 25°C** [a]

	Log K_a	
Compound	Na$^+$	K$^+$
Tetraglyme	1.28	1.72
Hexaglyme	1.60	2.55
Heptaglyme	1.67	2.87
1,2-$C_6H_4(O(CH_2CH_2O)_2CH_3)_2$	1.44	2.15
1,2-$C_6H_4(O(CH_2CH_2O)_3CH_3)_2$	1.61	2.83
1,2-$C_6H_4(O(CH_2CH_2O)_4CH_3)_2$	1.74	3.30

[a] From Chaput *et al.* (1975).

monoglyme is about $5 \times 10^{-4}\,M$ at $-50°C$ and about $6 \times 10^{-4}\,M$ at $-60°C$. More concentrated and apparently more kinetically stable solutions can be obtained when polyethers such as tetraglyme or octaglyme are present in the solution. The association constants for potassium ion in methanol provide a measure of the relative effectiveness of the coordination of ethers such as hep-

$$K^+ + Ether \underset{CH_3OH}{\overset{}{\rightleftarrows}} K^+(Ether)$$

taglyme and the dihydroxybenzene derivative (**1**) (Table I).

1

Crown ethers such as dicyclohexyl-18-crown-6 (**2**), and cryptate compounds such as **3** also promote the solution of alkali metals in ethers and amines (Dye *et al.*, 1970; Kaempf *et al.*, 1974; Komarynsky and Weissman, 1975). Although a discussion of the arguments is beyond the scope of this

2

3

review, it is pertinent that Dye and his associates conclude that the concentration of Na$^-$ is much greater than the concentration of solvated electrons in the presence of **2**. The role of Na⁻ in the chemistry of metal reductions has not been examined as yet.

When the crown ether or the cryptate is dissolved in benzene or toluene in the presence of a potassium mirror, the alkali metal dissolves to form a blue solution at ambient temperature. Benzene(-1) and toluene(-1) form much more readily in the presence of these complexing agents than in their absence (Kaempf *et al.*, 1974; Komarynsky and Weissman, 1975). In addition, the rate constant for the electron transfer reaction is about 10^4 smaller than the rate constant expected for the diffusion limited reaction that is usually observed in

$$K^{+} \cdot 2, C_6H_6^{\, \bar{\cdot}} + C_6H_6 \rightleftharpoons K^{+} \cdot 2, C_6H_6 + C_6H_6^{\, \bar{\cdot}}$$

other solvents (Komarynsky and Weissman, 1975). Nevertheless, the benzene and toluene ion pairs are exceedingly reactive by virtue of their high energy content.

The reactions of organic molecules with reducing agents in ether differ from the reactions of the same molecules with reducing agents in ammoniacal solvents. The subsequent chemical reactions of the reduced compounds with other organic reagents also differ. Consequently, the outcome of the reduction reaction and the reductive alkylation reaction depends upon the manner in which the reaction is initiated. The nature of the coal sample also plays a very important role in the outcome of the reaction. The aromatic molecules separated from the coal by fragmentation reactions and the molecules freed from the coal by the collapse of the solid matrix serve to transport electrons from the insoluble metal to the remaining portions of the coal. Thus, while the reactions between the coal and the metal may be limited by the rate of electron transfer in the initial states of the reduction, the coal molecules that are extruded from the solid coal increase the rate at which the reduction occurs, and, more important, these coal molecules level the differences in the original reaction conditions as the very rapid electron exchange reactions occur between solvated electrons and the soluble coal molecules or between the anions and dianions of naphthalene and these soluble molecules. These soluble entities become important for the transport of electrons from the solid metal to the reactive acceptors in the solid coal.

Etheral solutions of alkali cations and the anions and dianions derived from aromatic hydrocarbons exhibit a complex series of equilibria involving both monomeric and dimeric diamagnetic and paramagnetic species virtually all of which are highly associated in concentrated solutions (Szwarc, 1969; Garst, 1969; Screttas and Micha-Screttas, 1981). The principal acceptors in the coal are the aromatic fragments, Coal—ArH, which are reduced to radical anions and dianions as illustrated for the reactions with naphthalene(-1) and naphthalene(-2). The extent of the reaction depends upon the reduction

TABLE II

Half-Wave Reduction Potentials for Representative Aromatic Hydrocarbons[a]

Compound	Half-wave reduction potential, $-e_{1/2}$ vs. sce (V)		
	In 75% dioxane–water	In 96% dioxane–water	In acetonitrile[b]
Biphenyl	2.70		2.70
Naphthalene	2.50	2.60	2.63
Phenanthrene	2.46		
Styrene	2.35		
Pyrene	2.12	2.10	2.19
			2.64
Anthracene	1.96	1.98	2.07
			2.52
Perylene	1.67	1.43	1.73
			2.21
			2.70
Tetracene	1.58	1.58	

[a] Abstracted from the information presented by Holy (1974). The data are drawn from Gough and Peover (1964) and Hoijtink (1970).

[b] The second and third values refer to the two and three electron reduction processes.

$$\text{Coal—ArH} + \text{C}_{10}\text{H}_8^{\cdot-} \rightleftharpoons \text{Coal—ArH}^{\cdot-} + \text{C}_{10}\text{H}_8$$

$$\text{Coal—ArH}^{\cdot-} + \text{C}_{10}\text{H}_8^{\cdot-} \rightleftharpoons \text{Coal—ArH}^{2-} + \text{C}_{10}\text{H}_8$$

$$\text{Coal—ArH}^{\cdot-} + \text{C}_{10}\text{H}_8^{2-} \rightleftharpoons \text{Coal—ArH}^{2-} + \text{C}_{10}\text{H}_8^{\cdot-}$$

potential of the aromatic residues within the coal. The polarographic half-wave reduction potentials for several common aromatic compounds are presented in Table II for convenient comparison.

The polarographic half-wave reduction potentials, electron affinities, and the energy content of the lowest unoccupied molecular orbital are closely related quantities which measure the relative thermodynamic stabilities of the hydrocarbon anion radicals and dianions. The half-wave potentials for this representative group range from 1.6 V for tetracene to 2.7 V for biphenyl. The introduction of a heteroatom into these compounds enhances the reduction reaction. Many heterocyclic compounds are reduced readily. Pyridine and

4

quinoline react with lithium in the absence of a proton donor to yield dimers (4) in rapid reactions (Ward, 1961; Chaudhuri *et al.*, 1968). Quinoline, carbazole, and isoquinoline are reduced to dihydro compounds by metals in liquid ammonia (Smith, 1963). Dibenzofuran and dibenzothiophene are also readily reduced and yield dihydro compounds upon treatment with sodium in liquid ammonia (Smith, 1963). Thus, the use of biphenyl and naphthalene as the electron transfer agents in the reduction reaction of coal virtually assures that all the aromatic residues in the structure except the less reactive phenyl derivatives will be reduced.

The outcome of the reaction depends, of course, on (1) the metal-to-hydrocarbon mole ratio; (2) the reduction potentials of the metal, the electron transfer agent, and its negative ion; and (3) the concentration of the ions in the solution. With one equivalent of the metal, biphenyl and naphthalene are readily converted to anion radicals. With two equivalents of the metal, dianions are formed from hydrocarbons such as naphthalene, anthracene, and phenanthrene. However, the equilibrium constant for disproportionation of the anion radical is very small (Szwarc, 1969). Consequently, the dianions are not present in appreciable concentration until sufficient reducing agent has

$$2 \text{ Hydrocarbon}(-1) \rightleftharpoons \text{Hydrocarbon} + \text{Hydrocarbon}(-2)$$

been provided. The anion radicals and the dianions form several different kinds of ion pairs and ion aggregates in the etheral solvents. Dimerization of the reduction products in solution presents a further complication. Naphthalene and diphenylethene and other related compounds form dianionic dimers in concentrated solution.

$$2 \text{ Naphthalene}(-1) \longrightarrow$$

The reaction media employed for the reduction of coal are strongly basic solutions. Consequently, the acidic protons in phenolic and thiophenolic residues are rapidly displaced to yield phenolates and thiophenolates. The radical anions and dianions, whether arising from the electron transfer agent or the coal, are also strongly basic molecules. Holy has noted that the anion radicals of compounds such as biphenyl and naphthalene effectively abstract protons from compounds with pK_a values lower than 33. For example, biphenyl(-1) abstracts the acidic proton of triphenylmethane completely; it does not react with toluene (Eisch and Kaska, 1962). Biphenyl(-2) and naphthalene(-2) are much more basic than the corresponding anion radicals. Indeed, naphthalene(-2) is apparently more basic than the benzyl carbanion (Brooks *et al.*, 1972). The intermediate (5) formed by proton abstraction is intermediate in basicity between the dianion and the anion

Naphthalene(-2) + $C_6H_5CH_3$ ⟶ $C_6H_5CH_2^-$ +

5

radical. Accordingly, these reagents accomplish a variety of proton transfer reactions in the coal molecules. The extent of these proton transfer reactions depends, of course, on the nature and the number of polycyclic aromatic hydrocarbon fragments in the coal because these compounds form less basic anions and, accordingly, moderate the reaction. Considerable evidence (Lindow *et al.*, 1972; Rabideau and Burkholder, 1978) is now available that compounds such as 5 play a very significant role in the alkylation reactions of biphenyl, naphthalene, and anthracene in mixtures of ammonia and ether.

Aldehydes and ketones are not present in abundance in most coals. However, small additional quantities of aldehydes and ketones may be formed during the degradation reactions of ethers as discussed subsequently. It is, of course, well known that metals reduce many compounds of this kind to ketyls and that the ketyls exhibit a diverse chemistry. In concentrated solu-

$$R_2CO + Na \rightarrow R_2CO^{\cdot-} + Na^+$$

tion, coupling reactions occur quite readily. In dilute solution, these anions are more stable and can be studied spectroscopically. Further reduction to the alcohol proceeds slowly in the absence of a proton donor (House, 1972).

$$R_2CO^{\cdot-} \underset{B}{\overset{H^+}{\rightleftharpoons}} R_2\dot{C}OH \underset{}{\overset{Na}{\rightleftharpoons}} R_2\bar{C}OH \underset{B}{\overset{H^+}{\rightleftharpoons}} R_2CHOH$$

The nature of the reactions of the carbonyl compounds in a solution rich in phenolates, carbanions, aromatic anion radicals, and aromatic dianions is more difficult to ascertain. However, the electron transfer reactions are reversible and rapid. Accordingly, the carbonyl compounds are in mobile equilibrium with the ketyls. The carbonyl compounds react rapidly with the carbanions to form adducts. For example, Holy (1974) has pointed out that aliphatic ketones and aldehydes react with anion radicals and dianions to form addition products. In appropriate circumstances the adducts can be obtained in appreciable yield as shown for the reaction between anthracene dianion and 3-pentanone (Walker, 1961). Consequently, the aldehydes and

$$2\text{Li}^+, \text{Anthracene}(-2) \ + \ (C_2H_5)_2C{=}O \ \xrightarrow{\ 70\%\ }$$

(product: 9,10-dihydroanthracene bearing $C(C_2H_5)_2OLi$ groups at the 9- and 10-positions)

ketones are converted to alcoholates, ketyls, and adducts with the carbanions and aromatic dianions. The exact product distribution depends upon the duration of the reduction reaction because the product distribution obtained under the conditions of kinetic control differs from the product distribution obtained under the conditions of equilibrium control. Only the more stable unreactive ketyls will not be converted to adducts.

The carboxylic acids present in the coal are rapidly converted to carboxylates by the basic agents. All the evidence suggests that their alkali metal salts are stable under the experimental conditions.

B. The Cleavage Reaction

1. Carbon–Oxygen Bond Cleavage

Certain of the anionic and dianionic fragments of the coal structure are unstable. The negatively charged fragments may decompose by carbon–oxygen or carbon–carbon bond cleavage to yield new compounds. Sometimes these cleavages reduce the molecular weight. Often the intermediate species produced in the cleavage reactions of the anionic intermediates are quite reactive radicals or carbanions and initiate other secondary reactions.

The earliest users of the reductive alkylation procedure emphasized that ether cleavage reactions were important. It is well known that diaryl ethers are quite readily cleaved under the experimental conditions. The dianions of these compounds undergo decomposition to yield an aryl oxide and an aryl carbanion (Gilman and Dietrich, 1957; Eargle, 1963; Eisch, 1963; Screttas, 1972).

$$\text{ArOAr} + \text{ArH}^{\cdot-} \rightarrow \text{ArOAr}^{\cdot-} + \text{ArH}$$

$$\text{ArOAr}^{\cdot-} + \text{ArH}^{\cdot-} \rightarrow \text{ArOAr}^{2-} + \text{ArH}$$

$$\text{ArOAr}^{2-} \rightarrow \text{ArO}^- + \text{Ar}^-$$

A somewhat different scheme has been proposed for aralkyl ethers (Screttas, 1972). The aralkyl ethers cleave much less readily than the diaryl ethers (Eisch, 1963).

$$\text{ArOR} + \text{ArH}^{\cdot-} \rightarrow \text{ArOR}^{\cdot-} + \text{ArH}$$

$$\text{ArOR}^{\cdot-} \rightarrow \text{ArO}^- + \text{R}\cdot$$

Allylic and benzylic ethers are readily decomposed. Allyl phenyl ether rapidly cleaves at $-15°C$ (Eisch and Jacobs, 1963). Benzyl ethers also easily cleave at either the radical anion or dianion stage to give a benzyl radical or a benzyl carbanion, respectively, together with the alkoxide, aryloxide, or ben-zyloxide (Gilman *et al.,* 1958; Cram *et al.,* 1959b). Reduction of the benzyl radical to a carbanion can occur (Cram *et al.,* 1959b). Several examples il-lustrate these reactions (Eargle, 1963; Cram *et al.,* 1959b; Angelo, 1966).

Even though dialkyl ethers usually cleave extremely slowly, methyl 2-(1-naphthyl)ethyl ether decomposes under basic conditions quite readily (Cram and Dalton, 1963).

Another mode of ether cleavage has been realized with dibenzylic ethers that can form stable carbanions.

We recently compared the reactivities of common ethers under the conditions generally employed for the preparation of the coal polyanions (A. Reed and L. M. Stock, unpublished results, 1981). The results are shown in Table III.

Potassium phenolate, a principal product of several of the cleavage reactions, is stable under the experimental conditions. It is also pertinent that bibenzyl rather than toluene is formed as the dominant product in the absence of an effective hydrogen atom donor in ammonia. Under the conditions of the reduction reactions of coal, the benzylic ethers presumably will produce benzylic radicals that abstract hydrogen atoms from other coal molecules. The differences in the nature of the ether cleavage reactions in ammonia and in ether are apparent. In particular, phenyl butyl ether is cleaved in the reaction with potassium in ether but not in the reaction in liquid ammonia.

TABLE III

Ether Cleavage Reactions in Liquid Ammonia and in Tetrahydrofuran

	Conversion (%)	
Compound	K, NH_3, $-33°C$ 6 hr	K, $C_{10}H_8$, THF, 25°C 100 hr
$C_6H_5OC_6H_5$	100	100
$C_6H_5OCH_2C_6H_5$	90 Phenol, bibenzyl	—
$C_6H_5CH_2OC_4H_9$-n	20 1-Butanol, bibenzyl	—
$C_6H_5OC_4H_9$-n	0	100
n-$C_4H_9OC_4H_9$-n	0	0

Diaryl ethers decompose to yield strongly basic aryl carbanions, as shown above. Carbanions of this kind are also formed in the metallation reactions aralkyl ethers (Letsinger and Pollart, 1956; Gilman and Dietrich, 1957; Screttas, 1972). Such strongly basic reagents produced within the confined regions of the coal matrix could engage in a variety of other reactions. Proton abstraction reactions leading to the formation of more stable carbanions would probably be the dominant reaction. However, the more reactive, more nucleophilic aryl carbanions may induce ether cleavage reactions by substitution reactions or by β-elimination reactions as illustrated in the general equations.

$$Ar^- + -CH_2CH_2OCH_2CH_2R \rightarrow -CH_2CH_2Ar + {}^-OCH_2CH_2R$$

$$Ar^- + -CH_2CH_2OCH_2CH_2R \rightarrow ArH + -CH{=}CH_2 + {}^-OCH_2CH_2R$$

In addition, the basic carbanions and other basic reagents in the solution may also abstract protons from carbon atoms adjacent to etheral oxygen atoms to yield another class of carbanions. This reaction is important because benzylic ethers are apparently a relatively abundant constituent in certain

$$Ar^- + -CH_2CH_2OCH_2C_6H_5 \rightarrow ArH + -CH_2CH_2O\bar{C}HC_6H_5$$

bituminous and subbituminous coals. The rearrangment reactions have been thoroughly studied (March, 1977). The reactive anion forms a radical pair in a reversible reaction. Recombination of the radical pair in the other manner yields the Wittig rearrangement product as shown for benzyl 2-butyl ether (6). The intermediate benzylic carbanion (7) may also decompose in another pro-

$$C_6H_5CH_2OCH(CH_3)C_2H_5 \xrightarrow{n\text{-}C_4H_9Li} C_6H_5\bar{C}HOCH(CH_3)C_2H_5$$

$$\textbf{6} \qquad\qquad\qquad\qquad\qquad \textbf{7}$$

$$C_6H_5\bar{C}HOCH(CH_3)C_2H_5 \rightleftharpoons C_6H_5CHO^{\cdot -} + \cdot CH(CH_3)C_2H_5$$

$$C_6H_5CHO^{\cdot -} + \cdot CH(CH_3)C_2H_5 \rightarrow \underset{\overset{|}{CH(CH_3)C_2H_5}}{C_6H_5CHO^-} \rightarrow \underset{\overset{|}{CH(CH_3)C_2H_5}}{C_6H_5CHOH}$$

cess to yield an elimination product.

$$C_6H_5\bar{C}HOCH(CH_3)C_2H_5 \rightarrow C_6H_5CH_2OH + CH_3CH{=}CHCH_3$$

$$\textbf{7}$$

Ether cleavage reactions can be initiated in other ways. For example, the methoxy group is eliminated during the reduction of the dimethoxynaphthalene (8) with lithium in ammonia (Weinstein and Fenselau, 1964; Marshall and Andersen, 1965). Although the exact course of the reaction is not known, intermediates such as 11 and 12 are presumably involved in the formation of 9 and 10.

Although most of the alkoxides produced in the ether cleavage reactions are stable under the conditions of the reactions, it is pertinent that Cram and Benkeser and their students found that appropriately constituted alkoxides, which can fragment to yield stable carbanions, decompose to give hydrocarbons and carbonyl compounds (Cram *et al.,* 1959 ; Benkeser and Broxterman, 1969).

$$CH_3CH_2C(CH_3)(C_6H_5)C(CH_3)_2OH \xrightarrow[\substack{25°C, 28 \text{ hr,} \\ 90\%}]{KOC(CH_3)_3} CH_3CH_2CH(CH_3)C_6H_5 + (CH_3)_2C{=}O$$

This brief synopsis of the chemistry of ethers in basic solution indicates the rich chemistry exhibited by these compounds. Coals such as Illinois No. 6 with a significant content of etheral oxygen atoms have an especially rich chemistry in the reductive alkylation reactions.

2. Carbon–Sulfur Bond Cleavage

Diaryl, aralkyl, and benzyl thioethers undergo cleavage reactions more readily than the corresponding ethers because the thioethers have much more favorable reduction potentials and because the thiophenolates and mercaptides are somewhat better leaving groups from the anionic intermediates (Gilman and Dietrich, 1957; T. Ignasiak *et al.,* 1977; B. S. Ignasiak *et al.,* 1978b; Screttas and Micha-Screttas, 1978, 1979).

The dialkyl thioethers are considerably more reactive than the comparatively unreactive oxygen analogs (T. Ignasiak *et al.,* 1977; B. S. Ignasiak *et al.,* 1978b). Moreover, the reactions apparently proceed equally well with sodium in liquid ammonia as with metals in ether (Smith, 1963). The reaction of cyclopentyl n-heptyl thioether is typical (Truce and Frank, 1967).

$$n\text{-}C_7H_{15}S \text{—} \underset{\text{(2) } CH_3OH}{\overset{\text{(1) Li, } CH_3NH_2}{\longrightarrow}} n\text{-}C_7H_{15}SH + \text{—} SH$$

$$\qquad\qquad 87\% \qquad\qquad 87\% \qquad 13\%$$

However, if the radical anion has an aromatic group with a sufficiently large electron affinity, the thioether does not cleave. Thus, 1-(1-naphthyl)-ethyl and 1-(2-naphthyl)ethyl phenyl thioether (13) are not decomposed by anion radicals (Screttas and Micha-Screttas, 1979).

$$\text{CH(CH}_3)\text{SC}_6\text{H}_5$$

13

Thioethers, like ethers, also react with strong bases to produce carbanions. It should be noted, however, that the carbanions derived from thioethers preferentially undergo elimination reactions rather than rearrangement reactions (Biellmann et al., 1979).

$$CH_3(CH_2)_3SCH_2C_6H_5 \underset{\substack{20°C, \\ \text{hexane}}}{\overset{BuLi}{\longrightarrow}} CH_3CH_2CH{=}CH_2 + HSCH_2C_6H_5$$

The course of the reactions of heterocycles such as thiophene, benzothiophene, and dibenzothiophene depends upon the reaction conditions. These compounds are reduced and cleaved when treated with sodium and an alcohol in liquid ammonia. At low concentration of the reducing agent, the reactions apparently yield relatively stable anions and dianions, that is, dibenzothiophene(-1) and dibenzothiophene(-2). Thus, the desulfurization reactions proceed slowly under the conditions of the reductive alkylation reactions in liquid ammonia (Smith, 1963).

The results obtained by T. Ignasiak and her co-workers (1977) for the reaction of potassium and naphthalene with the thioethers in tetrahydrofuran are summarized in Table IV.

Under the conditions of these experiments the thioethers cleave to give thiols and hydrocarbons. In some instances the thiols and the hydrocarbons are reduced further. For example, dibenzothiophene gives only biphenyl and 2-naphthalenethiol yields dihydronaphthalene and tetralin in addition to naphthalene. These observations indicate that only a few unreactive sulfur compounds such as n-alkanethiols and benzenethiols will survive the reduction reactions usually employed in reductive alkylation intact.

TABLE IV

Reaction Products Obtained in the Reactions of Thioethers
with Potassium and Naphthalene after 168 hr at 25°C[a]

Thioether	Products on protonation	Yield[b]	Products on octylation[c]	Yield[b]
Diphenylsulfide	Benzene	—	Benzene	91
	Benzenethiol	63	Phenyloctylsulfide	96
	Biphenyl	11	Biphenyl	16
Dibenzylsulfide	Toluene	72	Toluene	86
	2-Benzyltoluene	51	2-Benzyltoluene	11
	Bibenzyl	Trace	Bibenzyl	Trace
Dioctylsulfide[d]	Octane	—		
	Octanethiol	25		
	Dioctyldisulfide[e]	22		
	Dioctylsulfide (recovered)	45		
Phenyloctylsulfide	Benzenethiol	69		
	Octane	67		
Dibenzothiophene	Biphenyl	52	Biphenyl	62
2-Naphthalene-thiol[f]	Dihydronaphthalene	36		
	Naphthalene	11		
	Tetralin	7		
	Unidentified dimer	15		

[a] From T. Ignasiak et al. (1977).
[b] Moles of product per 100 moles of starting material.
[c] Dioctylsulfide was also produced when desulfurization occurred.
[d] This compound was reduced for 3 days only.
[e] This compound was formed during the workup of the reaction.
[f] No naphthalene was used, the reaction was carried at the reflux temperature for 21 hr.

3. Carbon–Carbon Bond Cleavage

The structures that have been proposed for coals contain molecular fragments which when converted to anions possess labile carbon–carbon bonds. It is well established that the carbon–carbon bonds in appropriately constituted polyaryl ethanes cleave (Eisch, 1963; Gilman and Gaj, 1963; Lagendijk and Szwarc, 1971; Gerson et al., 1976; Elschenbroich et al., 1977; Collins et al., 1980). The course of these reactions is dictated by the relative values of the reduction potentials of the diarylethane, $ArCH_2CH_2Ar$, and the initiators, ArH^- and ArH^{2-}, present in the solution. The cleavage reactions apparently occur with both the monoanion and the dianions. Walsh and Megremis (1981) recently presented rather secure evidence for the view that the dianion of 9,9-diarylfluorene selectively undergoes cleavage to yield the phenyl anion as one reaction product.

$$ArCH_2-CH_2Ar + ArH^{\cdot-} \rightleftharpoons ArCH_2-CH_2Ar^{\cdot-} + ArH$$

$$ArCH_2-CH_2Ar^{\cdot-} \rightarrow ArCH_2{}^{\cdot} + ArCH_2{}^{-}$$

$$ArCH_2-CH_2Ar^{\cdot-} + ArH^{2-} \rightleftharpoons ArCH_2-CH_2Ar^{2-} + ArH^{\cdot-}$$

$$ArCH_2-CH_2Ar^{2-} \rightarrow 2\,ArCH_2{}^{-}$$

Aryl groups are important for the effective stabilization of the radicals and anions produced in these reactions. For example, the reaction proceeds much more readily with 1,1,2,2-tetraphenylethane than with 1,1,1-triphenylethane (Eisch, 1963). However, many 1,2-diarylethanes undergo the reaction under

$$(C_2H_5)_2CHCH(C_6H_5)_2 \xrightarrow[25°C,\ 2\ hr]{Li_2(C_6H_5C_6H_5)} \xrightarrow{CO_2} \underset{83\%}{(C_6H_5)_2CHCO_2H}$$

$$(C_6H_5)_3CCH_3 \xrightarrow[Reflux,\ 6\ hr]{Li_2(C_6H_5C_6H_5)} \xrightarrow{CO_2} \underset{2\%}{(C_6H_5)_3CCO_2H}$$

relatively mild conditions. 1,2-Diphenylethane and 1,2-di-(1-naphthyl)-ethane, as well as strained compounds such as [2.2](2,7)-naphthalene (14), cleave at the sp^3-sp^3 carbon–carbon bond to yield toluene, 1-methylnaphthalene, and 2,7-dimethylnaphthalene (15) as the primary products.

The reductive cleavage of sp^2-sp^3 carbon–carbon bonds has also been observed in modestly strained hydrocarbons. Triptycene (16) reacts with potassium or sodium–potassium alloy to give 9-phenyl-9,10-dihydroanthracene (17) but does not react with sodium (Theilacker and Möllhoff, 1962;

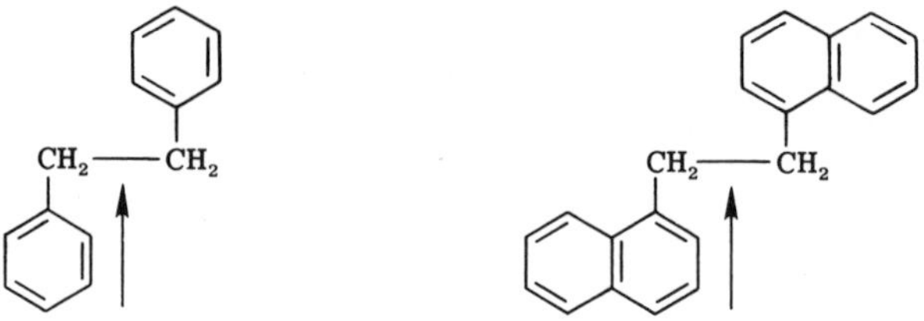

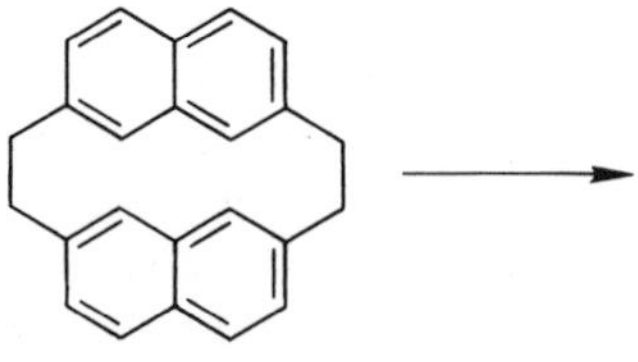

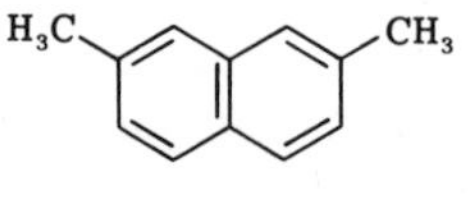

14

15

Walsh and Ross, 1968). However, sodium slowly cleaves 2-phenyltriptycene at room temperature to give 2,9-diphenyl-9,10-dihydroanthracene (Walsh

16   **17**

and Ross, 1968), and lithium reacts with 1,2-benzotriptycene in liquid ammonia to give 9-(2-naphthyl)-9,10-dihydroanthracene (Rabideau *et al.*, 1979). Potassium in liquid ammonia also cleaves a 2,4a,9,10-tetrahydrophenanthrene derivative **(18)** to the corresponding bibenzyl derivative **(19)** (Razdan *et al.*, 1979). Collins and his co-workers (1980) found that sodium–potassium alloy cleaves sp^2–sp^3 bonds and abstracts benzylic protons

18   **19**

from 1,2-diphenylethane and diphenylmethane. It is notable that the aro-

$$C_6H_5CH_2CH_2C_6H_5 \xrightarrow[\substack{triglyme,\\0°C}]{NaK} \xrightarrow{CH_3I} \underset{37\%}{C_6H_5CH_3} + \underset{42\%}{C_6H_5CH_2CH_3} + \underset{21\%}{C_6H_5CH(CH_3)CH_2C_6H_5}$$

matic rings are not reduced under these conditions. Collins and his co-workers (1981) have pointed out that there are two rational reaction pathways

$$C_6H_5CH_2CH_2C_6H_5 \xrightarrow{NaK} \xrightarrow{H_2O} C_6H_5CH_3$$

for the cleavage of the sp^2–sp^3 carbon–carbon bond in these molecules. One

$$C_6H_5CH_2C_6H_5 \xrightarrow[\substack{88\%\\triglyme,\\0°C}]{NaK} \xrightarrow{CH_3I} \underset{13\%}{C_6H_5CH_3} + \underset{54\%}{C_6H_5CH(CH_3)C_6H_5}$$

$$C_6H_5CH_2C_6H_5 \xrightarrow[triglyme]{NaK} \xrightarrow{H_2O} \underset{43\%}{C_6H_5CH_3} + \underset{33\%}{3\text{-}C_6H_5CH_2C_6H_4C_6H_5}$$

involves the initial reduction of diphenylmethane to an anion radical and subsequent coupling of the intermediate with diphenylmethane to yield **20**, which decomposes to yield the benzyl anion and a radical.

$$C_6H_5CH_2C_6H_5 + K \rightarrow K^+ + C_6H_5CH_2C_6H_5^{\cdot-}$$

20a   **20b**

$$20a \longrightarrow$$

$$+ \; C_6H_5CH_2^-$$

$$3\text{-}C_6H_5CH_2C_6H_4C_6H_5 \qquad C_6H_5CH_3$$

The second reaction pathway was proposed by Grovenstein *et al.* (1977). Their proposal focuses on the dimerization of the radical anion to yield **21**, which decomposes to give the benzyl anion and another anion, which eventually releases the hydrogen perhaps as a hydrogen atom following another electron transfer reaction.

$$C_6H_5CH_2C_6H_5 \longrightarrow \qquad \longrightarrow \qquad C_6H_5CH_2^- \; + \;$$

21

These ipso aromatic substitution reactions provide pathways for the transformation of hydroaromatic fragments in coal molecules leading either to smaller or larger structures. As noted previously, these ipso substitution reactions apparently occur in preference to the reduction reactions of these molecules.

The reductive cleavage of sp^2–sp^2 carbon–carbon bonds has also been reported. Potassium in monoglyme converts 9,9-bianthryl to anthracene (Solodovnikov and Zaks, 1969). In contrast, potassium in this solvent transforms 1,1-binaphthyl into perylene (Solodovnikov *et al.,* 1968). Even

simple compounds such as biphenyl and terphenyl undergo reductive cleavage in low yield ($< 25\%$) when treated with lithium in hexamethylphosphoramide to produce phenylcyclohexadiene and cyclohexa-1,3-diene, as

C_{18} compounds

well as the reduced terphenyls (Kotlarek and Pacut, 1978). The concept that rather stable carbon–carbon linkages are broken under the conditions of reductive alkylation was verified in a rather direct way by Lazarov and Angelov (1980). They investigated the reduction of an oxygen-free polymer of phenanthrene and formaldehyde. The polymer contained 1.2 methylene units for each phenanthrene unit with not more than 0.2% oxygen by direct determination. They suggest that the methylene groups bridge predominantly through the 9 and 10 positions of phenanthrene (**22**). The polymer was reduced with potassium in tetrahydrofuran. The mixture of anions was treated with a proton donor, an alkyl halide or trimethylsilyl chloride to complete the

$$[(\text{Phenanthrene})_{1.0}(\text{CH}_2)_{1.2}]_n \xrightarrow[\substack{\text{THF,} \\ 25°C, 72\ hr}]{\text{K, C}_{10}\text{H}_8} \xrightarrow{\text{electrophile}} \text{Product mixture}$$

$[(\text{Phenanthrene})_{1.0}(\text{CH}_2)_{1.2}]_n$

22

reaction. The microanalytical data, enhanced solubility, reduced molecular weights, and spectroscopic properties of the reaction products strongly suggest that the phenanthrene molecules were reduced and, more important, that carbon–carbon bond cleavage reactions occurred under the conditions of the reduction reaction. The spectroscopic evidence favors the view that the reduction reactions occurred to yield 9,10-dihydrophenanthrene fragments as well as benzhydryl fragments, which were subsequently protonated, alkylated, or silylated. The results suggest that one carbon–carbon bond cleavage reaction occurs for each 8 or 10 phenanthrene units in the polymer.

The radicals and carbanions produced in the reduction reaction undergo two reactions, rearrangement and coupling, that further complicate the interpretation of the experimental results. Fortunately, the 1,2-alkyl rearrangements of carbanions are very slow. However, 1,2-phenyl shifts do occur. These reactions are particularly important when the rearrangement reaction leads to a more stable carbanion, as illustrated for the reaction of the phenyl anion with 1,1,1,2-tetraphenylethane (Grovenstein, 1977).

$$C_6H_5^- + (C_6H_5)_3CCH_2C_6H_5 \rightarrow C_6H_6 + (C_6H_5)_3C\bar{C}HC_6H_5$$

$$(C_6H_5)_3C\bar{C}HC_6H_5 \rightarrow (C_6H_5)_2\bar{C}CH(C_6H_5)_2$$

The related radicals exhibit a similar chemistry. The 1,2-phenyl shifts are also more important than 1,2-alkyl shifts in this case. The principal side reaction of the alkyl radicals is fragmentation via β-scission. This reaction is particularly important when the radical formed in the process is stabilized by delocalization and the alkene is conjugated with an aromatic nucleus.

$$(C_6H_5)_2CHCH_2\dot{C}HC_6H_5 \rightarrow (C_6H_5)_2CH\cdot + H_2C{=}CHC_6H_5$$

Coupling products as well as fragmentation products are also formed from the reactive intermediates initially produced in the reduction reactions. Several examples including the dimers formed from naphthalene and pyridine have already been mentioned. Such products are quite generally observed when the reduction reactions are carried out at high hydrocarbon concentration. However, lesser quantities of these compounds are formed in the

$$\begin{array}{c} C_6H_5 \\ \\ C_6H_5 \end{array} C=C \begin{array}{c} H \\ \\ H \end{array} \xrightarrow[NH_3]{Na} \xrightarrow{NH_4} (C_6H_5)_2CHCH_3 + ((C_6H_5)_2CHCH_2)_2$$

$$ 67\% 17\%$$

presence of good donors. It is relevant to note that many coals are excellent hydrogen atom donors.

In summary, the initial products obtained under the reaction conditions through electron transfer reactions, proton abstraction reactions, ether cleavage reactions, and carbon–carbon bond cleavage reactions equilibrate as the coal matrix collapses to yield a mixture of aromatic radical anions; aromatic dianions; resonance-stabilized radicals and carbanions; aryloxides, alkoxides, benzyloxides, and their sulfur analogs; ketyls; and neutral species such as arenes and alkylated arenes. Fortunately, only two types of rearrangement reactions appear to be important. Consequently, the structures of the products are, for the most part, closely related to the structures of the molecules originally present.

C. The Alkylation Reaction

When an alkylating agent is added to the solution, the equilibrium mixture of aromatic anions, relatively stable carbanions, aryloxides, alkoxides, benzyloxides, their sulfur analogs, basic nitrogen compounds, and carboxylates react.

The reactions of the oxygen, sulfur, and nitrogen nucleophiles with the alkylation reagent are straightforward and do not require special comment except to note that elimination reactions may compete with substitution reactions to consume a portion of the alkylating agent. For this reason, it is customary to employ an excess of the alkylation reagent.

The reduced aromatic compounds, which constitute a very important fraction of the reactive molecules in solution, have a more complex chemistry because electron transfer reactions and proton transfer reactions both occur in competition with the alkylation reaction. The available evidence suggests that anion radicals play only a minor role in these alkylation reactions. However, the competitive proton transfer reactions can exert an important influence on the outcome of the reaction. The influence of the reaction conditions, including such matters as the nature of the cation associated with the aromatic anion, the leaving group of the alkylating agent, and the order of addition of the reagents, is more fully resolved for the reactions in liquid ammonia than for the reactions in etheral solvents (Lindow *et al.,* 1972; Rabideau and Burkholder, 1978).

When the alkylating agent is used in large excess in ammonia or in ammonia–ether solutions, the aromatic anions in solution yield highly alkylated products. For example, the reductive methylation of biphenyl in a 1:2 (by volume) mixture of tetrahydrofuran and ammonia at $-33°C$ yields the

mono-, di-, and trialkylated products (Lindow *et al.*, 1972). Anthracene generally yields a small quantity of the tetraalkylated product, 9,9,10,10-tetramethyl-9,10-dihydroanthracene, under these conditions. The significance of the order of addition of the reagents is well illustrated by the results obtained in the reductive ethylation of naphthalene. This reaction yields 25% of the monoethyl compound (23) and 75% of a mixture of *E*- and *Z*-1,4-diethyl-1,4-dihydronaphthalene (24 and 25) when ethyl iodide is added to the reaction mixture, the normal method of addition. On the other hand, when

the solution of sodium naphthalene anions is added to ethyl iodide, the inverse addition procedure, the monoethyl compound (23) constitutes 90% of the product mixture (Rabideau and Burkholder, 1978). Under the conditions of the alkylation reactions in ammoniacal solvents, the principal entity in solution is the reduced mononegative anion (5). This compound readily undergoes alkylation and in the presence of a strong base further alkylation occurs via proton abstraction reactions of the initial alkylation product (26).

5

26

Less information is available concerning the nature of these alkylation reactions in etheral solvents. The character of the ion pairs and ion aggregates present in these solutions influences the course of the alkylation reaction. As already discussed, the unfavorable equilibrium constants for the disproportionation of the anion radicals virtually assure that the hydrocarbons will be present in solution as anion radicals in the absence of excess reducing agent.

$$2C_{10}H_8^{\cdot-} \rightleftharpoons C_{10}H_8 + C_{10}H_8^{2-}$$

However, the Sternberg alkylation reactions are generally performed with excess metal. In this situation, the alkylation reactions of the anion radicals are not the dominant processes. The reactions of the anion radicals can generally be distinguished from the reactions of the dianions by the reaction rate and the product distribution. For example, when compounds such as sodium naphthalene(-1) react with alkylating agents a mixture of products is obtained (Sargent and Lux, 1968). The product distribution highlights the fact that the

$$Na^+,C_{10}H_8^{\cdot-} + n\text{-}C_5H_{11}I \rightarrow C_{10}H_8 + C_{10}H_8(C_2H_5)_2 + C_{10}H_{22} + C_5H_{12} + C_5H_{10}$$
$$\phantom{Na^+,C_{10}H_8^{\cdot-} + n\text{-}C_5H_{11}I \rightarrow}\quad 35\% \qquad\quad 46\% \qquad\quad 16\% \qquad 3\%$$

aromatic radical anions are poor nucleophiles but efficient electron transfer agents. These species very effectively convert alkyl halides to alkyl radicals (Garst, 1971; Bank and Juckett, 1976).[3] The rate at which alkyl radicals are

$$Na^+,C_{10}H_8^{\cdot-} + n\text{-}C_5H_{11}I \rightarrow NaI + C_{10}H_8 + C_5H_{11}{}^\cdot$$

formed depends only on the rate of the electron transfer from the aromatic radical anion to the alkyl halide. Since the larger halogen atoms have the most favorable reduction potentials, the electron transfer reactions are most important for the alkyl iodides (Sargent *et al.,* 1966; Garst, 1971). Indeed, the reactions of anion radicals with alkyl iodides frequently yield free radicals in 50–70% yield (Garst, 1971; Holy, 1974). The radical intermediates obtained in such reactions usually dimerize and disproportionate as shown in the equation (Bank and Juckett, 1976; Bergbreiter and Killough, 1978; Savoia *et al.,* 1978). Under the conditions of a reaction with coal, these radicals would be produced in an isolated environment. Accordingly, reactions of the radicals with reactive components in the coal would probably be predominant in these circumstances and recombination reactions with aromatic anions to yield alkylated aromatic compounds would be important. The radicals also react with aromatic radical anions (Sargent and Lux, 1968) to yield the same products as would be obtained in the ionic reaction as illustrated for sodium

naphthalene(-1) and *n*-butyl iodide. Clearly, the presence of butyl radicals would enlarge the chemistry of the Sternberg alkylation procedure.

[3] Ketyls also generate alkyl radicals from alkyl halides.

$$R_2CO^{\cdot-} + n\text{-}C_5H_{11}I \rightarrow R_2CO + n\text{-}C_5H_{11}{}^\cdot + I^-$$

Several lines of evidence favor the view that radical reactions play only a minor role in the alkylation reaction. The reactions of the polyanion obtained from Illinois No. 6 coal with methyl, n-butyl, and n-octyl iodide were compared in tetrahydrofuran at 25°C (Alemany *et al.*, 1979). The reaction could be monitored quite readily by the rate at which potassium iodide precipitated from solution. It was established that the reaction is not unusually rapid and that the rate of the reaction with methyl iodide is clearly greater than the rate of the reaction with n-butyl or n-octyl iodide under the same conditions. These observations, of course, suggest that the S_N2 reactions of the coal polyanion with the alkylating agent are more significant than the electron transfer reactions. Although there is a clear distinction in the reaction rate, the extent of the alkylation reaction after 48 hr is the same for methylation, butylation, and octylation with more than 10 alkyl groups added per 100 carbon atoms.

The observation that the reactions of the coal polyanion with n-butyl chloride and n-butyl bromide proceed much less efficiently than the reaction with n-butyl iodide also strongly supports the view that S_N2 reactions rather than electron transfer reactions are primarily responsible for the production of soluble materials. This interpretation is based on the fact that the reactivities of nucleophiles with n-butyl iodide, bromide, and chloride are in the approximate order 80:40:1 and that these substitution reactions are all slow relative to the electron transfer reactions of the butyl halides with anion radicals. To illustrate, the rate constants for the electron transfer reactions of primary alkyl iodides with anion radicals of the kind formed under the conditions of these experiments are about 10^4 to 10^8 M^{-1} sec^{-1} (Bank and Juckett, 1976). The rate constants for the reactions of primary alkyl iodides with nucleophiles are less, only about 10^4 M^{-1} sec^{-1} in the fastest processes such as the reaction of hexyl iodide with diphenylmethyllithium or 9-lithio-9,10-dihydroanthracene in tetrahydrofuran at 20°C (Bank *et al.*, 1977). These considerations suggest that the butyl halides could undergo rapid electron transfer reactions to produce butyl radicals only during the initial stages of the coal alkylation process and that such processes are not generally important for the formation of soluble products. Rather the available information supports the view that the alkylation reactions which occur during the reductive alkylation process involve carbon, oxygen, nitrogen and sulfur nucleophiles such as relatively stable carbanions (27) and dianions (28), as well as phenolates (29) and carboxylates (30).

27

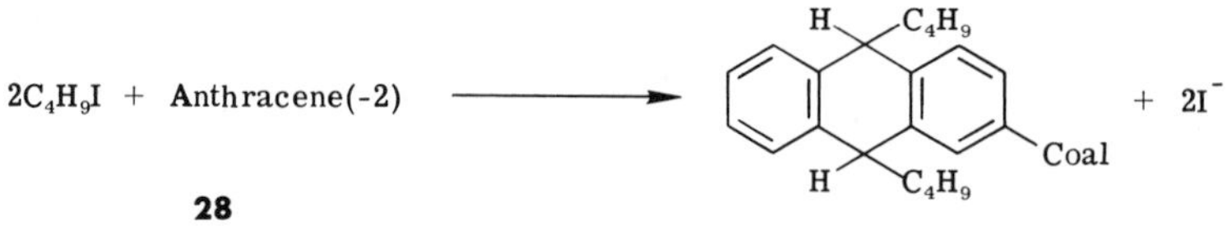

$2C_4H_9I$ + Anthracene(-2)

28

C_4H_9I +

29

C_4H_9I +

30

The observation that the precipitation of potassium iodide from the reaction of butyl iodide with the Illinois No. 6 coal polyanion is not complete after about 24 hr provides support for this interpretation. Thus, the much slower S_N2 reactions of the polyanion with the other halides would not be complete in 48 hr. The observations indicate that the alkylation reactions of even rather reactive alkyl halides require a considerable time at ambient temperature. The spectroscopic information discussed subsequently provides strong support for the view that hindered aryloxides and carboxylates and possibly tertiary alkoxides are present in the polyanions. The O-alkylation reactions of these molecules have substantial energy requirements thereby accounting for the slow reaction rate. The fact that the alkylation reactions of the polyanion are slow may be responsible for certain discrepancies in the experimental observations obtained in different laboratories because of differences in the time alloted for alkylation. Quite recent results suggest that it may be advantageous to complete the alkylation reaction by the treatment of the alkylation product obtained under aprotic conditions with additional alkyl iodide in the presence of a phase transfer catalyst (Liotta, 1979; C. I. Handy and L. M. Stock, unpublished results, 1981).

III. METHODS AND PROCEDURES

The factors that are important for the successful reductive alkylation reactions of coals are examined in this section. Generally, the reductive alkylation is carried out in two discrete steps. The reduction reaction can be performed, as already mentioned, with a metal in ammonia, with a metal in ether, or with a metal and an electron transfer agent in ether. Inasmuch as the latter approach has been more thoroughly studied, this discussion will necessarily emphasize the work on the reactions with electron transfer reagents. The less plentiful but equally significant contributions concerning the reactions in ether and ammonia without added electron transfer agents are incorporated into the discussions in Section III,A. Although the coal polyanion reacts readily with a variety of electrophilic agents, the alkylation reactions apparently proceed very successfully only with primary alkyl iodides. The research work on the alkylation reaction is discussed in Section III,B. The tactics used for the isolation of the reaction products from the reaction mixtures are as significant as the reactions themselves. Accordingly, the separation procedures and the methods used to test their effectiveness are discussed in Section III,C.

A. The Reduction Process

The reagents used in the reduction reactions in ether and in ammonia require special attention because the anion radicals, dianions, and carbanions that are generated in the reaction react readily with adventitious impurities. Accordingly, most workers have used carefully cleaned metals in redistilled ammonia or in ethers that have been distilled from potassium or a suitable reactive hydride. The electron transfer agents have similarly been purified prior to use. Although it may not be essential, most investigators have demineralized the coal prior to the reduction alkylation. Also, most investigators have dried the finely divided coals, -325 mesh, in vacuum and protected the dried products from the atmosphere prior to the reductive alkylation reactions.

In the first stage of the procedure, equilibrium is achieved between the metal and the solvent or between the metal, the electron transfer agent and the solvent. The coal is then added to the highly colored solution of aromatic anions. The color of the reaction solution changes almost immediately upon the addition of the coal. For example, the intense dark green color of the naphthalene anions is discharged in a few minutes and the color is only slowly restored about 1.5 hr later. The course of the reaction can be studied by the hydrolysis of aliquots of the reaction mixture and the determination of the quantity of hydroxide ion formed in the aqueous solution (Sternberg *et al.*, 1970, 1971; Wachowska *et al.*, 1974; Alemany *et al.*, 1979). Typical results for the reduction of Illinois No. 6 coal by potassium in tetrahydrofuran in the presence of biphenyl are shown in Fig. 1. Quite similar results were obtained with naphthalene, anthracene, 9,10-diphenylanthracene, and pyrene with Illinois No. 6 coal and with potassium and naphthalene and Pocahontas coal. Unfortunately, there are no comparable rate data for the reduction of coal by solvated electrons in ammonia or in etheral solvents.

The negligible rate of reaction of potassium with tetrahydrofuran is shown in curve a of the figure. The much more rapid reaction of potassium with biphenyl is sketched in curve b. This electron transfer agent undergoes reduction relatively rapidly to the dianion. Further reduction occurs much more slowly. After 5 days, the hydrolyzed solution contains about 2.4 equivalents

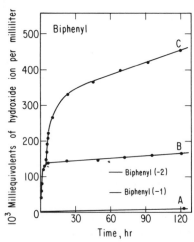

Fig. 1.   The titration curves for the reactions of potassium with tetrahydrofuran (a), with biphenyl (b) and with biphenyl and Illinois No. 6 coal (c). The equivalence levels for the conversion of biphenyl to biphenyl(-2) and to biphenyl(-1) are also noted by the straight lines.

TABLE V

Reactivity Data for the Reduction of Illinois No. 6 Coal by Potassium
in Tetrahydrofuran in the Presence of Various Electron Transfer Agents at 25°C

Compound	10^3 meq of hydroxide ion required for the coal polyanion per ml of solvent[a,b]		
	3 hr	45 hr	120 hr[c]
Biphenyl	78	220	260
9,10-Diphenylanthracene	—	153	246
Naphthalene	78	115	209
Anthracene	74	100	136

[a] Based upon the titer observed for the reaction solution containing coal and the electron transfer agent minus the titer observed for the solution of the electron transfer agent at the same time, please see Fig. 1.

[b] About 0.5 eq of potassium is present initially.

[c] The titers of aliquots withdrawn from vigorously stirred reaction mixtures are about tenfold greater than the titers of aliquots withdrawn from reaction mixtures from which the coal particles have been allowed to precipitate for 1 hr. Hence, the polyanion of this coal is not significantly soluble in tetrahydrofuran at the end of the reduction reaction.

of potassium hydroxide per mole of biphenyl in the original solution. Other electron transfer agents such as naphthalene and anthracene exhibit similar behavior (P. Heimann and L. M. Stock, unpublished research, 1981). The coal reacts rapidly with the electron transfer agent as shown in curve c. Indeed, the Illinois No. 6 coal used in these experiments acquires about 15 negative charges per 100 carbon atoms during the first day of reaction. The reduction reaction in the heterogeneous solution slows to a nearly constant rate after about 12 hr and the coal molecules acquire about 0.1 negative charge per 100 carbon atoms per hour during the last four days of the reaction.

The reactivities of several electron transfer agents have been compared (P. Heimann and L. M. Stock, unpublished research, 1981). The titers obtained for several compounds at 3, 45, and 120 hr after the addition of Illinois No. 6 coal are presented in Table V.

The results obtained for aliquots of the reaction mixture withdrawn three hours after the coal was added to the reaction mixture reveal that there are only modest variations in the reactivity of the electron transfer agents. Presumably, many of the rapid initial reactions occur between the acidic protons of the coal and the basic anions as illustrated for biphenyl(-2). At these

Biphenyl(-2) + Coal—ArOH $\longrightarrow$ + Coal—ArO⁻

short reaction times, therefore, the small differences observed in the reactivities of the electron transfer agents are not meaningful.

The results obtained after 45 hr are much more significant. At this stage of the process, the reduction reaction initiated in the presence of biphenyl has proceeded to a much greater extent than the reactions initiated in the presence of 9,10-diphenylanthracene, naphthalene and anthracene. The reactivity order biphenyl > naphthalene > anthracene correlates very well with the reduction potentials for these hydrocarbons. This order is also observed at 120 hr. The rate data for these compounds also correlate well with the quantity of soluble product obtained in these reactions as discussed subsequently. These data strongly suggest that the reaction is kinetically controlled at 45 hr and that the less stable, more reactive anionic electron transfer agents are more effective for the reduction and depolymerization of the molecules in Illinois No. 6 coal as illustrated for ether cleavage.

$$ArH(-2) + Coal—O—Coal \rightarrow ArH + Coal(-1) + Coal—O^-$$

After 120 hr, the situation has changed significantly. At this time, the differences in most of the electron transfer agents have been appreciably leveled and only anthracene is significantly less reactive. The leveled reactivity at 120 hr may be understood, at least in part, on the basis of the influence of the reduced coal molecules on the progress of the reaction. The anions, anion radicals, and dianions produced from the coal molecules must exert an important influence on the behavior of the other electron transfer agents in solution. We postulate that at long reaction times in the absence of an excessive concentration of an added electron transfer agent the anionic coal molecules moderate the differences in reactivity of the added electron transfer agents and limit the extent of the reduction of the aromatic compounds and the extent of carbon–oxygen and carbon–carbon bond cleavage reactions in this coal and presumably in other subbituminous and bituminous coals.

The results for 9,10-dihydroanthracene, anthracene, and biphenyl suggest that steric factors do not play a major role. Indeed, 9,10-diphenylanthracene is more reactive than the unsubstituted compound. These results certainly suggest that the anionic electron transfer agents do not, of necessity, propagate the reaction through the small pores of the coal. Rather reaction models in which electron density is propagated from the electron transfer agent to an acceptor molecule on the surface of the coal and from this surface acceptor to another acceptor in the interior of the coal by an electron transfer process not involving the further intervention of the soluble electron transfer agent or models in which the molecular fragments of coal are peeled away from the solid as layers from an onion appear more compatible with these results. It is pertinent to note that the soluble products obtained in the first reductive alkylation reaction of Illinois No. 6 coal have properties that are quite similar to the soluble products obtained by the reductive alkylation of the insoluble residue from the first reaction. This finding is, of course, more compatible with the second interpretation.

Preliminary work is usually necessary to establish the appropriate quantities of the metal and the electron transfer agent because the amounts needed depend upon the rank of the coal. It has been recommended that the reduction reaction be continued until the analysis of aliquots of the mixture show that the coal is not being reduced further. Reaction times as long as 12 days have been employed. However, the results obtained by Sternberg and by Wachowska and their associates (Table VI) are more typical.

Generally, the lower rank coals consume a greater quantity of the reducing agent and the electron transfer agent per gram of the coal used in the reaction. This feature of the reaction was first noted by Sternberg, and it is particularly important for the reactions of subbituminous coals with potassium and naphthalene in ether (Sternberg and Delle Donne, 1974; Franz and Skiens, 1978). It is also very important in the reactions of subbituminous coals with solvated electrons in glyme solvents (D. A. Blain and L. M. Stock, unpublished research, 1981) and in liquid ammonia (C. I. Handy and L. M. Stock, unpublished research, 1981). As pointed out by Sternberg, the acidic protons and readily extractable hydrogen atoms of these coals react with the metal and reduce the electron transfer agents to compounds, for example, dihydronaphthalene, that are not effective transfer agents. The resulting decrease in the concentration of the active metal or the electron transfer agent adversely influences the rate of the reduction reaction.

TABLE VI

The Quantity of Naphthalene and the Reaction Times for the Reduction of Coals of Different Rank by Potassium in Tetrahydrofuran at 25°C

Coal	Carbon content (%, daf)	Naphthalene (g/g coal)	Reaction time (hr)	Negative charges per 100 C atoms
Subbituminous[a]	71	0.40	168	24
Bruceton, hvAb[a]	82	0.18	360	26
Pocahontas, lvB[a]	90	0.05	72	12
Anthracite[a]	96	0.05	72	11
K-I[b]	78	0.18	96	12
K-II[b]	81	0.18	120	10
K-III[b]	87	0.18	144	11
K-IV[b]	88	0.18	144	14
Anthracite[b]	93	0.05	96	8

[a] Sternberg and Delle Donne (1974).
[b] Wachowska and Pawlak (1977).

Much shorter reaction times have been employed for the reduction of coals by potassium in the absence of an electron transfer agent. Miyake and his co-workers (1980) found that potassium in refluxing tetrahydrofuran produced a significant amount of soluble product when the reaction was carried out between 2 and 6 hr. The reactions with metals in liquid ammonia at -78 to $-33°$C generally have been carried out for 6 hr (Lazarov and Angelova, 1968; B. S. Ignasiak and Gawlak, 1977), and the reactions with metals in the glyme solvents have been carried out between 2 and 16 hr at -30 to $-50°$C (Niemann and Hombach, 1979; Haenel *et al.,* 1980). The facility with which these reactions occur indicates that the reduction reactions of the coal molecules are quite rapid and that the use of a soluble reducing agent, the solvated electron, considerably shortens the reaction time.

Lazarov *et al.* (1978) worked out another method for the reductive alkylation of coal in the absence of an added electron transfer agent. In their approach, the finely divided coal is reacted directly with molten potassium at 120°C in a partial vacuum. The weight loss under these conditions, about 0.5%, is insignificant. The blue-tinted potassium coal adduct prepared in this way from a low-volatile bituminous Ruhr coal (89.2% C daf) is paramagnetic and resembles an intercalation compound. The adduct reacts with methyl iodide in tetrahydrofuran to yield the alkylation product. The analytical data obtained for this methylated coal and the methylated coal obtained by reductive alkylation using potassium and naphthalene to generate the coal polyanion in tetrahydrofuran in the usual way are very similar with about 12–13 methyl groups per 100 carbon atoms. Reductive methylation in the more

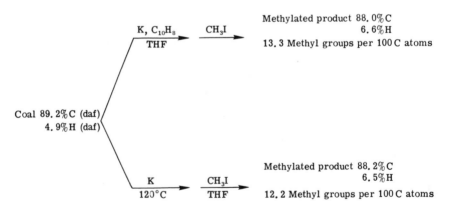

customary way yields a product that is 55% soluble in benzene; the product obtained by the direct reaction is somewhat more soluble, 61%, in this solvent. These initial results suggest that the molten potassium method yields a coal alkylate useful for spectroscopic study and other chemical investigations. However, more work will be necessary with low-rank coals to establish that the method is generally useful.

Several investigations of the reaction variables (i.e., the metal, the electron transfer agent, the solvent,and so forth)have been carried out to establish the conditions most favorable for the conversion of insoluble coal to a soluble alkylated product. Sternberg *et al.* (1970, 1971) examined the reductive ethylation of Pocahontas coal using naphthalene as the electron transfer agent to assess the effectiveness of lithium, sodium, and potassium. They found that potassium was appreciably more effective for the preparation of a

$$\text{Pocahontas coal} \xrightarrow[\text{C}_{10}\text{H}_8]{\text{metal}} \xrightarrow[\text{THF}]{\text{C}_2\text{H}_5\text{I}} \text{Alkylated products}$$

benzene-soluble coal alkylate. Similar experiments have been carried out with Illinois No. 6 coal (Alemany *et al.,* 1979) employing the reductive butylation reaction. Potassium was also found to be the most effective reagent for the preparation of an alkylated product that was soluble in tetrahydrofuran. The

$$\text{Illinois No. 6 coal} \xrightarrow[\text{C}_{10}\text{H}_8]{\text{metal}} \xrightarrow[\text{THF}]{\text{C}_4\text{H}_9\text{I}} \text{Alkylated products}$$

quantitative results obtained in these studies are summarized in Table VII.

Two factors contribute to the specific effectiveness of potassium. First, from the viewpoint of thermodynamics, the conversion of the coal molecules to polyanions may be more favorable when the counter ion is the potassium ion. Second, from the standpoint of kinetics, the evidence favors the view that the potassium salts are more ionic and less stable and undergo fragmentation reactions more readily. In the case of the Pocahontas coal, for example, lithium and sodium enable the introduction of 7 ethyl groups per 100 carbon atoms, with a resultant 55% solubility in benzene. Potassium enables the introduction of about 9 ethyl groups per 100 carbon atoms; this product is 95% soluble in benzene. The results for the Illinois No. 6 coal are quantitatively

TABLE VII

Effect of Alkali Metal on the Reductive Alkylation Reaction in Tetrahydrofuran at 25°C

	Conversion (%)	
Metal	Pocahontas coal[a] ethylation solubility in benzene	Illinois No. 6 coal[b] butylation solubility in THF
Lithium	58	5
Sodium	54	37
Potassium	95	52

[a] Sternberg *et al.* (1970, 1971).

[b] Alemany *et al.* (1979).

TABLE VIII

**The Influence of Electron Transfer Agents on the Butylation
of Illinois No. 6 Coal[a]**

Electron transfer agent	Reduction potential (V)[b]	Conversion (% solubility in THF)
Biphenyl	2.70	58
Naphthalene	2.50	52
Anthracene	1.90	41

[a] From Alemany *et al.* (1979).

[b] From Table II.

different; potassium enables the formation of a butylated product that is ten-fold more soluble in tetrahydrofuran than the alkylated product from the reaction with lithium.

The influence of the electron transfer agent on the course of the reductive butylation of Illinois No. 6 coal has also been examined. In these experiments,

$$\text{Illinois No. 6 coal} \xrightarrow[\substack{\text{Hydrocarbon,} \\ \text{THF}}]{\text{K} \qquad \text{C}_4\text{H}_9\text{I}} \text{Alkylated product}$$

potassium was employed as the reducing agent and the effectiveness of biphenyl, naphthalene, and anthracene as reagents for the formation of soluble butylated coal products was determined. The results are summarized in Table VIII.

Although the differences in the conversion of the coal to soluble alkylated products is not large, the differences do appear to be significant with biphenyl and naphthalene more effective than anthracene as electron transfer reagents in the heterogeneous reduction reaction of this coal. These observations are in accord with the results obtained in the study of the rate of the reactions of these hydrocarbons with Illinois No. 6 coal (Table V). The rate data and the conversion data indicate that the less thermodynamically stable biphenyl anions are much more effective reagents than the more thermodynamically stable anions of anthracene. Presumably, the biphenyl anions are more effective for the transfer of electron density to the acceptors in the coal molecules. The greater concentration of negative charge in the coal molecules would, of course, increase the concentration of aromatic anions available for alkylation and, perhaps even more importantly, increase the rates of the fragmentation reactions of ethers and hydrocarbons.

Complementary information is available for the reductive butylation reaction of Illinois No. 6 coal in liquid ammonia (C. I. Handy and L. M. Stock, unpublished research, 1981). The metal, its concentration, the reaction time, and the reaction temperature influence the yield of tetrahydrofuran-soluble

$$\text{Illinois No. 6 coal} \xrightarrow[\text{NH}_3,\ T^\circ\text{C}]{\text{metal} \quad \text{C}_4\text{H}_9\text{I}} \text{Butylated product}$$

butylated products. The principal observations are summarized in Tables IX–XI.

Preliminary work indicated that the quantity of soluble alkylated product obtained in replicate experiments varied by about 5%. Within the limits of this uncertainty, the experimental results indicate that potassium is the most effective reducing agent and that both potassium and lithium are more satisfactory reagents for the reaction in liquid ammonia than sodium. Presumably, both kinetic and thermodynamic factors are important in this heterogeneous reaction. Thermodynamic considerations suggest that lithium should be the better reducing agent in liquid ammonia (Jolly, 1956). However, the potassium salts of the anionic coal molecules appear to be more

TABLE IX

The Influence of the Reducing Agent on the Reductive Butylation
of Illinois No. 6 Coal in Ammonia

Metal[a]	$T\,(^\circ\text{C})$	Conversion (% solubility in THF)
Li	-33	42
Na	-78	32
K	-33	52
K	-78	49

[a] The same quantity of the metal, 30 mmole, per gram of coal was used in each experiment. The reduction reaction was carried out for 6 hr.

TABLE X

The Influence of the Quantity of the Reducing Agent
on the Reductive Butylation of Illinois No. 6 Coal in Ammonia

Metal[a]	$T\,(^\circ\text{C})$	Quantity (mmole/g coal)	Conversion (% solubility in THF)
K	-33°	15	31
K	-33°	30	52
K	-78°	30	49
K	-33°	44	46

[a] The reduction reaction was carried out for 6 hr.

TABLE XI

The Influence of Reaction Temperature and Reaction Time
on the Reductive Butylation of Illinois No. 6 Coal
with Potassium in Liquid Ammonia

Reaction temperature (°C)	Reaction time (hr)	Conversion (% solubility in THF)
−78	6	49
−78	24	38
−33	6	52
25	20	45

[a] The same quantity of potassium, 30 mmole, per gram of coal was used in each experiment.

reactive in a kinetic sense as mentioned in the discussion of the reactions in etheral solvents in the presence of added electron transfer agents. The experimental data also suggest that about 30 mmole of potassium per gram of coal are necessary for the reduction reaction but that additional quantities of the reducing agent do not increase the yield of tetrahydrofuran soluble products. The finding that the reaction proceeds equally well at −78, −33, and 25°C suggests that the reaction temperature is not a critical factor. Long reaction times do not offer a major advantage for the preparation of soluble products. Indeed, the quantity of soluble product obtained after a 24 hr reduction reaction is somewhat less than the quantity of soluble product obtained after a 6 hr reaction. However, the weight gain realized in the 24 hr reaction is about 25% greater than the weight gained in the 6 hr reduction. More effort will be necessary to distinguish between two alternative explanations. On the one hand, it is possible that the longer reaction time does yield a more negatively charged coal polyanion which is subsequently alkylated to a greater degree but that these additional butyl groups do not enhance the solubility of the product. On the other hand, the coal polyanion may undergo coupling reactions during the longer reaction time which cross-link the structure and limit the effectiveness of the butylation reaction for the formation of soluble products. Finally, the results indicate that the yields of soluble alkylated products obtained under the most favorable conditions for the reductive butylation reactions in liquid ammonia remain about 10–15% less than the yields obtained in the reductive butylation reaction of the same coal in the presence of an electron transfer agent in tetrahydrofuran.

Only a modest amount of information is now available concerning the reductive alkylation of coals in etheral solvents in the absence of added electron transfer agents. Niemann and Homback (1979) report that the reduction of a medium volatile bituminous coal proceeds about equally well with potassium or sodium potassium alloy in three different solvents including

$$\text{Robert coal} \xrightarrow[\substack{\text{monoglyme,} \\ \text{2 hr, } -70°C}]{K} \xrightarrow{(CH_3)_2CHOH} \text{Reduction product}$$

monoglyme at $-70°C$, monoglyme and triglyme (1:1) at $-20°C$, and triglyme and hexamethylphosphoramide (1:1) at $-20°C$. Unfortunately, hexamethylphosphoramide is incorporated into the reaction product.

The reductive alkylation of this coal has also been described (Haenel et al., 1980). About 75% of the methylated coal could be extracted into pyridine

$$\text{Robert coal} \xrightarrow[\substack{\text{monoglyme,} \\ \text{16 hr, } -30°C}]{K} \xrightarrow{CH_3I} \text{Alkylated product}$$

using a Soxhlet apparatus. When an equivalent procedure was adopted for the reductive methylation of Illinois No. 6 coal and Colorado subbituminous coal the alkylated products were 24 and 5% soluble, respectively, in tetrahydro-

$$\text{Illinois No. 6 coal} \xrightarrow[\substack{\text{monoglyme–} \\ \text{triglyme,} \\ -20°C, \text{ 16 hr}}]{K} \xrightarrow{CH_3I} \text{Alkylated product}$$

furan (D. A. Blain and L. M. Stock, unpublished research, 1981). Although the yields of soluble products obtained in other reductive methylation reactions of Illinois No. 6 coal (30% for the reaction in ammonia and 51% for the reaction in ether with an electron transfer agent) are somewhat greater, the simplicity of the method justifies further study to increase its effectiveness for the preparation of soluble alkylated coals.

B.   The Alkylation Reaction

The coal polyanions have been reacted with many different alkyl halides. The results obtained with the polyanions prepared from bituminous and sub-bituminous coals indicate that rather long reaction times (48 hr) are desirable for the completion of the alkylation reaction. Unfortunately, rather short reaction times have been employed in some studies. This is regrettable because, as already discussed, many of the reactions of the coal polyanion with the alkylating agent are slow. Apparently, the hindered phenolate anions and the residual carboxylate groups undergo alkylation quite slowly, and it is necessary to alkylate these structural units to achieve the maximum solubility of the reductively alkylated products. These slow reactions can apparently be accelerated by phase transfer catalysts such as tetrabutylammonium hydroxide (Liotta, 1979).

The impact of variations in the nature of the leaving group has been investi-

$$\text{Illinois No. 6 coal} \xrightarrow[\substack{\text{THF,} \\ 25°C}]{\text{K, } C_{10}H_8} \xrightarrow[\substack{\text{THF,} \\ 48 \text{ hr, } 25°C}]{C_4H_9X} \text{Coal alkylate}$$

gated. The results for the butylation of Illinois No. 6 coal are summarized in Table XII.

The reactions of the potassium coal polyanions with these butylation reagents differ markedly. Butyl chloride and butyl bromide are distinctly less effective than the iodide for the formation of soluble reaction products. The butyl sulfonate esters are somewhat more reactive. The mesylate is as effective as the iodide in its reaction with this polyanion. However, the triflate is much too reactive for practical work; it induces the polymerization of tetrahydrofuran to a viscous gel. The observation that butyl iodide is much more effective than the other butyl halides in the alkylation reaction provides further support for the view presented previously (Section II,C) that the alkylation process proceeds by a bimolecular nucleophilic substitution reaction rather than by an electron transfer process, followed by recombination of the components of a radical pair.

Several research groups have investigated the influence of the alkyl group on the outcome of the alkylation reaction. Representative results for the reductive methylation, ethylation, butylation, octylation, and benzylation of three North American coals and for the methylation, ethylation, butylation, and octylation of five Polish coals are summarized in Tables XIII and XIV, respectively. In each case, the reduction reaction was performed with potassium and naphthalene in tetrahydrofuran.

Although the data from different research groups may not be comparable because different methods have been used to remove naphthalene and its reductive alkylation products from the reductively alkylated coal, the trends observed in the experimental results are unmistakably the same. First,

TABLE XII

The Influence of the Leaving Group
on the Reductive Alkylation Reaction
of Illinois No. 6 Coal[a]

Reagent	Conversion (% solubility in THF)
Butyl chloride	23
Butyl bromide	51
Butyl iodide	62
Butyl mesylate	64
Butyl triflate	Polymer

[a] From Alemany et al. (1979).

TABLE XIII

The Solubility Achieved in the Reductive Alkylation of Three North American Coals
Using Several Different Alkyl Iodides

Reaction	Pocahontas[a] in benzene	Coal alkylate (% solubility) Balmer 10[b]		Illinois No. 6[c] in THF
		In benzene	In pyridine	
Methylation	48	63	80	51
Ethylation	95	—	—	—
Butylation	93	84	84	52
Octylation	—	85	87	54
Benzylation	—	71	82	—

[a] Sternberg *et al.* (1970, 1971).
[b] B. S. Ignasiak and Gawlak (1977).
[c] Alemany *et al.* (1979).

benzene is a poor solvent for the methylated coals, but a satisfactory solvent for the ethylated, butylated, octylated, or benzylated coals. More polar solvents such as tetrahydrofuran and pyridine are quite effective for the dissolution of all the alkylated coals. Second, the larger primary alkyl groups generally yield a greater quantity of soluble coal alkylate. However, it is necessary to consider the impact of the added weight of the alkyl groups on the results. This point is illustrated for coal K–III in Table XV.

In this case, reductive ethylation, butylation, and octylation enable a greater conversion of the coal molecules into alkylated products that are soluble in benzene than does reductive methylation. Generally, significant gains are realized in the solubility of the alkylate in hydrocarbons such as heptane or benzene when a primary alkyl group other than a methyl group is introduced into the coal. However, the differences in the effectiveness of the alkylating agents is considerably lessened when more polar solvents such as pyridine or tetrahydrofuran are examined.

Quite similar observations have been made in studies of the reductive alkylation reaction in liquid ammonia. Typical results for the octylation, butylation, and methylation of Illinois No. 6 coal are summarized in Table XVI.

The reductive butylation reaction of the Illinois No. 6 coal in liquid ammonia is more effective for the preparation of soluble alkylated coal than the reductive methylation reaction in this solvent. In addition, butyl iodide is a more effective reagent than butyl bromide or butyl chloride for the alkylation reaction. Thus, the results obtained for the coal polyanion prepared from Illinois No. 6 coal in ammonia parallel in a regular way the results obtained for the reduction reaction in tetrahydrofuran with an electron transfer catalyst.

TABLE XIV

The Extractability Achieved in the Reductive Alkylation of Five Polish Coals
Using Several Different Alkyl Iodides [a]

| Coal | Reaction | Coal alkylate properties | |
		Alkyl groups per 100 C atoms	Extractability (% in benzene)
K-1	Demineralized fresh	—	0.6
78.2% C (daf)	Reduced	—	10.0
	Methylation	10.5	38.6
	Ethylation	9.6	44.5
	Butylation	7.2	48.2
	Octylation	6.3	49.5
K-II	Demineralized fresh	—	0.4
81.1% C (daf)	Reduced	—	9.2
	Methylation	9.6	34.4
	Ethylation	8.0	44.5
	Butylation	6.2	52.4
	Octylation	5.4	58.1
K-III	Demineralized fresh	—	0.8
86.9% C (daf)	Reduced	—	10.1
	Methylation	9.7	50.3
	Ethylation	8.9	69.9
	Butylation	7.1	73.6
	Octylation	6.8	78.4
K-IV	Demineralized fresh	—	0.4
87.9% C (daf)	Reduced	—	3.5
	Methylation	9.9	46.0
	Ethylation	8.6	57.9
	Butylation	8.7	60.3
	Octylation	7.4	75.0
Anthracite	Demineralized fresh	—	0.1
92.6% C (daf)	Reduced	—	1.5
	Methylation	3.2	1.2
	Ethylation	1.8	1.5
	Butylation	1.3	2.9
	Octylation	1.3	5.4

[a] From Wachowska (1979).

All the results are in accord with the view that the alkylation reactions of the coal polyanion proceed by conventional ionic substitution reactions. Although no definitive data are now available concerning the behavior of the coal polyanions prepared in etheral solvents in the absence of electron transfer agents, it seems reasonable to postulate that the alkylation reactions of these anions will proceed in the same way as for the polyanions prepared in other ways.

TABLE XV

An Analysis of the Yield and Solubility Data for the Reductive Alkylation of Polish K–III Coal[a,b]

Reaction	Yield of products (wt. % of coal)	Solubility of product in benzene (%)	Wt. of soluble product (g)	Wt. of added alkyl group (g)	Wt. of coal in soluble product (g)
Methylation	121	50.3	0.61	0.21	.50
Ethylation	133	69.9	0.93	0.33	.70
Butylation	150	73.6	1.10	0.50	.73
Octylation	179	78.4	1.40	0.79	.78

[a] From Wachowska (1979).
[b] Based on the assumption that the added alkyl groups are proportioned equally between the soluble and the insoluble products.

TABLE XVI

The Influence of the Alkylating Agent on the Reductive Alkylation
of Illinois No. 6 Coal in Liquid Ammonia[a]

$$\text{Illinois No. 6 coal} \xrightarrow[\text{NH}_3, 6\text{ hr}]{\text{K}} \xrightarrow[\text{THF, 48 hr}]{\text{alkyl halide}} \text{Alkylated product}$$

Metal	Reaction temperature (°C)	Alkyl halide	Conversion (% solubility in THF)
K	−33	Methyl iodide	30
K	−33	Butyl chloride	21
K	−33	Butyl bromide	34
K	25	Butyl bromide	36
K	−33	Butyl iodide	52
K	−78	Butyl iodide	49
Na	−78	Octyl iodide	45

[a] The reduction reactions were carried out with 40 mmole of the metal per gram of the coal for about 6 hr.

C.   The Isolation of the Products

The alkylated coal products can be readily isolated when the reactions have been carried out in ammonia or in ether in the absence of electron transfer catalysts. Generally, the reaction mixtures are hydrolyzed and the aqueous phase is then acidified to precipitate all the coal products which are collected and washed with suitable solvents to remove the residual alkylation agents, reaction solvents, and salts.

Two quite different procedures have been used to separate the alkylated coal products from the residual naphthalene and its alkylation products in the reactions employing electron transfer reagents. The original procedure (Method A in Fig. 2) used by Sternberg et al. (1970, 1971, 1974) depended upon solvent extraction techniques for the selective removal of the compounds derived from the electron transfer agent from the desired coal products. The alternate procedure (Method B in Fig. 2) was developed by Burk and Sun (1975). This procedure employs chromatography on silica gel to separate the naphthalene derivatives from the alkylated coal products. Control experiments demonstrate that the naphthalenic compounds can be quantitatively removed from mixtures of the alkylated coal products obtained from Illinois No. 6 coal (Burk and Sun, 1975; Larsen and Urban, 1979; Alemany, 1980). Specifically, Larsen and Urban (1979) carefully examined the ethylation of Illinois No. 6 coal using ^{14}C-labeled naphthalene and ^{14}C-labeled tetrahydrofuran to probe for the incorporation of these compounds into the reaction products. They used Method B for the separation of naphthalene and

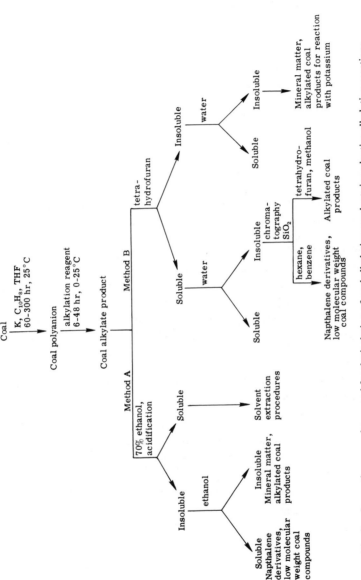

Fig. 2. Separation procedures used for the isolation of coal alkylation products in reductive alkylation reactions.

TABLE XVII

The Weight Percent of Naphthalene, Tetrahydrofuran, and Ethyl Groups in the
Chromatographic Fractions Obtained from the Reductive Ethylation of Illinois No. 6 Coal[a]

Fraction	Eluent	Fraction (wt. %) derived from		
		Naphthalene	THF	Ethyl
1	Hexanes	45.2	0.9	6.1
2	Transition between hexanes and C_6H_6/ hexanes	58.3	0.7	2.2
3	Benzene/hexanes (1:1)	46.3	1.8	8.2
4	THF	1.6	4.8	13.3
5	Left on column	0.8	0.8	11.0
6	Insolubles	0.6	0.7	10.8

[a] From Larsen and Urban (1979).

its reduction and alkylation products from the coal alkylation products. Their results are summarized in Tables XVII and XVIII.

The desired coal products are contained in fractions 4–6. The modest quantities of labeled naphthalene in these fractions indicate that the products of the reductive ethylation reaction of this coal can be cleanly separated from the products of the alkylation of naphthalene by this chromatographic method. The experimental results presented in Tables XVII and XVIII indicate that a somewhat greater quantity of tetrahydrofuran is incorporated into the alkylated coal products. However, it is important to note that most of the tetrahydrofuran is physically bound. The presence of physically entrapped tetrahydrofuran can be readily discerned by NMR spectroscopy, and Larsen and Urban point out that not less than 40% of the labeled tetrahydrofuran in fraction 4 can be removed by solvent extraction. Moreover, physical binding

TABLE XVIII

The Amounts of Naphthalene, Tetrahydrofuran,
and Ethyl Groups Incorporated into
Illinois No. 6 Coal in Reductive Ethylation[a]

Fraction	Molecular units per 100 C atoms of coal		
	Naphthalene	THF	Ethyl
1	5.3	0.2	5.1
2	6.8	0.1	1.2
3	5.2	0.4	6.2
4	0.2	1.0	11.0
5	0.1	0.2	9.7
6	0.1	0.2	9.9

[a] From Larsen and Urban (1979).

of this ether is apparently more important in reductive methylation and ethylation than in reductive butylation. In summary, the results obtained in several laboratories indicate that the desired coal alkylation products can often be separated from the reductive alkylation products of naphthalene and from the solvent by chromatographic method B (Fig. 2). However, it should not be assumed that this desirable result can always be obtained because many coals may present special problems.

It is less clear that the solvent extraction procedure is equally suitable. Investigators who have employed this method for the separation of the alkylated coal molecules from the reaction solvents and the electron transfer agents have noted large weight gains, unusually high oxygen contents, and other related anomalies. Such observations have prompted these investigators to suggest that tetrahydrofuran and its decomposition products and naphthalene and its alkylation products may be incorporated into the product mixture (Bimer, 1976; T. Ignasiak *et al.*, 1977; B. S. Ignasiak *et al.*, 1978a). Franz and Skiens (1978) studied the reductive ethylation of subbituminous Kaiparowitz coal to examine this feature. They found that the unseparated product mixture contained a considerable quantity of reduced and alkylated naphthalenes. The volatile products were identified (Table XIX).

There is evidence that the solvent extraction method does not remove all of these compounds from the coal alkylate. Franz and Skiens (1978) show that

TABLE XIX

**Volatile By-Products from the Reductive Ethylation
of Kaiparowitz (hvB) Coal**

Product	Yield (g)[b]
Tetralin	0.096
Naphthalene	0.030
Ethyldihydronaphthalene (2 isomers)	0.128
Ethyltetralin	0.018
Ethylnaphthalene	0.012
Diethyldihydronaphthalene (4 isomers)	0.049
Ethylbitetralyl	0.012
Tetralyldihydronaphthalene	0.050
Tetralylethyldihydronaphthalene (2 isomers)	0.038
Bitetralyl	0.490
Tetralyldihydronaphthalene (2 isomers)	0.182
Ethylbitetralyl	0.058
Total volatile products:	1.2

[a] From Franz and Skiens (1978).

[b] The reaction mixture contained 2.3 g naphthalene and 7 g of the coal.

the most prominent signals in the carbon NMR spectrum of ethylated sub-bituminous Kaiparowitz coal arise from bitetralyl rather than from a product of the coal. It is pertinent that the principal signals in the carbon NMR spectrum of an ethylated lignite (67.2% C, daf) and an ethylated bituminous coal (89.0% C, daf) obtained by Dogru and his associates (1978) also arise from bitetralyl (31). All these observations point to the presence of such com-

31

pounds in the coal alkylate obtained from reductive ethylation using naphthalene as an electron transfer agent. Presumably the solvent extraction procedures have failed to remove these contaminants from the alkylation products. As a consequence, it must be acknowledged that the solubility data and the spectroscopic properties of the samples isolated by Method A may be biased. The alkylated coals separated from the reaction mixture by Method B are apparently free of these contaminants. Further separations based upon gel permeation chromatography may be necessary in some instances to remove the naphthalene residues completely. Clearly, such procedures may be essential in some circumstances to avoid confusion in the interpretation of the experimental results.

The oxygen content of the products of the reductive alkylation reactions is sometimes greater than the oxygen content of the coal employed in the reaction (Sternberg et al., 1971; Bimer, 1976; T. Ignasiak et al., 1977; B. S. Ignasiak and Gawlak, 1977). Such observations have prompted several investigators to propose that tetrahydrofuran undergoes cleavage reactions, which lead to the chemical incorporation of the solvent into the coal. Alternative explanations can be based upon the absorption of oxygen by the alkylated coal product (B. S. Ignasiak and Gawlak, 1977) or the physical entrapment of the solvent within the solidified alkylated product.

Although it is clear that potassium does not react to any appreciable extent with tetrahydrofuran at ambient temperature in five days, as shown in Fig. 1, it is necessary to consider the reactions which might occur under the conditions of the reduction reaction because some of the carbanions or aromatic anions formed from the coal molecules might be more reactive. Indeed, it is well known that most alkyl ethers are slowly cleaved by strongly basic reagents of the kind present in the reduction reactions. Eisch (1963) reported that lithium biphenyl(-2) reacts with tetrahydrofuran at reflux for 8 hr to yield about 3% by weight of a crude product in which 1-butanol is the principal component. Fujita *et al.* (1972) found that naphthalene and lithium react in refluxing tetrahydrofuran to give a 50% yield of two alcohols. Other

(naphthalene) + (tetrahydrofuran)

$\xrightarrow[\substack{65°C,\ 8\ hr \\ 50\%}]{Li}$

H, CH$_2$CH$_2$CH$_2$CH$_2$OH structure (8%) + H, CH$_2$CH$_2$CH$_2$CH$_2$OH structure (92%)

hydrocarbons such as anthracene, biphenyl, and the methylnaphthalenes also react in this way. Gilman and Gaj (1963) reported that lithium reacts with 1,1,2,2-tetraphenylethane in refluxing tetrahydrofuran to yield 12% diphenylmethane and 62% 5,5-diphenyl-1-pentanol after 7 days. In further work, Bates *et al.* (1972) observed that butyllithium reacts with this ether to form

(tetrahydrofuran) + BuLi $\longrightarrow$ (tetrahydrofuran anion) $\longrightarrow$ CH$_2$CH$_2$ + CH$_2$=CHO$^-$

ethylene and the enolate of acetaldehyde in a rather rapid reaction ($t_{1/2} = 10$ min at 35°C). In equally pertinent investigations Carpenter *et al.* (1972) and Köbrich and Baumann (1973) observed that triphenylmethyllithium and 1,1-diphenylhexyllithium reacted slowly with tetrahydrofuran to yield cleavage products. If such reactions occurred with the carbanions formed from coal molecules, the —(CH$_2$)$_4$O— fragment would be incorporated into

$(C_6H_5)_3C^-, Li^+$ + (tetrahydrofuran) $\xrightarrow{95°C}$ $(C_6H_5)_3C(CH_2)_3CH_2O^-, Li^+$

the reaction products. However, the rate constant determined for the first-order reaction between triphenylmethyllithium and tetrahydrofuran at 30°C is 4.7×10^{-8} sec^{-1}, which corresponds to a half-life of about 170 days for this organolithium compound in this ether. The observation that the solution of

$$CH_3(CH_2)_4C(C_6H_5)_2{}^-, Li^+ \; + \; \underset{O}{\square} \quad \xrightarrow[\text{20 hr, } 70°C]{\text{120 hr, } 25°C}$$

$$CH_3(CH_2)_4C(C_6H_5)_2H \; + \; CH_3(CH_2)_4C(C_6H_5)_2(CH_2)_3CH_2OH$$

$$16\% \qquad\qquad\qquad\qquad 84\%$$

1,1-diphenylhexyllithium must be heated at 70°C to complete the reaction with tetrahydrofuran also indicates that this displacement reaction is slow under the conditions of the reductive alkylation reaction. The observations of Franz and Skiens (1978) and Miyake *et al.* (1980) are in accord with the results obtained in the study of the tertiary alkyllithiums. Neither of these research groups were able to detect coupling products, (e.g., 32) or other ring-opened derivatives. Miyake, for example, carried out the reduction reaction with

H   H

H   CH$_2$CH$_2$CH$_2$CH$_2$OH

32

potassium and naphthalene for 2–6 hr in tetrahydrofuran at 66°C without the formation of adducts. They also report that the Yūbari coal (86% C, daf) can be reduced by potassium in refluxing tetrahydrofuran without any increase in the oxygen content of the alkylated product. Thus, undesirable changes in the oxygen content of the coal were not observed even though the reaction conditions were somewhat more severe than those employed in the conventional reductive alkylation reactions.

Clearly, when the oxygen content of the products is unusually large, additional work must be done to establish that tetrahydrofuran or one of its decomposition products is not chemically bonded or physically adsorbed into the alkylated coal. The available data including the observation that ^{14}C-labeled tetrahydrofuran cannot be readily washed from the coal alkylate (Larsen and Urban, 1979) strongly suggest that physical incorporation of the ether is often more significant than chemical incorporation. Nevertheless, it is not possible to discount the high reactivity of the carbanions which may be formed during the reduction reactions.

Carson and Ignasiak (1979) reported that hydrogen was evolved during the alkylation of a Cretaceous vitrinite (89.1% C, daf) and other vitrinites after treatment of the coal with potassium and naphthalene in tetrahydrofuran. The hydrogen evolution during the octylation of a reaction mixture of Balmer

10 coal and naphthalene in tetrahydrofuran-d_8 consisted of 10% H_2, 22% HD, and 68% D_2. Thus, almost 80% of the hydrogen evolved originated from the tetrahydrofuran-d_8. Carson and Ignasiak also found that the vitrinite was required for the evolution of hydrogen and that chars and other more aromatic coal products were not reactive under the same conditions. The observations suggest that the alkylating agents serve as oxidizing agents for the aromatization of certain hydroaromatic compounds in the reaction mixture. The shuttling of hydrogen atoms between tetrahydrofuran-d_8, naphthalene, and the coal apparently occurs to a limited degree. Walsh and Dabestani (1981) observed that naphthalene is a more important donor of hydrogen atoms than tetrahydrofuran. The point is illustrated by the deuterium content of the E- and Z-2-hexenes formed in the presence and absence of the labeled compounds. Naphthalene appears to be a far more effective donor than the more abundant solvent. Walsh and Dabestani infer that an alkyldi-

$$\text{Na,Naphthalene} + CH_2{=}CH(CH_2)_4F \xrightarrow[\substack{7\ hr,\\25°C}]{THF\text{-}d_8} Z\text{-2-hexene} + E\text{-2-hexene}$$

	29%	13%
	8% D	8% D

$$\text{Na,Naphthalene-}d_8 + CH_2{=}CH(CH_2)_4F \xrightarrow[\substack{7\ hr,\\25°C}]{THF} Z\text{-2-hexene} + E\text{-2-hexene}$$

	39%	16%
	48% D	39% D

hydronaphthalene anion, $C_{10}H_8R^-$, may be the actual proton transfer reagent. The requirement that coals or coal products be present in the reactions leading to hydrogen evolution also suggests that the naphthalene anions (e.g., 5) play a significant role in the hydrogen atom transfer reactions. Unfortunately, no

5

one has yet established the rate at which the anions of naphthalene exchange with tetrahydrofuran and more research will be required to sort out the factors governing these processes. However, it has been known for some time that compounds such as 1,4-dihydronaphthalene can react with organometallic compounds such as phenyllithium to give naphthalene in 85% yield (Gilman, 1954). The reactions of 1,4-dihydronaphthalene and 1,4-dihydrodibenzofuran with phenyllithium yield appreciable quantities of the oxidized aromatic compounds under vigorous reaction conditions. Lithium hydride is the formal product of the reaction. Harvey and Cho (1974) have also observed that the dehydrogenation reactions of dihydrophenanthrene occur readily. The occurrence of equivalent reactions during the reductive alkylation reaction might be responsible for the formation of hydrogen. It is notable that the low-rank coals are the most effective reduction reagents for naphthalene and that it is the reduced naphthalenes which react with organometallic compounds to yield substances capable of hydrogen formation. In accord with Gilman's proposal that there is a significant driving force for conjugation in long-duration experiments at ambient temperatures, the conversion of tetrahydrofuran to furan may be another important reaction. Such reactions would, of course, account for the significant quantity D_2 observed by Ignasiak and his co-workers.

D. Comparison of the Methods

Several procedures for the reduction and for the reductive alkylation of coal have been described in the previous sections. In this section, we compare the relative effectiveness of these various methods for the formation of soluble products and contrast the results of the reductive alkylation reactions with experiments designed to reduce or to alkylate the coal or to effect elimination reactions. In this survey, attention is focused on a typical Polish coal and on Illinois No. 6 coal with special reference to their conversion to compounds soluble in polar organic solvents.

Bimer and Witt (1978) compared the effectiveness of several reagents for the reduction of an intermediate rank coal with about 80% carbon. The experiments were performed with the fresh coal and with the demineralized coal. The results obtained in this investigation are summarized in Table XX.

Curiously, the reduction reactions of this coal depend in a significant way upon the mineral content of the coal. The sample with the lesser mineral content acquires more hydrogen atoms per 100 carbon atoms and contains fewer hydroxyl groups and more volatile matter. As discussed subsequently, the results for lithium in ethylenediamine cannot be taken too literally because of the incorporation of the reaction solvent into the products. Similar problems may exist for the reduction products obtained with potassium naphthalene(-1) in tetrahydrofuran. However, the analytical results obtained by Bimer and Witt for the reduction with this reagent indicate that only six hydrogen atoms are added per 100 carbon atoms and that there is no large increase in the oxygen content of the product. Hence, the incorporation of the solvent is not an important factor with this coal.

The effectiveness of the reduction in liquid ammonia is not enhanced appreciably by the addition of an alcohol to the reaction mixture in spite of the fact that the added alcohol should enable the reduction of the benzene

TABLE XX

A Comparison of Several Methods for the Reduction of a Representative Polish Bituminous Coal[a]

Reaction	Reduction (H added per 100 C atoms)	Solubility (% in pyridine)	Hydroxyl groups (%)	Volatile matter (%)
Fresh coal (79.91% C, daf)				
Starting coal	—	12.1	4.49	33.8
Sodium, NH_3, −33°C	3	12.9	5.27	35.5
Sodium, t-BuOH, NH_3, −33°C	3	13.5	5.11	34.5
Lithium, $H_2NCH_2CH_2NH_2$, 80–90°C	2	19.2	5.65	38.5
Lithium, t-BuOH, $H_2NCH_2CH_2NH_2$, 80–90°C	2	17.2	6.53	38.5
Demineralized coal (80.14% C, daf)				
Starting coal	—	13.8	4.18	35.0
Sodium, NH_3, −33°C	5	14.1	5.27	35.0
Sodium, t-BuOH, NH_3, −33°C	5	13.9	5.21	33.0
Lithium, $H_2NCH_2CH_2NH_2$, 80–90°C	12	21.9	6.15	40.0
Lithium, t-BuOH, $H_2NCH_2CH_2NH_2$, 80–90°C	15	20.2	4.33	39.0
K^+, $C_{10}H_8^{-}$, THF, 25°C	6	21.9	4.69	43.0

[a] From Bimer and Witt (1978).

derivatives under the experimental conditions. The results obtained in this study suggest that this bituminous coal is not more significantly reduced in the presence of the alcohol than in its absence. This finding is in accord with other work, which suggests that the reduction reactions of monocyclic aromatic compounds in the coal molecules do not occur at all readily in liquid ammonia solution.

Several other methods for the reduction of coal have been described more recently in the literature, but they have not yet been studied comparatively. These methods are discussed in subsequent sections of this review.

The reductive alkylation reactions of Illinois No. 6 coal have been carried out in ethers in the presence or absence of electron transfer agents such as naphthalene or biphenyl and in liquid ammonia. As already discussed, these reactions appear to proceed most effectively with potassium and butyl iodide. The results obtained under the five different sets of experimental conditions that are generally used are summarized in Table XXI.

The procedures used for the alkylation of this coal correspond closely with the methods described by other investigators. On the one hand, the reduction reactions of the coal in the solutions containing solvated electrons proceed quite rapidly and rather short (less than 24 hr) reaction times were used in these reduction reactions. On the other hand, the reduction reactions in the presence of an electron transfer agent were generally carried out for a long reaction time, as is customary. The butylation reaction was performed insofar as possible at the same temperature and for the same time in each experiment. The results indicate that the procedure first advanced by Sternberg and his associates (1970, 1971; Sternberg and Delle Donne, 1974) using an electron transfer agent is more successful for the conversion of Illinois No. 6 coal to soluble compounds than any of the reactions carried out without electron transfer agents. The fact that 52% of the coal is converted to soluble products by the reaction in ammonia indicates that the reaction system is also quite suitable for the formation of soluble alkylated products. It should be noted that an increase in the time alloted for this reduction reaction does not increase the extent of the conversion of the solid coal to soluble alkylated products. The reaction in glyme solvents without an electron transfer agent is not as successful for the preparation of soluble product from the Illinois No. 6 coal. Whether the reaction will be more successful at longer reaction times remains to be determined. The addition of a proton donor (e.g., water) to the reaction in ammonia considerably reduces the yield of soluble products. This

TABLE XXI

A Comparison of the Reductive Butylation Reactions of Illinois No. 6 Coal Using Potassium and Butyl Iodide

Experiment	Reduction[a] Solvent	Addenda	Time (hr)	Temperature (°C)	Alkylation[b] solvent	Conversion (% solubility in THF)
1[c]	NH_3	None	6	−33	Tetrahydrofuran	52
2[c]	NH_3	H_2O	6	−33	Tetrahydrofuran	27
3[d]	Monoglyme–triglyme	None	15	−50	Monoglyme–triglyme	30
4[e]	Monoglyme	Naphthalene	120	Ambient	Monoglyme	67
5[e]	Tetrahydrofuran	Naphthalene	120	Ambient	Tetrahydrofuran	67

[a] The reduction reactions were performed by the general methods outlined in the literature.

[b] The alkylation reactions were performed for 48 hr at ambient temperature in each experiment.

[c] C. I. Handy and L. M. Stock (unpublished research, 1981), using the approach of Lazarov and Angelova (1968).

[d] D. A. Blain and L. M. Stock (unpublished research, 1981), using the approach of Niemann and Hombach (1979).

[e] Alemany (1980), using the approach of Sternberg et al. (1971).

observation suggests that the reduction reactions of aromatic structures in the coal should be avoided if the maximum degree of solubilization is to be achieved. Presumably, the less stable, more reactive anions formed transiently in ammonia are reduced in the presence of the proton donor, and the concentration of carbanions in the reaction mixture is less than in the absence of the proton donor. In this situation, the reaction yields more reduced and less alkylated aromatic compounds. This interpretation is supported by the fact that fewer alkyl groups are added to the coal when a proton donor is present in the initial reaction mixture.

The conversion of the Illinois No. 6 coal to a polyanion and its subsequent protonation with ammonium chloride yields a product that is not appreciably soluble in tetrahydrofuran (Table XXII). Indeed, the material seems to be less soluble than the untreated coal. Very similar results are obtained when the reduction capacity of the system is increased by the incorporation of proton donors such as ethanol, *t*-butyl alcohol, or water. Although many monocyclic aromatic compounds should be reduced to dihydro compounds under these

conditions, the experimental results obtained in this laboratory and the results presented by Bimer and Witt (1978) suggest that these reactions do not pro-

TABLE XXII

**The Conversion of Illinois No. 6 Coal to Soluble Products
Using Reduction, Elimination and Alkylation Reactions** [a]

Experiment	Reaction conditions	Conversion (% solubility in THF)
6	Reduction, K in NH_3 without an additive for 6 hr at $-33°C$; NH_4Cl [b]	3
7	Reduction, K in NH_3 with ethanol present for 6 hr at $-33°C$; NH_4Cl	2
8	Elimination, KNH_2 in NH_3 for 6 hr at $-33°C$; NH_4Cl	5
9	Alkylation, butyl iodide in aqueous THF with tetrabutylammonium hydroxide [c]	23
10	Alkylation, butyl iodide in NH_3 with KNH_2 [d]	28
5	Reductive alkylation as described in Table XXI	67

[a] From A. Reed and L. M. Stock (unpublished results, 1981).
[b] The procedure of Lazarov and Angelova (1968) was used.
[c] The procedure of Liotta (1979) was used.
[d] The procedure of B. S. Ignasiak *et al.* (1979) was used.

TABLE XXIII

TABLE XXIII

The Butylation of Reduction Products Obtained from Illinois No. 6 Coal[a]

Experiment	Source of coal product (Table XXII)	Reaction conditions	Conversion (% solubility in THF)
11	Experiment 6, K, NH_3/NH_4Cl	Butylation in THF	25
12	Experiment 7, K, C_2H_5OH, NH_3/NH_4Cl	Butylation in THF	17

[a] From A. Reed and L. M. Stock (unpublished results, 1981).

ceed readily with bituminous coals in ammonia. The reaction was also examined in the presence of a strong base, potassium amide, capable of the initiation of elimination reactions of benzyl ethers, as already discussed previously (Section II,A,1). In this instance, the reaction product was only slightly more soluble than the products obtained in the reduction reactions (Table XXII). Thus, ether cleavage reactions alone are inadequate to provide a large quantity of soluble coal products. Alkylation reactions are much more successful than reduction or elimination reactions. The alkylation of this coal using the methods of Liotta and Ignasiak and their associates (Liotta, 1979; B. S. Ignasiak *et al.,* 1979) provided much more soluble product. However, the quantity of the soluble product obtained in these reactions was only about half of the amount of material obtained in the reductive alkylation reactions in the same solvents (Table XXII).

The experimental observations summarized in Table XXIII reveal that reduction followed by butylation yields a considerably greater quantity of soluble product. As in the other cases, the more reduced product obtained in the presence of a proton donor yields a lesser quantity of soluble products. These observations establish that the reduction reactions which occur in liquid ammonia do not fragment the coal molecules to a sufficient extent to render a large quantity of the product of the subsequent alkylation reaction soluble in tetrahydrofuran. These results strongly imply that the alkylation reactions of carbanions and aromatic anions are necessary for the conversion of more than 50% of the coal to a soluble product.

The behavior of the insoluble reaction products in subsequent reactions was also examined, as outlined in Table XXIV. The insoluble residue obtained in the reductive alkylation reaction of the Illinois No. 6 coal with potassium and naphthalene in tetrahydrofuran could be reductively alkylated with potassium and naphthalene in tetrahydrofuran to yield products that were about 60% soluble in tetrahydrofuran. The residue from the reaction in liquid ammonia was also treated with potassium and naphthalene to yield a product that was also about 60% soluble in tetrahydrofuran. The insoluble residue from the reductive alkylation reaction with potassium and naphthalene in tetrahydrofuran was 48% converted to a soluble product by

TABLE XXIV

The Reductive Butylation of the Insoluble Reaction Products Obtained in the Initial Reductive Butylation Reactions[a]

Experiment	THF-insoluble residue from reductive butylation (Table XXI)	Reaction conditions	Conversion (% solubility in THF)
13[a]	Experiment 1, K, NH_3/C_4H_9I	The same reaction was repeated	6
14[b]	Experiment 1, K, NH_3/C_4H_9I	Reductive alkylation	59
		K, $C_{10}H_8$, THF/C_4H_9I (Experiment 5)	
15[b]	Experiment 5, K, $C_{10}H_8$, THF/C_4H_9I	The same reaction was repeated	66
16[b]	Experiment 5, K, $C_{10}H_8$, THF/C_4H_9I	Reductive alkylation	48
		K, NH_3/C_4H_9I (Experiment 1)	

[a] From C. I. Handy and L. M. Stock (unpublished research, 1981).

[b] A. Reed and L. M. Stock (unpublished results, 1981).

treatment with potassium in liquid ammonia. However, the insoluble reaction product obtained from the reaction in liquid ammonia could not be reductively alkylated successfully by further reaction in liquid ammonia. The results obtained in these subsequent reactions indicate that the reaction in liquid ammonia follows a somewhat different course than the reaction in ether. It is striking that the insoluble residue from the reductive butylation reactions in tetrahydrofuran undergoes a second reductive butylation reaction in

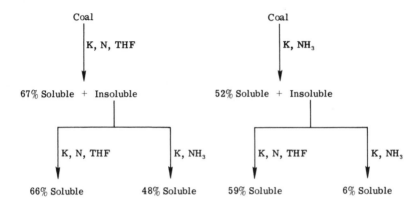

tetrahydrofuran or in ammonia to yield an important quantity of soluble alkylated products and that the insoluble residue from the reductive butylation reaction in ammonia only reacts effectively in the ether. Although it is difficult to present a completely adequate explanation for these observations on the basis of the information now available, the differences in the rate of ether cleavage reactions in the two reaction systems and the differences in the mode of the electron transfer reactions may be responsible for these results. On the one hand, ether cleavage reactions with potassium in tetrahydrofuran proceed much more effectively than the corresponding reactions in liquid ammonia. Presumably, the ethers retained in the insoluble reaction products from the reductive butylation in ammonia are cleaved in the second reaction in tetrahydrofuran and that reaction is successful. On the other hand, the solvated electron is presumably a much more effective reagent for the reduction of coal than potassium naphthalene(-1) or potassium naphthalene(-2). The solvated electron is especially effective in the sense that it is able to propagate through the micropores of the coal and, thereby, is more effective for the reduction of the structural elements deep within the coal matrix. Such reactions may be quite effective for the fragmentation of the coal molecules in most circumstances; however, in other cases, weak carbon–carbon or carbon–oxygen bonds may break in the interior of the coal, and, in the absence of an effective proton or hydrogen atom donor, recombination reactions may occur to yield new, more stable structures, which are more resistant to reductive cleavage reactions. To illustrate, relatively labile benzyl phenyl ethers cleave to give bibenzyl and phenol (Table III). In the interior of a coal such reactions could cross-link rather than fragment the structure to give a much less soluble material.

IV. APPLICATIONS OF REDUCTION, ALKYLATION, AND REDUCTIVE ALKYLATION

A. The Reduction of Coal

In the years following the Second World War, several investigators used alkali metals in ammoniacal solution to reduce cellulose, lignin, and the humic acids. Reactions of this kind were applied for the study of coal in 1958 when two groups (Given et al., 1958; Reggel et al., 1958) reported the successful reduction of coals using lithium in ethylamine and in ethylenediamine, respectively. Given and his associates found that lithium in ethylamine reacted with several different coals ranging from 82 to 92% carbon (daf). The number of hydrogen atoms added per 100 carbon atoms of the coal ranged from about 15 for the coal with 82% carbon to about 35 for the coal with 89% carbon. The proportion of the reduced coal product that is soluble in cold pyridine also depends upon the carbon content. Only about 15% of the reduction product of the coal with 82% carbon is soluble in cold pyridine, whereas almost 80% of the reduction product of the coal with 89% carbon is soluble in this solvent. Coals with a greater carbon content are not reduced to the same extent and the products are not as soluble in cold pyridine. Because solutions of lithium in ethylamine are very effective reducing agents for aromatic compounds, Given and his associates suggested that the enhanced solubility of the coal was a consequence of structural deformations resulting from the decreased rigidity of the coal structure resulting from the extensive reduction of the planar aromatic compounds.

In another study, Given et al. (1960) investigated the reduction reactions of macerals by lithium in ethylamine. The results for a group of vitrinites and exinites are presented in Table XXV. Generally, the vitrinites appear to be more readily reduced than the exinites. The lower rank coal provides a notable exception to this generalization. The reduction products of the vitrinites are also much more soluble in cold pyridine than the reduction products of the exinites. It is also significant that about 40% of the oxygen atoms in the exinite of the Cannock Wood coal are removed in the reduction reaction. Little or no change in the oxygen atom content of the other macerals occurred in this reaction. Given and his co-workers suggest that deoxygenation reactions of the

TABLE XXV

The Reduction of Vitrinites and Exinites of Several Coals by Lithium in Ethylamine[a]

Coal	Carbon content (%, daf)		Reduction (H added per 100 C atoms)		Solubility (% in pyridine)	
	Vitrinite	Exinite	Vitrinite[b]	Exinite	Vitrinite[b]	Exinite
Cannock Wood, Shallow	79.5	80.0	6	26	—	27
Markham Main, Barnsley	82.1	81.3	14	3	20	7
Dinnington Main, Barnsley	84.9	84.2	23	6	40	—
Aldwarke Main, Silkstone	86.8	86.9	32	4	50	9
Chislet, No. 5	88.5	89.5	37	23	70	29

[a] From Given et al. (1960).

[b] These results are estimated from the relationships between the degree of reduction of the coal and the solubility in pyridine and the carbon content of the coal as reported by Given et al. (1960).

kind observed in the electrolytic reduction of anthraquinone and xanthone occur with this coal (Given and Peover, 1959; Given *et al.,* 1960).

The application of lithium in ethylenediamine as a reducing agent for coal was also reported in 1958. Reggel and his associates found that a variety of coals including subbituminous, bituminous, and anthracite could be effectively reduced by this reagent at the boiling point in 2.5 hr (Reggel *et al.,* 1957, 1958, 1961). Under these vigorous conditions, the hydrogen atom content of the coals increases appreciably and many of the coals are converted to compounds that are soluble in cold pyridine. Unfortunately, the reduction products of the lower rank materials, the lignite and the subbituminous coals, contain appreciable quantities of ethylenediamine and water that cannot be removed. However, the reduction products of the bituminous coals contain less than one molecule of the diamine per 100 carbon atoms. The results presented in Table XXVI have been corrected for the presence of the diamine.

The observations suggest that this reduction reaction is most effective for coals with about 89% carbon as also found by Given *et al.* (1960). However, the more vigorous conditions used by Reggel and his co-workers generally provide a greater degree of reduction than obtained with lithium in ethylamine. Indeed, when the product of the initial reduction reaction of Pocahontas No. 3 coal with lithium in ethylenediamine is reduced in a second experiment an additional 17 hydrogen atoms per 100 carbon atoms can be added to the coal. The absorption bands associated with aliphatic C–H vibra-

TABLE XXVI

The Reduction of Vitrinites of Several Coals by Lithium in Ethylenediamine[a]

Coal	Carbon content (%, daf)	Reduction (H added per 100 C atoms)[b]	Solubility (% in pyridine)
Harmatten	78.4	−5	—
Bruceton	82.5	17	31
Pond Creek	84.5	26	38
Lower Banner	88.7	28	97
Pocahontas No. 3	89.8	45	63
Dorrance	93.1	18	0
Graphite	99.6	20	—

[a] From Reggel *et al.* (1961).

[b] A correction has been made for the ethylenediamine incorporated into the reduction product.

tions in the infrared spectrum exhibit increased intensity in all of the reduced coals, whereas virtually all of the other absorption bands including carbonyl absorptions and the absorptions normally identified with aromatic structures exhibit reduced intensity. Elemental analyses indicate that sulfur compounds are removed in this reduction reaction. The solubility data obtained by Reggel and his co-workers (Table XXVI) are qualitatively similar to the results obtained by Given and his associates (Table XXV). In particular, both groups found that the reduction products of the coals with about 89% carbon were more soluble than the reduction products obtained from coals with either lower or higher carbon contents. This intriguing observation has not been completely explained. The factor that seems to be most important is the significant increase in the hydrogen content of these coals with as many as 50 hydrogen atoms added per 100 carbon atoms of the original coal after two reduction reactions. Clearly, the aromatic compounds in these coals must be appreciably reduced with a concomitant disruption of the van der Waals interactions between these molecules. Ether and thioether cleavage reactions also occur under the conditions of these reactions. Although the hydroxyl groups are, for the most part, retained the mercaptan groups are eliminated from the structure. Given and Reggel and their associates note that the extensive reduction reactions of the aromatic compounds will cause structural distortions, which also reduce intermolecular interactions in the solid and enhance the solubility of the reduction product. It is also reasonable to postulate that the carbon–carbon bond cleavage reactions, which also occur under these experimental conditions, reduce the molecular weight and enhance the solubility of the product. Just why these factors are most important for coals with about 89% carbon is not clear.

The somewhat simpler reduction of coals using alkali metals in liquid ammonia was carried out by Halleux and his associates (1961; Halleux and de Greef, 1963). They studied the reduction of a pyridine extract of vitrinite obtained from Bruceton coal using sodium in ammonia and obtained evidence for the cleavage of ether bonds by determination of the hydroxyl content before and after the reduction reaction. Kröger and his associates (1965) investigated the possibility that alkali metals in liquid ammonia could be used for the quantitative determination of ethers in subbituminous coals but without conclusive results. Lazarov and Angelova reported on the reduction reactions of whole coals with sodium in liquid ammonia in 1968. The results obtained for seven eastern European coals ranging from 66 to 90% carbon are shown in Table XXVII.

This mild reduction reaction increased the hydrogen content of these coals to a lesser degree than had been realized in the work with lithium in amine solvents. Sodium in liquid ammonia led to the addition of only 15 hydrogen atoms per 100 carbon atoms of the Kachulka coal compared to the approx-

TABLE XXVII

The Reduction Reactions of Seven Eastern European Coals with Sodium in Liquid Ammonia[a]

$$\text{Coal} + \text{Na} \xrightarrow[-78°C, 6 \text{ hr}]{\text{NH}_3, (\text{C}_2\text{H}_5)_2\text{O}} \xrightarrow{\text{NH}_4\text{Cl}} \text{Reduction product}$$

Coal	Carbon content ($\%$, daf)	Reduction (H added per 100 C atoms)	Volatile matter ($\%$, daf)	
			Before	After
Maritza–Iztak (Bulgaria)[b]	65.7	6	65.0	66.3
Pernik (Bulgaria)[c]	75.1	2	49.9	52.5
Irkutsk (USSR)[d]	76.5	18	52.6	55.0
Donbass (USSR)	82.2	5	39.8	45.9
Zelenigrad (Bulgaria)	83.1	15	36.8	44.0
Kachulka (Bulgaria)	87.6	15	26.4	43.4
Chumerna (Bulgaria)	89.9	13	16.6	33.9

[a] From Lazarov and Angelova (1968).
[b] A brown coal.
[c] Lignite.
[d] This coal contains 5.4% (daf) organic sulfur.

imately 30 hydrogen atoms which were added per 100 carbon atoms of the coals of equivalent rank by lithium in the ethylamine or ethylenediamine. In addition, the extent of the reduction reaction in liquid ammonia is not as dependent upon the composition of the coal as are the reactions with lithium in amines. One clear advantage of the reaction of sodium in liquid ammonia arises in the fact that the reaction solvent is not incorporated into the product in any significant degree. The Donbass coal and the Kachulka coal gain 0.1 and 0.2% nitrogen, respectively, in the reduction reaction.

Studies of the reductive alkylation reaction in ammonia suggest that potassium may be a better reducing agent than sodium. Unfortunately, the reactions of these coals and other coals with potassium have not been investigated.

The properties of the reduction products of reactions with sodium in liquid ammonia have been examined. The initial studies of Lazarov and Angelova (1968) were supplemented by further work from the same laboratory in 1971 (Lazarov et al., 1971). These contributions describe the significant changes that have occurred in the physicochemical properties of the coals as a consequence of the reduction. The extractability into chloroform, the free swelling indices, the density and the reflectance of two representative coals are presented in Table XXVIII.

In the second contribution, Lazarov et al. (1971) compared the products of reduction, acetylation, and thermolysis of the same two coals. The reduction reaction was carried out as already described. The acetylation reaction was performed by the well-known method of Blom et al. (1957). The thermolyses

TABLE XXVIII

The Properties of Donbass and Kachulka Coal before and after Reduction in Ammonia[a]

Property	Donbass coal (82.2% C, daf)		Kachulka coal (87.6% C, daf)	
	Before	After	Before	After
Volatile matter (%, daf)	39.8	45.9	26.4	43.4
Free swelling index	2.5	3.0	5.0	9.0
Chloroform extraction (%)[b]	0.8	3.2	0.9	11.2
Density, g ml⁻¹ (in methanol)	1.35	1.35	1.30	1.27
Reflectance in air, R (%)	6.10	7.25	5.33	8.24

[a] From Lazarov and Angelova (1968).
[b] The Soxhlet method was used for this determination.

were carried out by heating the sample at a constant rate of 3.3°C min⁻¹, to 400 or 450°C and then rapidly cooling the sample. The average molecular weight, the fraction of aromatic carbon atoms and the reflectance of the products were compared. The results are summarized in Table XXIX.

These results indicate that significant increases occur in the volatile matter, the free swelling index, and the reflectance, and that the densities are little

TABLE XXIX

Properties of Chloroform Extracts of the Reaction Products of Donbass and Kachulka Coals[a]

Reaction	H/C atomic ratio	Physicochemical property		
		Average molecular weight[b]	f_a^c	R^d
Donbass coal (82.2% C, daf)				
None[e]	0.987	690	0.62	10.6
Thermolysis, 400°C	1.054	430	0.62	6.5
Acetylation[f]	0.978	780	0.61	12.2
Reduction, Na, NH₃	1.208	460	0.46	6.4
Kachulka coal (87.6% C, daf)				
None[e]	0.928	740	0.72	10.4
Thermolysis, 450°C	0.926	620	0.71	9.3
Acetylation[f]	0.906	770	0.70	12.0
Reduction, Na, NH₃	0.984	670	0.62	10.7

[a] From Lazarov *et al.* (1971).
[b] An ebulliometric method was employed.
[c] The fraction of aromatic carbon atoms was estimated from the proton NMR spectra.
[d] The reflectance in air, %.
[e] The coal was extracted with pyridine and the chloroform-soluble part of the extract was investigated.
[f] Acetylation was carried out by the method of Blom *et al.* (1957).

changed by reduction. The change in the extractability into chloroform is appreciably different for the two coals. Unfortunately, data are not available concerning the solubility of these reduction products in cold pyridine. However, it seems safe to conclude that the reduction product of the Kachulka coal is appreciably soluble in this solvent. The changes in the properties of the higher rank coal are, in general, more pronounced than the changes in the properties of the lower rank coal. This difference corresponds, of course, to the greater degree of the reduction of the higher rank coal. The molecular weight data obtained for the reduction products are more similar to the data for the thermolysis reaction than to the data for acetylation reaction products or untreated extracts. This observation prompted Lazarov and his co-workers to conclude that bond cleavage reactions were significant in the reduction reaction. Even though the f_a values may not be quantitatively correct, the f_a data suggest that the concentration of the aromatic compounds in the reduction products is considerably smaller than in the untreated coals or the products of thermolysis or acetylation. This observation is in good accord with the idea that the aromatic structures in these coals are reduced in the course of this reaction. As already discussed, this reduction reaction is more selective than the reactions with lithium and the reduction of the polycyclic aromatic compounds would be expected to occur in preference to the reduction of benzene derivatives.

The extent of the reduction reaction and the quantity of volatile matter are considerably larger for the higher rank coals than for the lower rank brown coal and lignite. Lazarov and Angelova suggest that the increased quantity of volatile matter and the increased solubility in chloroform are indicative of ether cleavage reactions as well as reduction reactions. To test this idea they determined the number of hydroxyl groups present in the coal before and after the reduction reaction; their observations are summarized in Table XXX.

As already mentioned, the quantity of hydrogen taken up by these coals strongly suggests that some of the aromatic compounds in the coal are reduced. The analytical data presented in Table XXX support this view with only 0.6 and 4.2 hydrogen atoms used in the reductive ether cleavage reactions of the Donbass and the Kachulka coal, respectively. Moreover, these results imply that there are significant differences in the reactivities of the oxygen atoms in these two coals. The oxygen atoms which cannot be acetylated in the original Donbass coal are not converted to oxygen atoms which can be acetylated in the reduced product. On the other hand, the unreactive oxygen atoms in the Kachulka coal are rendered reactive by the reduction reaction. The results for these two coals are compared with the results for several other coals in Table XXXI.

The results suggest that there is a larger variation in the nature of the oxygen

TABLE XXX

The Composition of Donbass and Kachulka Coal before and after Reduction in Liquid Ammonia[a]

Coal sample	Composition (%, daf)							
	C	H	S	N	O^b	$O_{OH}{}^c$	$O_{CO}{}^d$	$O_{NR}{}^e$
Donbass before	82.2	5.9	0.6	2.7	8.4	3.9	0.9	3.6
Donbass after	82.0	6.1	0.5	2.8	8.3	4.2	0.8	3.3
Kachulka before	87.6	5.3	1.4	1.3	4.1	0.8	0.5	2.8
Kachulka after	86.7	6.4	0.8	1.5	4.0	2.9	0.5	0.6

[a] From Lazarov and Angelova (1968).
[b] The total oxygen content determined by difference.
[c] The hydroxyl group content was determined by the method of Blom et al. (1957).
[d] The carbonyl group content was determined by a method developed by Angelova.
[e] The quantity of unreactive oxygen was determined by difference.

atoms in these coals. The experimental observations strongly imply that the ether cleavage reactions are most effectively accomplished in the higher rank coals even though the ether linkages are more numerous in the lower rank coals. Two factors may be responsible for this difference. First, sodium in liquid ammonia effectively cleaves aryl ethers but not aryl alkyl ethers. Second, phenolate anions resist reductive cleavage. Thus, molecular fragments such as the partially etherified 1,2-dihydroxybenzenes, which would be relatively abundant in the lower rank coals, would be converted to anions that strongly resist further reduction, thereby preventing the ether cleavage reaction. The

TABLE XXXI

Distribution of Reactive Hydroxyl and Carbonyl Groups, Cleavable Ethers, and Noncleavable Ethers in Several Coals[a]

Coal	Type of oxygen atom (% of total O)		
	$O_R{}^b$	$O_{Na}{}^c$	$O_{NR}{}^d$
Maritza	60.5	10.2	29.3
Pernik	54.0	5.0	41.0
Donbass	57.1	5.2	37.7
Zelenigrad	54.9	26.4	18.7
Kachulka	31.7	51.4	16.9
Chumerna	37.9	46.4	15.7

[a] From Lazarov and Angelova (1968).
[b] The quantity of reactive oxygen atoms in the whole coal.
[c] The quantity of unreactive oxygen atoms in the whole coal that are rendered active by reduction in ammonia.
[d] The quantity of unreactive oxygen atoms in the whole coal that are not rendered active by reduction in ammonia.

$$\text{Coal—C}_6\text{H}_3(\text{O—Alkyl})(\text{OH}) \longrightarrow \text{Coal—C}_6\text{H}_3(\text{O—Alkyl})(\text{O}^-)$$

finding that proportionally more ethers are cleaved in the higher rank coals is, accordingly, compatible with a decrease in the dihydroxybenzene content and an increase in the diaryl ether content of the higher rank coals.

Montgomery and Wachowska and their associates investigated the use of potassium in the presence of naphthalene as a reagent for the reduction of coal macerals, coals, and their oxidation products in tetrahydrofuran under the conditions used by Sternberg for the preparation of the coal polyanion. This

$$\text{Coal} \xrightarrow[\text{96–144 hr, 25°C}]{\text{K, C}_{10}\text{H}_8} \text{Coal polyanion} \xrightarrow{\text{H}_2\text{O}} \text{Reduction product}$$

research effort was focused from the outset upon the structural factors which were important for the definition of the coking properties of coal, weathered coal, and the coal macerals. They were especially concerned about the influence of hydroxyl and etheral functional groups upon the dilatation, contraction, softening points, and plasticity indices. In the course of this study it was established that a variety of Canadian and Polish coals ranging from brown coal to anthracite could be reduced by the reaction with potassium in the presence of the electron transfer agent and that ether cleavage reactions occurred in coals of all ranks. The reaction times varied from 96 to 144 hr with between 10 and 13 negative charges introduced per 100 carbon atoms in the Polish coals and about 8 negative charges introduced per 100 carbon atoms in the anthracite. The Canadian coals were reduced in a similar fashion with between 9 and 11 negative charges introduced per 100 carbon atoms. The results for the whole coals are summarized in Table XXXII. These investigators also determined the hydroxyl, carboxyl, and carbonyl content of the coals and their reduction products (Table XXXIII).

The reduction reaction under these conditions provides a coal product with up to 20 additional hydrogen atoms per 100 carbon atoms. As in the other reduction reactions, potassium naphthalene(-1) is most effective for coals with between 85 and 89% carbon. The reduction products of the low-rank coals are modestly soluble in benzene. In contrast the products of the higher rank coals are only slightly soluble in benzene even though the reduction reaction has proceeded to a greater extent.

The number of hydroxyl, carboxyl, and carbonyl groups in the coals and in the reduction products were determined by the usual methods of analysis. Wachowska and Pawlak estimated the number of reactive ether groups (sixth column) in the coals from the increase in the number of hydroxyl groups in the

TABLE XXXII

The Reduction Reactions of Several Canadian and Polish Coals
with Potassium and Naphthalene in Tetrahydrofuran[a-c]

Sample	Carbon content (%, daf)	Reduction (H added per 100 C)	Solubility (% in benzene) Before reduction	After reduction
Polish coals[a,b]				
Brown coal	72.8	10	5.2	28.8
K-I	68.2	15	0.6	10.0
K-II	76.4	9	0.4	9.2
K-III	82.5	14	0.8	10.1
K-IV	86.0	20	0.4	3.5
Anthracite	90.6	13	0.1	1.5
Canadian coals[c]				
Moss 3	81.3	3		
Balmer 10	88.0	10		
Balmer 10[d]	86.5	8		

[a] Wachowska and Andrzejak (1976).
[b] Wachowska and Pawlak (1977).
[c] Wachowska et al. (1974).
[d] The Balmer 10 coal was oxidized in air at 100°C to destroy the swelling characteristics of the coal completely.

product compared to the starting material. They described the remaining oxygen atoms as heterocyclic and unreactive. There are two principal problems in this analysis. First, it is unfortunate that it is not possible to present a fully balanced analysis for the reduction products because the total oxygen content of the products shows some disturbing fluctuations. Coals K-I and K-II lose oxygen during the reaction, whereas all the Polish and Canadian coals of higher rank gain oxygen during the process. Moreover, the oxygen atoms that are acquired appear in the unreactive category. Although it is known that oxygen can be adsorbed by the anionic coal molecules, the precautions taken in this research and the fact that such oxygen compounds would be reactive negate an interpretation of this kind and suggest that tetrahydrofuran is either physically or chemically incorporated into the reduction products of the higher rank coals. The second problem concerns the analytical data and the estimation of the number of ether bonds cleaved during the reaction. As already noted, the number of reactive ethers was estimated by the difference between the number of hydroxyl groups in the product and in the coal. This estimate neglects the fate of the carboxyl groups and the carbonyl groups in the coal that the data (Table XXXIII) indicate have been reduced. If the carboxyl groups are reduced to hydroxyl groups, then the number of ether groups

TABLE XXXIII

The Distribution of Oxygen Functional Groups before and after the Reduction Reaction (Oxygen Atoms per 100 Carbon Atoms)[a,b]

	Before reduction					As reactive ether	After reduction				
	Total	As OH	As CO_2H	As CO	Unreactive		Total	As OH	As CO_2H	As CO	Unreactive
K-I	14.52	2.39	0.67	0.55	10.91	1.15	12.26	3.54	0.00	0.47	8.25
K-II	10.79	1.19	0.43	0.45	8.72	0.49	9.16	1.62	0.00	0.45	7.06
K-III	5.14	—	0.33	0.21	4.60	1.27	6.10	1.27	0.00	0.21	4.62
K-IV	3.85	—	0.31	0.32	3.22	2.35	5.47	2.35	0.00	0.31	2.81
Anthracite	3.41	—	—	0.16	3.25	0.96	4.93	0.96	0.00	0.14	3.83
Moss 3		0.85		0.23		2.06		2.92		0.20	
Balmer 10		0.92		0.30		1.76		2.69		0.29	
Balmer 10		0.98		0.30		2.00		2.98		0.30	

[a] Wachowska and Pawlak (1977).

[b] Wachowska et al. (1974).

[c] The results for the oxidized coal are given; please see Table XXXII.

cleaved per 100 carbon atoms is considerably smaller than that estimated by Wachowska and Pawlak. While such a literal interpretation of the analytical data can be made, we are reluctant to do so because there is another puzzling feature. Other chemical evidence suggests that carboxylic acids are not readily reduced under the mild conditions of these experiments. Indeed, work in our laboratory has established that the reductive butylation of Illinois No. 6 coal yields a considerable quantity of butyl carboxylates (L. M. Stock and R. S. Willis, unpublished results, 1981). These considerations suggest that the quantity of carboxyl groups in the reduction product recorded in Table XXXIII is much too small. Thus, the actual number of ethers cleaved in these reduction reactions is probably only moderately smaller than the value presented in the original article (Wachowska and Pawlak, 1977). However, it is evident that additional work will be necessary to resolve the origin of the discrepancy.

Although there are no obvious trends in the data, the results strongly suggest that ether cleavage reactions occur in all the coals. Ether cleavage appears to be most extensive with the low-rank brown coal, with about 3 ether cleavage reactions per 100 carbon atoms in the single reduction reaction. The extent of the reaction with the higher rank Polish K–IV coal and the Canadian coals is equally impressive, with about 2 ethers cleaved per 100 carbon atoms. Somewhat lower values are realized for the unweathered coal of intermediate rank. This feature of the results for the Balmer 10 coal is illustrated by the similarity of the dilatation, contraction, and other thermal properties of the reduction products shown in the "after reduction" columns in Table XXXIV. Subsequent reactions of the reduced materials with barium hydroxide, diazomethane, or hexamethyldisilazane to convert the hydroxyl groups to less reactive functions reduce the dilatation but do not significantly alter the softening temperature. Wachowska et al. (1974) suggest that this difference arises because of the prevention of dehydration reactions and steam formation at the temperature of the dilatation measurements. In subsequent work, Wachowska and Pawlak (1977) compared the coking properties of the three intermediate-rank Polish coals, K–II, K–III, and K–IV, and their reduction products. The results obtained with these coals were somewhat different. Only the observations for K–III are presented in Table XXXIV for comparison with the observations made on the Canadian Balmer 10 coal.

The influence of the reduction reaction on the coking properties of the three Canadian coals and the Polish K–II, K–III, and K–IV coals have been examined (Wachowska et al., 1974; Wachowska and Pawlak, 1977). The high-volatile Moss 3 coal and the medium-volatile Balmer 10 coal were oxidized in air at 85 and at 100°C to reduce or eliminate their swelling characteristics. These artificially weathered coals were subsequently treated to remove or to block the reactive oxygen functional groups introduced during the weather-

TABLE XXXIV

The Properties of Balmer 10 and Polish K-III Coals in the Ruhr Dilatation Analysis

	Balmer 10[a]			Oxidized Balmer 10[a]			K-III[b]		
							Before Reduction		After
	Before reduction	After reduction	After Ba(OH)$_2$	Before reduction	After reduction	After Ba(OH)$_2$	Before demineralization	After demineralization	demineralization and reduction
Dilatation (%)	80	260	155	Nil	254	157	75	Nil	192
Contraction (%)	27	15	16	12	19	14	29	31	25
Softening point (°C)	421	375	378	447	378	375	381	382	281
Temperature, maximum dilatation (°C)	510	525	495	Nil	510	498	465	Nil	452
Temperature, maximum contraction (°C)	468	405	439	525	420	441	426	454	356
Plasticity index	0.57	0.50	0.26	0.15	0.45	0.21	—	—	—

[a] Wachowska et al. (1974).
[b] Wachowska and Pawlak (1977).

ing. Wachowska and her associates showed that neither the removal of these functional groups nor the alkylation of the hydroxyl groups in the oxidized coals with barium hydroxide or diazomethane had any appreciable influence on the restoration of the swelling properties of these coals. On the other hand, when the oxidized Balmer 10 coal was reduced with potassium in the presence of naphthalene in tetrahydrofuran, the swelling properties of the product were very similar to the properties of the reduction product.

Because Balmer 10 coal contains an especially high proportion, about 50%, of semifusinite, these workers also studied the structural factors governing the coking process by an investigation of the coking properties of macerals from several coals including Balmer 10 (Nandi *et al.,* 1974; Wachowska *et al.,* 1979). They examined vitrinite, semifusinite, and fusinite and their reduction products. They note, however, that fusinite is not reduced by potassium naphthalene(− 1) in tetrahydrofuran (Nandi *et al.,* 1974). The photographs of the semicoke produced in a Ruhr dilatometer at 550°C reveal that there are major differences in the materials prepared from the macerals and their reduction products. Special attention was given to the nature of the oxygen atoms in the coal macerals and in their reduction products to gain perspective on the impact of changes in oxygen functionality on volatility, dilatation, contraction, and related properties of the macerals. The results obtained in the reduction reactions of coal fractions greatly enriched in the maceral content are shown in Tables XXXV and XXXVI and the coking properties of these substances are summarized in Table XXXVII.

Vitrinite, the oxidized vitrinite and semifusinite undergo reduction in a similar fashion. However, fusinite is much less reactive than the other macerals. The fresh vitrinite and the oxidized vitrinite have somewhat different swelling properties, whereas the products of their reduction reactions have virtually the same properties. There is clearly a different relationship between the properties of the fresh semifusinite and its reduction product. The experimental results obtained in the reduction reaction were sup-

TABLE XXXV

**The Reduction of Macerals from Balmer 10 Coal
with Potassium and Naphthalene in Tetrahydrofuran**[a]

Maceral	Carbon content (%, daf)	Reduction (H added per 100 C atoms)
Vitrinite[b]	89.4	8
Oxidized vitrinite	87.5	10
Semifusinite[c]	89.4	10

[a] From Wachowska *et al.* (1979).

[b] This sample contains 80.6% vitrinite.

[c] This sample contains 62.2% semifusinite, 14.4% fusinite, 19.2% vitrinite.

TABLE XXXVI

The Oxygen Content of the Macerals of Balmer 10 Coal
and Their Reduction Products[a]

	Oxygen content (O atoms per 100 C atoms)					
	Before reduction				After reduction	
Maceral	As OH[b]	As CO[c]	As ether[d]	Other[e]	As OH[b]	As CO[c]
Vitrinite	0.78	0.25	1.57(2.0)	0.68	2.35	0.25
Oxidized vitrinite	0.84	0.26	1.78(2.6)	2.35	2.62	0.26
Semifusinite	0.71	0.21	1.54(3.8)	1.57	2.25	0.23
Fusinite	1.53	0.31	— (1.2)	1.20	—	—

[a] From Wachowska et al. (1979).

[b] The OH content was determined by the method of Knotherus (1956).

[c] The CO content was determined by the method of Kröger et al. (1965).

[d] The reactive ether content of the original coal was estimated from the change in the number of hydroxyl groups in the maceral and its reduction product. The values shown in parentheses are estimated from octylation–deoctylation experiments.

[e] The other kinds of oxygen were determined by difference. The coals were free of carboxylic acids as judged by ion exchange with calcium acetate using the method of Ihnatowicz (1952).

plemented by a study of the properties of the products obtained by the reductive octylation of the macerals in an effort to obtain new information concerning the impact of structural changes on coking properties. A full discussion of this work is somewhat beyond the scope of this review.

Niemann and Hombach (1979) used solutions of potassium in glyme solvents at low temperatures to reduce a medium volatile bituminous coal (89.4% C, daf) from the Robert seam of the Westerholt colliery. The reduction of this coal proceeded in a straightforward manner to yield a product with about 21 additional hydrogen atoms per 100 carbon atoms. Subsequent reduction reactions yielded even more hydrogen-rich products, as shown in the equation, until about 70–80 hydrogen atoms were introduced after six

$$C_{100}H_{111}O_{2.0} \longrightarrow C_{100}H_{144}O_{1.8}$$

Pyridine soluble
50%

$$C_{100}H_{63}O_{3.2} \xrightarrow[\text{glymes}]{K} C_{100}H_{83.1}O_{2.6}$$

$$C_{100}H_{64.5}O_{2.9} \longrightarrow C_{100}H_{91.3}O_{3.4}$$

Pyridine insoluble
50%

TABLE XXXVII

The Properties of the Balmer 10 Macerals in the Ruhr Dilatometer Analysis[a]

Property	Vitrinite		Oxidized vitrinite		Semifusinite	
	Before reduction	After reduction	Before reduction	After reduction	Before reduction	After reduction
Dilatation (%)	70	260	Nil	254	Nil	Nil
Contraction (%)	27	16	12	19	3	26
Softening point (°C)	421	375	447	378	480	381
Temp. max. dilatation (°C)	480	525	Nil	510	Nil	Nil
Temp. max. contraction (°C)	468	405	525	420	528	477

[a] From Wachowska (1979).

reactions. The products are light gray (81.2% C, 9.2% H, 5.2% O) and are about 80% soluble in pyridine. Niemann and Hombach note that the reduction products are free of the solvent, but the analytical data presented indicate that certain products possess more oxygen atoms per 100 carbon atoms than the starting material. For example, the formula of the product of the third reduction reaction is $C_{100}H_{120}O_{5.1}$ compared with $C_{100}H_{63}O_{3.2}$ for the raw coal or $C_{100}H_{83.1}O_{2.6}$ for the first reduction product. Whether such solvent incorporation occurs is important because the glyme ethers used in these experiments have a large molecular weight and the incorporation of even one such molecule per 100 carbon atoms would have an appreciable influence on the analytical results and their interpretation.

The addition of excess potassium to the reaction mixture does not increase the degree of the reduction of this coal. Niemann and Hombach postulate that the limited reduction reaction occurs because the chemical potential of the blue solution is not adequate to overcome the repulsive forces in the polyanion. However, the coal product recovered from the reaction mixture is readily reduced, as shown in Fig. 3.

The course of the reaction was studied by quantitative infrared spectroscopy using the bands at 2920 and 2850 cm^{-1} for the aliphatic C—H stretching frequency, at 1600 cm^{-1} for aromatic C—C stretching frequency, at 1375 cm^{-1} for C—H symmetrical bending mode in methyl groups, and at 810 and 745 cm^{-1} for the out-of-plane, aromatic C—H deformation frequency. The experimental observations are presented in Figs. 4, 5, and 6.

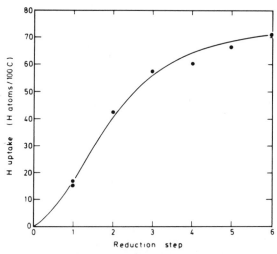

Fig. 3. The relationship between hydrogen uptake of the Robert coal and the number of the reduction reactions. [Adapted from Niemann and Hombach (1979) and used with the permission of IPC Science and Technology Press, Ltd.]

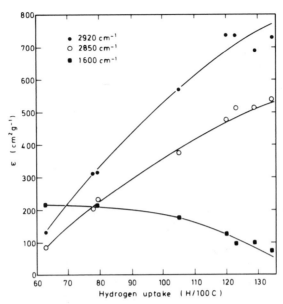

Fig. 4. The relationship between the intensities of characteristic aliphatic absorptions at 2920 cm^{-1} (●), 2850 cm^{-1} (○), and 1600 cm^{-1} (■) in the IR spectra of the reduction products obtained from the Robert coal and the number of the reduction reactions. [Adapted from Niemann and Hombach (1979) and used with the permission of IPC Science and Technology Press, Ltd.]

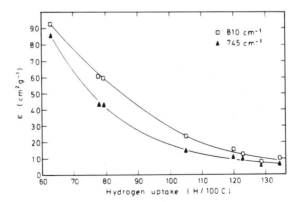

Fig. 5. The relationship between the intensities of the absorption bands of aromatic compounds at 810 cm^{-1} (□) and at 745 cm^{-1} (▲) and the degree of the reduction reaction. [Adapted from Niemann and Hombach (1979) and used with the permission of IPC Science and Technology Press, Ltd.]

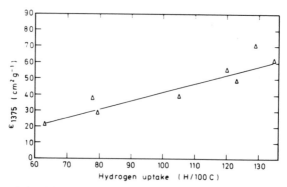

Fig. 6. The relationship between the absorption intensity of the methyl group at 1375 cm^{-1} ($\triangle$) and the extent of the reduction reaction. [Adapted from Niemann and Hombach (1979) and used with the permission of IPC Science and Technology Press, Ltd.]

The trends shown in these figures unmistakably point to the decreasing quantity of aromatic carbon–hydrogen bonds and the increasing quantity of aliphatic carbon–hydrogen bonds. These same trends have been observed by the other investigators who have studied the infrared spectra of the reduction products of coals. However, it is particularly pertinent that the relatively characteristic frequency of the methyl group at 1375 cm^{-1} also shows a distinct increase. Niemann and Hombach point out that this observation is compatible with the cleavage of carbon–carbon bonds in diphenylmethane or 1,2-diphenylethane structures under the conditions of this reduction reaction.

Ouchi *et al.* (1980) reduced eight coals with excess sodium and *t*-butyl alcohol in hexamethylphosphoramide (HMPA). This reduction reaction is

$$\text{Coal} \xrightarrow[\substack{\text{HMPA, 20 hr,} \\ \text{ambient} \\ \text{temperature}}]{\text{Na, C}_4\text{H}_9\text{OH}} \xrightarrow{\text{NH}_4\text{Cl}} \text{Reduction product}$$

similar to the reductions using lithium in ammoniacal solution in the sense that isolated carbon–carbon double bonds are quite readily reduced under the experimental conditions. Fortunately, hexamethylphosphoramide is not incorporated into the pyridine-soluble reaction products. The results obtained for these eight representative coals are summarized in Table XXXVIII.

The results obtained in this reduction reaction are similar in many respects to the results obtained with lithium in ethylamine and in ethylenediamine. All three of these reactions proceed with the substantial incorporation of hydrogen atoms into the coal product and are most effective for the formation of pyridine-soluble products with the coals which have about 89% carbon content. In addition, the reduced coals contain more oxygen atoms than the starting materials. Repetitive reduction reactions with sodium and *t*-butyl alcohol as do many of the other reduction reactions increase the yield of

TABLE XXXVIII

**The Reduction Reactions of Eight Coals with Sodium and *t*-Butyl Alcohol
in Hexamethylphosphoramide at Ambient Temperature[a]**

Coal	Carbon content (%, daf)	Reduction[b] (H added per 100 C atoms)	Weight gain (%)	Extraction in pyridine	
				Coal (%)	Product (%)
Bayswater	83.56	—	101.7	9.3	16.1
Miike	83.88	35	100.4	21.8	45.9
Daiyon	84.04	—	100	12.5	18.8
Shin-Yūbari	87.46	61	110.4	13.0	88.1
Indian Ridge	87.40	66	117.2	5.2	88.9
Goonyella	87.93	58	113.6	9.9	91.8
Balmer	89.37	66	112.8	3.8	79.1
Beatrice	91.51	66	114.9	2.9	65.6
Preasphaltene	85.60	39	—	—	—

[a] From Ouchi *et al.* (1980).

[b] These results are based upon the analytical data for the portion of the coal product that was soluble in pyridine.

pyridine soluble products. For example, three reactions of the Miike coal yield a product that is 71% soluble in pyridine. The infrared spectra and the nuclear magnetic resonance spectra of the reduction products indicate that the reduction products are highly aliphatic substances. The aliphatic absorptions at 2920, 2840, 1450, and 1380 cm^{-1} in the infrared spectra are significantly enhanced, whereas the aromatic absorptions at 3030, 1600, and between 860 and 740 cm^{-1} are significantly decreased in intensity. The pyridine-soluble extracts have little absorption in the aromatic region of the nuclear magnetic resonance spectrum but have intense absorptions ranging from 1 to 6 ppm. The appearance of the signals between 4 and 6 ppm suggests that the products are relatively rich in cycloalkene and methine hydrogen atoms as might result from the reduction of a naphthalene fragment in the coal structure.

Ouchi and his associates studied the molecular weights of the pyridine soluble reduction products. The reduced coals were partitioned by chromatography using pyridine as the eluent on two BioBead® (Bio-Rad Laboratory) gels with exclusion limits of 14,000 and 1400, respectively. Several chromato-

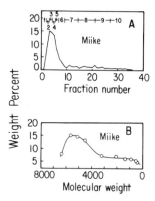

Fig. 7. (A) The elution profile observed for the gel permeation chromatography of the product from the Miike coal. (B) The molecular weight distribution observed for the ten combined fractions shown in the upper part of A. [Adapted from Ouchi *et al.* (1980) and used with the permission of IPC Science and Technology Press, Ltd.]

graphic fractions were combined and the molecular weights of the products were determined by vapor pressure osmometry. The results for Miike coal (83.9% C, daf) and Balmer 10 coal (89.4% C, daf) are illustrated in Figs. 7 and 8.

The first five fractions in each chromatographic experiment contained about 40% of the total quantity of the reduced coal eluted with pyridine. The fractions were combined as shown at the top of Figs. 7A and 8A and the molecular weights of the 10 fractions were determined. Generally, the molecular weight distribution was bimodal as shown in Figs. 7B and 8B, and for some of the coals examined in this work the molecular weights observed in

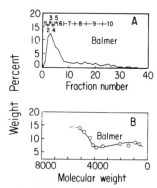

Fig. 8. (A) The elution profile observed for the gel permeation chromatography of the product from the Balmer coal. (B) The molecular weight distribution observed for the 10 combined fractions shown in the upper part of A. [Adapted from Ouchi *et al.* (1980) and used with the permission of IPC Science and Technology Press, Ltd.]

the initial fraction exceeded 7000 daltons. Ouchi and his co-workers point out that the maximum and average molecular weights observed for the higher rank coals seem to be smaller than the corresponding values for the lower rank coals (Fig. 9). This finding and the information on the coking properties of these coals and their chemical reactions prompted these workers to suggest that the data are most compatible with a structural model for coal in which the higher rank coal

"has highly condensed aromatic rings but a minimum extent of polymerization. With coals of lower rank the size of the aromatic structures becomes smaller but there is an increase in molecular weight and the extent of polymerization, i.e., the coalification process is a type of depolymerization reaction . . ." (Ouchi *et al.*, 1980, p. 756).

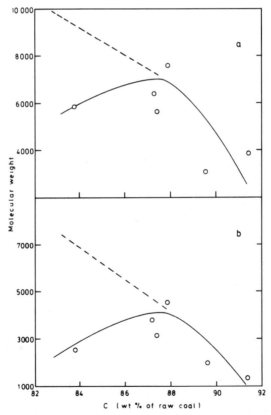

Fig. 9.   The relationship between the maximum molecular weight observed for the reduction product (a) and the average molecular weight observed for the reduction product (b) and the percentage by weight of carbon in the original coal. [Adapted from Ouchi *et al.* (1980) and used with the permission of IPC Science and Technology Press, Ltd.]

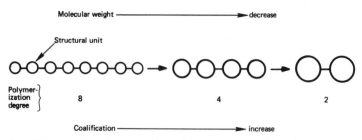

Fig. 10. The trends in molecular weight and the dimensions of the polymer units as a function of the degree of coalification. [Adapted from Ouchi *et al.* (1980) and used with the permission of IPC Science and Technology Press, Ltd.]

Their idea is illustrated in Fig. 10. This view certainly requires careful consideration, but it must be pointed out that it is very difficult to ascribe special significance to the molecular weight data obtained for reduction products.

Most of the research workers who have examined the reductive alkylation reactions of coal have also studied the reduction reactions of the same coals to assess the relative importance of the addition of hydrogen atoms and alkyl groups. The results presented in the previous discussion are quite representative of the pattern of results that have been obtained in these investigations. For example, Bimer (1974) reported that the reductive butylation of a Polish coal with potassium and naphthalene yielded a product that was 52% soluble in benzene, whereas the reduction of the same coal gave a product that was only 5% soluble in the same solvent. B. S. Ignasiak and Gawlak (1977) found that the reduction of Balmer 10 coal with sodium in liquid ammonia or with potassium and naphthalene in tetrahydrofuran produced only a modest amount of soluble products. Similar results were realized in this laboratory using either potassium in liquid ammonia or potassium and naphthalene in tetrahydrofuran at ambient temperature for the reduction of Illinois No. 6 coal and Colorado subbituminous coal (C. I. Handy and L. M. Stock, unpublished research, 1981; D. A. Blain and L. M. Stock, unpublished research, 1981). In related work, Miyake *et al.* (1980) found that about 11% of the reduction product was soluble in benzene when Yūbari coal (86.0% C, daf) was reduced with potassium in refluxing tetrahydrofuran for 2 hr and treated with isopropyl alcohol. When these coals were subjected to reductive alkylation, the products were considerably more soluble. To illustrate, the butylation product obtained from the Yūbari coal is 75% soluble in benzene. Thus, much of the research work indicates that the reduction reaction is not as effective as the alkylation reaction for the conversion of the coal molecules to substances that are readily soluble in conventional organic solvents. Nevertheless, many coals with about 89% carbon can be readily reduced and repetitive reduction reactions enable a high conversion of the coal to soluble products rich in hydrogen.

In summary, the reduction reactions of many coals have been investigated by a variety of chemical methods. Vigorous methods, for example, sodium and t-butyl alcohol in hexamethylphosphoramide, potassium in glyme solvents, or lithium in amines, very effectively reduce many coals. These reactions are most effective for the reduction of coals with about 89% carbon. Coals with lower and higher carbon content are not as effectively reduced in the sense that a lesser number of hydrogen atoms are added and in the sense that the solubility properties are not altered as significantly. Milder agents, for example, sodium in liquid ammonia, are generally less effective for the formation of soluble reduction products. Whereas the change in the solubility of the coals reflects the effectiveness of the reduction reactions for the disruption of the structure, the products obtained in the milder, more selective reduction reaction may be quite valuable for the assessment of the influence of modest changes in structure upon the reactivity of the coal via studies of the physical and chemical properties of the products. For example, Montgomery and Wachowska and their associates exploited the relatively mild reaction with potassium and naphthalene to study the role of coal structure upon coking properties.

The infrared spectra of the reduction products obtained in the various reactions uniformly indicate that the reduction reactions decrease the number of aromatic molecules in the coals with a corresponding increase in the aliphatic content of the product. The degree of the reduction reaction depends, of course, upon the reagent employed. With potassium in glyme solvents, the reduction of the Robert seam coal proceeds with a significant increase in the absorption at 1375 cm^{-1}, a frequency often assigned to the methyl group; this observation suggests that relatively strong carbon-carbon bonds such as those in diphenylmethane and diphenylethane cleave under these reaction conditions (Niemann and Hombach, 1979; Collins *et al.*, 1981).

The nuclear magnetic resonance spectra provide confirmatory information for many of the conclusions reached in the examination of the infrared spectra. Lazarov and Angelov (1968) established that the soluble reduction product had decreased aromatic character, and Ouchi *et al.* (1980) found that alkene resonances were present in the highly reduced reaction products of a variety of coals.

Analyses of the coals and their reduction products have proven very informative. However, it must be acknowledged that there are significant uncertainties in the analytical measurements of the oxygen atom content in the coals and in the reduction products and in the distribution of the oxygen atoms among the various functional groups in these substances. Major experimental uncertainties arise in the determination of the oxygen atom content and its distribution. Major uncertainties also arise because the behavior of the coal and its polyanion are not always well understood. Two of the complications arise because the reaction solvent, often tetrahydrofuran, can be incorporated either chemically or physically into the reaction product and because oxygen in trace amounts in the protective atmospheres can also be incorporated into the coal. In spite of these difficulties the experimental results strongly suggest that ether cleavage reactions occur under the conditions of all these reduction reactions and that there are about as many ether cleavage reactions per 100 carbon atoms in the high-rank coals with 89% carbon as in the lower rank coals. The lesser reactivity of the ethers in the lower rank coals may stem, at least in part, from the presence of ethers derived from unreactive aromatic compounds with two or more hydroxyl groups such as lignin-like fragments derived from coniferryl and sinapyl alcohol (33, 34).

The molecular weights of the pyridine soluble fractions shown in Figs. 7–9 are impressively large. Inasmuch as relatively narrow chromatographic bands were analyzed in this study (Ouchi *et al.,* 1981), the results provide a better indication of the true molecular weight distribution than the data obtained by the study of the entire quantity of the reduction product.

In summary, the reduction reaction has been exploited by several research groups to achieve a better understanding of the structure and the reactivity of a variety of coals. The work on this reaction and other complementary work on these coals has increased our knowledge of these materials. New research on the reduction reactions and the reaction products will certainly lead to further gains.

B. The Alkylation of Coal

Attempts to transform coal into a more soluble substance by the conversion of the more reactive functional groups into ethers or esters have not generally been successful. However, two new alkylation procedures have been described which appear to be very useful and which will provide some new opportunities for the investigation of coal chemistry.

An effective mild method for alkylation has been devised by Liotta (1979) for the selective O-methylation of lower rank subbituminous and bituminous coals with reactive hydroxylic and carboxylic acid groups which can be alkylated in the presence of tetrabutylammonium hydroxide. This compound serves both as a base to activate the hydroxylic functional groups and as a phase transfer catalyst. In another approach, the Alberta Research Council group has demonstrated that sodium and potassium amide in liquid ammonia can be used in conjunction with an alkylating agent such as ethyl bromide for the alkylation of different coals (B. S. Ignasiak *et al., 1979*; Gawlak *et al., 1979*). The products of the C- and O-alkylation reactions which occur under these conditions are sometimes surprisingly soluble.

The method developed by Liotta enables the selective alkylation of the weakly acidic phenolic and carboxylic acid groups in bituminous and sub-bituminous coals (Liotta, 1979; Liotta *et al., 1981*). In this procedure, tetrabutylammonium hydroxide, a compound expected to be a swelling agent for the coal (Dryden, 1952), is employed as a phase transfer catalyst and a base for the activation of the hydroxyl groups. The alkylating agents are added in a second step. Under these experimental conditions alcohols, phenols, car-

$$C_{100}H_{84}O_{11}S_{1.8}N_{1.4} \xrightarrow[\substack{\text{THF-H}_2\text{O}, \\ 2\,\text{hr}}]{(C_4H_9)_4N^+,OH^-} \xrightarrow[\substack{\text{THF-H}_2\text{O}, \\ 16\,\text{hr}}]{CH_3I} C_{105}H_{94}O_{11}S_{1.8}N_{1.4}$$

Illinois No. 6 coal   Alkylated product

boxylic acids and other nucleophilic groups readily undergo alkylation. The rate of reaction of the coals is surprisingly rapid when it is recognized that the coal has not been subjected to either reduction or fragmentation reactions.

Study of the infrared spectrum of the starting material and the product revealed that the strong, broad OH absorption from 3300 to 3600 cm^{-1} in the original coal was eliminated in the methylated product. In addition, a new band at 1720 cm^{-1} indicative of methyl esters appeared in the spectrum of the product. The proton NMR spectrum of a soluble extract of the methylated coal exhibited a relatively strong resonance from 3.5 to 4 ppm. This resonance was not present in extracts of the original coal or in extracts of the coal alkylated with methyl-d_3 iodide. The carbon NMR spectrum of extracts of the coal methylated with methyl-^{13}C iodide had two relatively strong bands in the methyl ether region from 54 to 61 ppm and one band in the methyl ester region from 49 to 52 ppm. All these results point strongly to the virtually complete methylation of all the hydroxylic and carboxylic groups in this bituminous coal. The quantitative microanalytical data for the methylated coal and the coal methylated with methyl iodide-d_3 indicate that five methyl groups have been added per 100 carbon atoms, as shown in the equation. The work on the extracts, which represent about 17% of the coal, suggests that 10% of the methyl groups are introduced at the carboxylic acid sites and the remainder are introduced at other phenolic and hydroxylic sites.

TABLE XXXIX
Extractability of Coals and O-Methylated Coals[a]

Substance	Extractability (wt. %)[b]		
	In THF	In C_6H_6	In $CHCl_3$
Illinois No. 6 coal	14.4	0.7	3.2
O-Methylated Illinois No. 6 coal	22.1	14.7	21.7
Rawhide coal	6.5	2.2	2.4
O-Methylated Rawhide coal	16.1	12.2	18.1

[a] From Liotta *et al.* (1981).

[b] The weight percent extracted on a dry-mineral matter-free basis by exhaustive Soxhlet extraction.

The reaction can be used for the introduction of other groups including butyl, heptyl, octadecyl, and benzyl fragments. About five of these alkyl groups per 100 carbon atoms are introduced into the Illinois No. 6 coal. The reaction also is suitable for the alkylation of subbituminous coals. The alkylated products obtained from Rawhide coal contain about 8 alkyl groups per 100 carbon atoms, and about 30% of these alkyl groups are bonded to carboxylic acid fragments.

Because these reactions can be accomplished on a relatively large scale, Liotta *et al.* (1981) have been able to examine several physical and chemical properties of the alkylated coals. The solvent extractability data summarized in Table XXXIX reveal that methylation reactions considerably enhance the solubility of the coals in the noninteractive solvents such as benzene and chloroform, which are better solvents for ethers and esters than for alcohols, phenols, or acids. The helium density measurements and the mercury porosity determinations suggest that the O-methylation reaction has considerably increased the microporosity of both the bituminous and the subbituminous coals. For example, the microporosity that is dependent upon short-range interactions such as hydrogen bonding and charge transfer associations of the O-methylated Illinois No. 6 coal and the O-methylated Rawhide coal is much greater than the microporosity of the untreated coal. Liotta *et al.* (1981) conclude that O-methylation has led to a more open structure because of the relaxation of short-range molecular interactions due to a decrease in the attractive forces between the polar fragments of the coal structure as a consequence of their etherification and between the nonpolar fragments of the coal structure as a consequence of the steric requirements of the alkyl groups introduced in the reaction.

Liotta *et al.* (1981) also studied the influence of O-alkylation on the caking properties of the coal. They found that O-methylation reduced the softening temperature and that the methylated coals and the other alkylated coals remained fluid over much broader temperature ranges than the untreated coal. The free swelling indices, contraction, and dilatation of the alkylated coals were all altered compared with the starting material. In general, the O-alkylation reaction enhances the agglomeration characteristics of the coal that is produced upon pyrolysis. As a result, the coke produced from O-methylated Illinois No. 6 coal is of a higher quality than the coke produced from the untreated coal. Similar results are realized for the Rawhide coal.

Nonreductive alkylation can also be accomplished using sodium or potassium amide in ammonia to form the polyanion and ethyl bromide or another halide as the alkylation reagent (B. S. Ignasiak *et al.,* 1979; Gawlak *et al.,* 1979). The product obtained under these somewhat more basic conditions is often surprisingly soluble in common organic solvents. These workers found that the alkylation reactions with indan and 1,2-diphenylethane proceed only to a small extent, but that the reactions of diphenylmethane, 9,10-dihydrophenanthrene, 9,10-dihydroanthracene, and fluorene proceed quite readily, as shown in the equations. The observation that polyalkylation reac-

$$(C_6H_5)_2CH_2 \xrightarrow[\text{NH}_3]{\text{NaNH}_2} \xrightarrow[(C_2H_5)_2O]{\text{C}_2\text{H}_5\text{Br}} (C_6H_5)_2CHC_2H_5 + (C_6H_5)_2C(C_2H_5)_2$$

$$94\% \qquad\qquad 6\%$$

24%

+

53%

+

22%

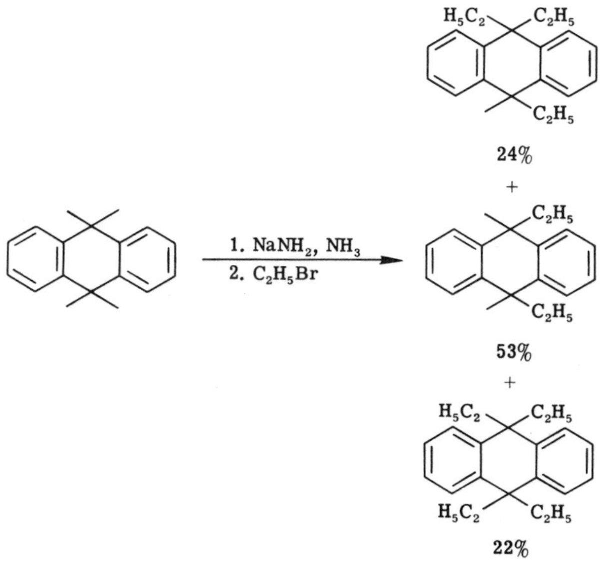

tions occur under these conditions is in accord with other observations concerning the reactions of carbanions in strongly basic etheral solutions as discussed previously (Section II,C). The experimental results for a group of Canadian Cretaceous coals and another group of American Carboniferous coals are summarized in Table XL.

After three repetitive alkylation reactions, the Cretaceous coals were about 25–35% soluble in chloroform. The ethylated West Virginia coal (83.2% C, daf) was somewhat more soluble in chloroform with about 39% in solution after three alkylations. Most surprising, however, was the finding that an ethylated Freeport seam coal (87.3% C, daf) was about 64% soluble in this solvent. The weight gained in the ethylation reactions and the % H in the ethylated products indicate that between 8 and 16 alkyl groups have been added per 100 carbon atoms to the different coals. However, these methods do not yield identical results concerning the number of alkyl groups that have been introduced and in some cases, the discrepancies are quite large. Although it may be premature to generalize from the limited observations presented in Table XL, the larger primary groups appear to be somewhat more effective for the formation of soluble products and the Carboniferous

TABLE XL

The Nonreductive Alkylation of Representative Coals Using Sodium Amide
in Liquid Ammonia for Formation to the Polyanion[a]

| | | Conversion (% | Alkyl groups per 100 C atoms | |
Coal sample (% C, daf)	Reaction[b]	solubility in chloroform)	on O(OH)[c]	on C[d]
Cretaceous coals				
Saskatchewan, 73.9	Ethylation	24.5	3.8(6.6)	9.5
Alberta, 77.7	Ethylation	33.2	3.8(5.5)	8.8
	n-Propylation	33.0	4.1(5.5)	9.6
	i-Propylation	30.7	2.7(5.5)	6.2
	n-Butylation	38.5	4.3(5.5)	8.1
	n-Hexylation	35.7	5.2(5.5)	10.7
British Columbia, 80.6	Ethylation	25.9	3.4(4.8)	5.5
Alberta, 86.8	Ethylation	28.0	0.7(1.1)	7.3
British Columbia, 88.8	Ethylation	28.9	0.0(0.3)	11.0
Carboniferous coals				
West Virginia, 83.2	Ethylation	39.0	1.4(2.3)	12.1
Freeport, 87.3	Ethylation	63.6	0.6(1.1)	7.2

[a] From Gawlak et al. (1979).

[b] Three repetitive alkylation reactions were carried out.

[c] Determined by analysis of the hydroxyl groups in the coal (shown in parenthesis) and the ethylated product.

[d] The average value was determined from change in weight and from the hydrogen content of the reaction product.

coals appear to be somewhat more readily converted to soluble compounds.

Preliminary molecular weight data suggest that the number average molecular weight of the soluble ethylated products is near 1250 and that about 30% of the soluble products have molecular weight greater than 5000.

The nuclear magnetic resonance spectra of the Freeport coal and its ethylated products provide some rather convincing evidence for the prominence of dihydroaromatic structures, specifically dihydrophenanthrene, in this coal. Gawlak and his co-workers established that the principal upfield bands in the C NMR spectrum at about 9, 12, and 14 ppm are associated with the methyl groups of the ethyl fragments introduced in the alkylation reaction by deuterium labeling experiments. The methylene resonances of the ethyl groups are broadly shifted and cannot be readily identified. One of the most prominent resonances in the NMR spectrum appears near 29 ppm. The authors assign this resonance to the 9 and 10 carbon atoms of 9,10-dihydrophenanthrene residues. They also note that this resonance appears to diminish when the coal is ethylated and that a new resonance near 46 ppm appears in the spectrum of the ethylated product. This interpretation is illustrated in the equation.

| Raw coal | First alkylation | Subsequent alkylation |

These results are very stimulating because they suggest that the soluble coal product obtained from this coal contains these molecular fragments in important abundance. Resonances near 29 ppm appear in many coals and coal products (Zilm et al., 1979) and it is difficult to discern whether these resonances are from the hydroaromatic structure in the coal or from a paraffin fragment.[4] Consequently, this nonreductive alkylation reaction may provide much needed information concerning this structural problem if the reaction can be controlled to yield mono- and dialkylation products selectively.

A substantial quantity of material with molecular weights below 1000 is produced in the reaction. It is not clear whether these products are merely

[4] It is also pertinent that the resonance of the one carbon atom of phenylethane appears at 29 ppm.

liberated from the solid in the course of the reaction or whether they are produced in fragmentation reactions of etheral fragments or strained hydro-

$$Coal-CH_2CH_2O-Coal + NaNH_2 \rightarrow Coal-CH=CH_2 + NaO-Coal + NH_3$$

carbon units in the structure.

It is apparent that additional work on these nonreductive alkylation reactions will provide important new data for the definition of the reactivity and
structure of coal.

C. The Reductive Alkylation of Coal

Both the reduction and the alkylation of coal are very useful for the
preparation of modified coals. However, neither of these reactions is generally suitable for the conversion of a variety of different coals to compounds
that are soluble to a significant extent in ordinary solvents, for example,
benzene, chloroform, tetrahydrofuran, or pyridine, such that the major portion of the original coal can be readily studied in homogeneous solution.
About 15 years ago Sternberg and his associates proposed that the reductive
alkylation reaction might better fulfill this need. With this objective in mind,
Sternberg *et al.* (1968, 1971; Sternberg and Delle Donne, 1974) studied the
reductive alkylation of four coals ranging in rank from a subbituminous Colorado coal to an anthracite. The coals were reduced with potassium in
tetrahydrofuran and the polyanions were alkylated with several different
primary alkyl iodides. They also found that the reaction products could be
partially dealkylated by treatment of the alkylated coal with additional

potassium and naphthalene in tetrahydrofuran. Their observations are summarized in Tables XLI and XLII.

All the coals are reduced by potassium and naphthalene in tetrahydrofuran and three of them can be successfully alkylated to yield products that are quite soluble. Sternberg and Delle Donne suggest that steric hindrance is probably responsible for the failure of this alkylation reaction of the anthracite polyanion. However, it seems more likely that this reaction fails because electron density is highly dispersed in these relatively stable anions and because many

TABLE XLI

The Reductive Alkylation of Four Representative Coals with Potassium, Naphthalene, and Butyl Iodide in Tetrahydrofuran [a,b]

Coal	Carbon content (%, daf)	Alkyl groups added per 100 C atoms [c]	Solubility		
			In hexane	In benzene	In pyridine
Subbituminous Colorado					
Unreacted	71		0	0.5	3
Ethylation		14.0	12	23	43
Second ethylation		14	30	41	73
Deethylation		14.2	34	40	89
Bruceton					
Unreacted	82		0	0.5	25
Ethylation		11	22	74	81
Second ethylation		16.2	37	88	95
Deethylation		9.8	10	68	91
Pocahontas					
Unreacted	90		Nil	0.5	3
Reduction			—	3	13
Methylation		8.1	3	48	95
Second methylation		12.6		92	
Demethylation		4.8		12	62
Ethylation		8.8	11	95	97
Deethylation		5.8			
Butylation		7.4	17	93	
Anthracite					
Unreacted	96			0	0
Ethylated		0.6		0	0

[a] Sternberg *et al.* (1971).

[b] Sternberg and Delle Donne (1974).

[c] The extent of the alkylation reaction was determined by assay of the ^{14}C content of the product, following the reaction with a labeled alkyl iodide.

TABLE XLII

The Oxygen Atom Content of the Coal and the Deethylation Reactions
of Its Ethylated Products[a]

Coal	Oxygen content (O atoms per 100 C Atoms)		Ethyl groups (groups per 100 C atoms)	
	Original	Ethylated	Ethylated	Deethylated
Colorado subbituminous	23.2	10.5	14.0	14.2
Bruceton	8.7	6.2	16.2	9.8
Pocahontas	3.0	2.5	8.8	5.8

[a] From Sternberg and Delle Donne (1974).

of the alkylation reactions only form products which are severely distorted as illustrated for the butylation of the central carbon atom of triphenylene for which the change in hybridization from sp^2 in **35** to sp^3 in **36** is stereochemically very unfavorable.

The three coals that can be alkylated successfully lose oxygen during the reductive alkylation reaction. The low-rank subbituminous coal experiences the greatest reduction in its oxygen content. Sternberg and Delle Donne postulate that oxygen atoms are lost during elimination reactions, which yield small soluble molecular fragments because about 20% of the coal was lost during the alkylation reaction, but the alternative explanation advanced by Given and his co-workers for the deoxygenation of other low-rank coals via reductive dehydration cannot be completely discounted (see Section IV,A). It is evident that structures proposed for subbituminous coals must take into consideration their rather facile loss of fragments that have a relatively high oxygen atom content.

The dealkylation reactions of the product yield interesting information about the nature of the ethyl–oxygen bonds in the coal alkylate (Table XLII). The results indicate that 0, 6.4, and 3.0 ethyl groups are lost from the Colorado, Bruceton, and Pocahontas coals, respectively. These observations suggest, as discussed by Sternberg and Delle Donne, that the alkylated products of these coals contain only a few reactive ethyl ethers (e.g., aryl ethyl ethers) that are readily susceptible to reductive dealkylation. The datum for the sub-bituminous coal is particularly puzzling because this coal is known to contain an abundant number of phenolic hydroxyl groups, which should have undergone the alkylation and dealkylation reactions. The data for the other two coals seem more reasonable. It is difficult to ascribe special significance to these results in the absence of independent information concerning the ratio of carbon to oxygen alkylation and the most cautious interpretation of the results is that the successful dealkylation reactions indicate the minimum number of readily cleaved aryl ethyl ethers in the product. The dealkylation reactions of many ethers may not proceed because the product of an initial dealkylation reaction may yield an unreactive intermediate (**38**), as illustrated in the conversion of **37** to **39**.

The molecular weights of the coal products were also investigated by vapor pressure osmometry and the molecular weight distribution was assessed by gel permeation chromatography. Exhaustive washing of the initial reaction products presumably removed all of the residual ethylated dihydronaphthalene but this washing procedure also removed about 20% of the coal. Consequently, the larger coal molecules contribute disproportionately to the number average molecular weight data presented in Table XLIII.

TABLE XLIII

The Number Average Molecular Weights of the Benzene-Soluble Fractions
of Several Ethylated Coals[a]

Coal	Solubility (% in benzene)	Number average molecular weight[b]
Subbituminous		
Colorado	41	700
Bruceton	86	2000
Pocahontas	95	2800 (3100)[c]

[a] From Sternberg and Delle Donne (1974).
[b] The molecular weight of the alkyl group free coal fragment is presented.
[c] The result for the butylation product is presented in parentheses.

Bimer (1974, 1976) examined the reductive alkylation of several representative coals. The reductive methylation of a bituminous coal (84.1% C, daf) from the Marcel mine was carried out with potassium and naphthalene in tetrahydrofuran for 24 hr and the methylation reaction was terminated after 2 hr (1976). The coal polyanion obtained from this coal was also treated with methanol to yield the reduction product. Bimer pointed out that repetitive washings of the product mixture with ethanol remove low molecular weight coal molecules as well as naphthalene and its reduction and alkylation products. To circumvent the loss of these coal molecules, he steam distilled the reaction mixture to remove the more volatile materials and to ensure the retention of the smaller coal molecules. Under these conditions, certain naphthalene derivatives are also retained. Thus, the mixtures obtained in the methylation and reduction reactions weigh 122 and 120%, respectively, more than the original coal and contain about 10% by weight of naphthenic compounds and some residual tetrahydrofuran. The solubility data and the analytical data for selected functional groups for these materials are summarized in Tables XLIV and XLV.

The material balances require that tetrahydrofuran as well as the reduced

TABLE XLIV

The Reduction and Reductive Methylation of Marcel Coal
with Potassium, Naphthalene, and Methyl Iodide in Tetrahydrofuran[a]

Reaction	Extractability (%)			
	In pentane	In ether	In benzene	Total
Unreacted coal	0.5	1.8	0.3	2.6
Reduction, CH_3OH	—	—	—	9.1
Methylation, CH_3I	6.6	12.6	23.4	42.6

[a] From Bimer (1976).

TABLE XLV

Analyses for Selected Functional Groups for the Marcel Coal
and Its Reduction and Methylation Products[a]

Material	Aromaticity f_a	Number of atoms or functional groups				
		C	H	O_{OH}	O_{OCH_3}	C_{CH_3}
Starting material	0.82	100	73	3.3	0.1	3.1
Reduced coal	0.72	120	—	4.3	—	—
Methylated coal	0.68	121	113	0.6	4.6	6.5

[a] From Bimer (1976).

and methylated naphthalenes be present in these product mixtures. Their presence considerably complicates the interpretation of the microanalytical data because, among other reasons, the distribution of the methyl groups between the naphthalene and the coal molecules is not known. The interpretation of the results obtained in the infrared studies, in the $C-CH_3$ determination by the Kuhn–Roth procedure, in the thermal gravimetric analyses, in the dehydrogenation reactions with phenanthrene, and in the oyxgen functional group analyses are also difficult. For example, the spectroscopic measurements and the results of the dehydrogenation reaction of the coal with phenanthrene clearly demonstrate that many aromatic molecules in the coal have been reduced, the Kuhn–Roth analysis suggests that only 3.4 methyl groups per 100 carbon atoms have been added to the reduced aromatic fragments of the coal. Functional group analyses (Table XLV) suggest that the hydroxyl and carboxyl content of the methylated coal is negligible, yet the OH stretching frequency is the most intense band in the infrared spectrum. These difficulties signal the need for caution in the quantitative interpretation of the analytical information. Qualitatively, however, the results indicate that ether cleavage reactions occur under the experimental conditions and that the alkylation of these free hydroxyl groups is necessary for the achievement of a high degree of solubility. This difference between reduction and reductive alkylation is also illustrated by the finding (Bimer, 1976) that methylation of the initial reduction product gave a much more soluble material.

$$\text{Marcel coal} \xrightarrow[\text{(2) } CH_3OH]{\text{(1) } K, C_{10}H_8, THF} \text{Reduced coal} \xrightarrow{(CH_3)_2SO_4} \begin{array}{l} \text{Alkylated coal} \\ \text{29\% soluble in} \\ \text{benzene} \end{array}$$

Bimer (1974) had previously studied the reductive alkylation of three European coals but focused attention on the Gliwice coal (88.3% C). As in the other study, only 24 hr were allowed for the reduction reaction with potassium and naphthalene in tetrahydrofuran and 2–3 hours were allotted for the alkylation reactions with methyl iodide and ethyl, n-butyl, and n-hexyl bromide. These times now are known to be too brief to achieve complete reduction and alkylation. Although the reaction conditions are not ideal, the reductive alkylation reactions yielded an appreciable quantity of coal products that could be extracted into conventional organic solvents, as shown in Table XLVI.

Bimer (1976) remarks that the presence of naphthalene and possibly tetrahydrofuran in the alkylated products complicates the interpretation of the data. Nevertheless, it is clear that he successfully alkylated three coals of different rank to yield appreciable quantities of soluble products. He concludes that the success of the alkylation reaction depends upon the number of alkyl groups introduced, upon the chain length, and, quite importantly, upon the nature of the coal. He also suggests that fragmentation and alkylation reaction are essential for the formation of soluble products.

Ignasiak and his co-workers employed the reductive alkylation reaction to obtain new information concerning the structural factors governing the in-

TABLE XLVI
The Reductive Alkylation of Three Coals with Potassium and Naphthalene in Tetrahydrofuran[a]

Reaction	Alkylation (alkyl groups per 100 C atoms[b])	Extractability (%)			
		In pentane	In ether	In benzene	Total
Gliwice coal					
Reduction,					
CH_3OH	—	5	4	5	14
Methylation,					
CH_3I	22	13	12	41	65
Ethylation,					
CH_3CH_2Br	13	15	19	51	84
Butylation,					
$CH_3(CH_2)_2CH_2Br$	13	26	19	52	97
Hexylation,					
$CH_3(CH_2)_4CH_2Br$	10	23	21	37	81
President coal					
Butylation,					
$CH_3(CH_2)_2CH_2Br$	12	13	11	4	27
Boleslaw Chrobry coal					
Butylation,					
$CH_3(CH_2)_2CH_2Br$	15	20	13	50	83

[a] From Bimer (1974).
[b] The extent of the alkylation is based upon the weight gained and microanalytical data.

fluence of ethers and thioethers on the reactivity of conventional and sulfur-rich coals (B. S. Ignasiak and Gawlak, 1977; B. S. Ignasiak et al., 1978a,b; Carson and Ignasiak, 1980). In their initial study, B. S. Ignasiak and Gawlak (1977) tested the proposal that essentially all the oxygen functional groups in bituminous coal exist as carbonyl and hydroxyl groups. They pointed out that the structures proposed for bituminous coals before 1977 assigned only a very minor role to ether functional groups but that there were several reasons to reconsider this view. Accordingly, they investigated the reductive alkylation of Balmer 10 coal (89.1% C, daf) using potassium and naphthalene in tetrahydrofuran and sodium in liquid ammonia as the reducing agents. The anion obtained in these reactions was then alkylated and the product was isolated using the extraction procedure discussed in Section III,C. Representative results are summarized in Table XLVII.

As in many other cases, the reductive alkylation of this coal proceeds smoothly to give reaction products that are appreciably soluble in benzene and in pyridine. The number average molecular weights of the benzene soluble portion of the products are uniformly higher than the molecular weights for the pyridine soluble portion of the products. The differences in the molecular weight values decrease as the size of the alkyl group increases. Ignasiak and Gawlak point out that this result suggests that associative interactions are more important in benzene than in pyridine. It is also pertinent that the molecular weight data obtained for the octylation in liquid ammonia are quite similar to the results obtained for the same reaction in tetrahydrofuran. Consequently, there is little reason to suspect that reduced derivatives of naphthalene or of tetrahydrofuran contribute in a significant way to the relatively low molecular weights observed for the reaction products obtained from the reaction in the ether.

TABLE XLVII
The Reductive Alkylation of Balmer 10 Coal[a]

Reaction	Alkylation (alkyl groups per 100 C atoms)	Solubility (%)		Number average molecular weight[b]	
		In benzene	In pyridine	In benzene	In pyridine
Methylation	11.6	63	80	1450	670
Butylation	10.8	84	84	1280	700
Octylation	10.9	85	87	670	680
Benzylation	11.2	71	82	680	550
Octylation[c]	7.6	59	75	740	590

[a] From B. S. Ignasiak and Gawlak (1977).
[b] The alkyl group free number average molecular weights were determined by vapor pressure osmometry.
[c] This reductive octylation reaction was carried out in liquid ammonia.

TABLE XLVIII

Extractabilities and Molecular Weights
of an Ethylated Carboniferous Vitrinite[a]

Extraction solvent	Extractability (%)	Number average molecular weight	
		In benzene	In pyridine
Ethanol	3.3	550	450
Pentane	2.1	810	710
Ethyl ether	16.5	1200	980
Benzene	51.3	8260	1900

[a] From B. S. Ignasiak et al. (1978a).

Ignasiak and Gawlak also found that reduced coal contains significantly more acetylable hydroxyl groups (2.9% OH) than the original Balmer 10 coal (0.7% OH). Because carbon–carbon bond cleavage reactions are not common, they conclude that ether cleavage reactions are principally responsible for the degradation of the structure of this Cretaceous coal to smaller structural units with a number average molecular weight of about 670.

In a subsequent communication, B. S. Ignasiak and his co-workers (1978a) showed that the molecular weight data obtained for extracts of an ethylated Carboniferous vitrinite (90.9% C, daf) by vapor pressure osmometry depend in a very significant way on the character of the extraction solvent and on the solvent used for the molecular weight measurement (Table XLVIII). Their observations show quite clearly that misleading data can be obtained when insufficient attention is given to the association of the coal molecules in solution. These associations are apparently minimized in pyridine and when larger alkyl groups such as the n-octyl group are introduced into the coal. In addition, the results establish that the less effective extraction solvents selectively extract the lower molecular weight coal molecules.

Carson and Ignasiak (1980) extended the investigation of the role of ether linkages in the chemistry of coal. They point out that the reactions of sodium in liquid ammonia generally fail to generate new hydroxyl groups from many low rank coals even though the oxygen content of such coals is quite high. They postulate that the low rank coals might contain phenoxyphenol fragments (40), which are not cleaved under the usual conditions used in reductive alkylation reactions. Indeed, they showed that 4-phenoxyphenol is quite unreactive with lithium or sodium in liquid ammonia but found that this compound does react quite readily with potassium in liquid ammonia to yield, after ethylation, 1,4-diethoxybenzene (**41**). To gauge the degree of depolymerization of a typical bituminous coal, they measured the molecular weight distribution of the products obtained from a West Virginia vitrinite containing 80.8% C and 11.6% O, following liquefaction reactions with tetralin and with tetrahydroquinoline at 385° C for several hours. The products of these donor solvent reactions were 55% soluble in benzene and 90% soluble in pyridine. The molecular weight distribution of the product revealed that the coal was highly fragmented and that more than 50% of the chloroform-soluble products had molecular weights below 500. The treatment of this vitrinite with either sodium or potassium in liquid ammonia yields a product that has the same hydroxyl content as the vitrinite (5% O_{OH}, 2.4% O_{CO}, 0.3% O_{CO_2H}, and 3.9% $O_{unreactive}$). Another sample of this vitrinite was nonreductively alkylated using sodium amide and ethyl bromide in liquid ammonia, and the molecular weight distribution of a chloroform extract of the product was determined. This extract was then reduced with a mixture of alkali metals in liquid ammonia and alkylated with ethyl bromide to yield the reductively ethylated material. The molecular weight distribution of this product was also determined. The molecular weight distributions of the chloroform extract of the reductively ethylated vitrinite and the nonreductively ethylated vitrinite are virtually identical with a broad spectrum of molecular weights ranging up to 6000. Their study of the degradation reactions of the hydroxydiphenyl ethers suggests that the molecular weights of the reaction products obtained

from the reductive ethylation of the ethylated coal should have been significantly lower than the molecular weights of the ethylated coal if hydroxydiphenyl ether fragments such as **43** were significant in the chemistry of the coal. Inasmuch as the molecular weight data for these two materials were very similar, Carson and Ignasiak infer that it is doubtful that phenoxyphenol ethers are present in low-rank Carboniferous vitrinites and that the molecular weight reduction observed for these vitrinites in hydrogen donor solvent processes must be ascribed in large part to carbon–carbon bond cleavage reactions rather than to ether cleavage reactions.

The group at the Alberta Research Council has also studied the reactions of sulfur-rich Rasa lignite (B. S. Ignasiak *et al.,* 1978b) and sulfur-rich Athabasca asphaltene (T. Ignasiak *et al.,* 1977). B. S. Ignasiak and his associates studied the reduction and the reductive alkylation of the lignite to establish the relationship between the structure of this material and its somewhat unusual coking properties. The lignite was reductively octylated with potassium and naphthalene in tetrahydrofuran and with sodium in liquid ammonia.

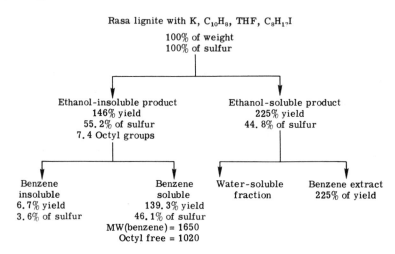

The reaction of this lignite with potassium and naphthalene in tetrahydrofuran yields a large quantity of hydrogen gas and a product which, in contrast to most alkylated coals, is appreciably soluble in ethanol. The quantity of the product requires that tetrahydrofuran be incorporated into it. The lignite undergoes extensive desulfurization to yield inorganic sulfides during the reduction reaction. These inorganic sulfides undergo alkylation to give dioctyl sulfide, which was identified as a major constituent in the product mixture. The low molecular weight observed for the ethanol soluble product is biased, of course, by the presence of dioctyl sulfide (5%) and hexadecane (2%) in the product mixture. Ignasiak and his co-workers suggest that the corrected number average molecular weight, 1020, of the ethanol-insoluble, benzene-soluble product probably is representative only of the higher molecular weight clusters.

The reductive octylation of this lignite with sodium in liquid ammonia provided quite different results. No hydrogen was evolved during the reduction in

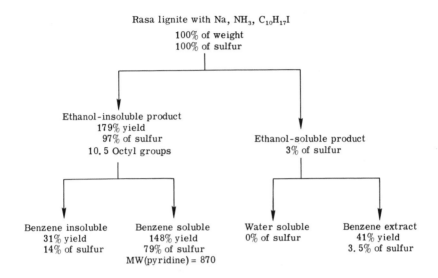

Rasa lignite with Na, NH_3, $C_{10}H_{17}I$
100% of weight
100% of sulfur

Ethanol-insoluble product
179% yield
97% of sulfur
10. 5 Octyl groups

Ethanol-soluble product
3% of sulfur

Benzene insoluble
31% yield
14% of sulfur

Benzene soluble
148% yield
79% of sulfur
MW(pyridine) = 870

Water soluble
0% of sulfur

Benzene extract
41% yield
3. 5% of sulfur

this solvent. The reaction in liquid ammonia yields a much smaller quantity of ethanol soluble products, and almost 97% of the sulfur is incorporated into the ethanol insoluble material. The difference in the extent of the desulfurization reactions by metals in ether and in ammonia is expected on the basis of the known reactions of sulfur compounds in these media (see Section II,B,2). The number average molecular weight of the products is about 500 when corrected for the octyl groups added during the reaction. About 36% of the sulfur in the ethanol-insoluble product occurs in the form of SC_8H_{17} units. Ignasiak and his co-workers suggest that these fragments are formed through the cleavage of thioethers that cross-link the lignite and conclude that the actual molecular weight of the lignite ranges between 1700 and 3300. Independent determinations of the distribution of the sulfur atoms among the possible functional groups in this lignite support the view that about 33% of the sulfur atoms are present as thioethers. Other lines of evidence discussed by B. S. Ignasiak et al. (1978a) suggest that the molecular clusters in the Rasa lignite have a quite uniform size. Thus, they conclude that the Rasa lignite should develop high fluidity on heating due to the cleavage of sulfur and oxygen ethers rather than to carbon–carbon bond cleavage reactions. Because the concentration of hydroxyl groups is quite low in this lignite, condensation reactions are not important processes in its thermal reactions. Consequently, this hydroaromatic material with labile carbon–sulfur linkages is highly thermoplastic.

Ignasiak and her associates studied the Athabasca asphaltenes using potassium naphthalene anions for the formation of the polyanion prior to protonation or octylation (T. Ignasiak *et al.*, 1977). The principal structural problem addressed in this work centered on whether this asphaltene was a carbon polymer with heteroatomic side-chain groups or a polymer joined by heteroatoms. Previous work had established that the asphaltene was readily reduced by lithium aluminum hydride and by lithium in ethylenediamine to yield much lower molecular weight compounds (Sawatzky and Montgomery, 1964). The large decrease in the average molecular weight from 6110 to 3550 in the reduction reaction was attributed by Sawatzky and Montgomery (1964) to the cleavage of sulfur and oxygen bonds. In the more recent study, Ignasiak and her co-workers employed potassium and naphthalene as the reducing agent in tetrahydrofuran with *n*-octyl iodide as the alkylation reagent. The polyanion produced in this reaction had about 18 negative charges per 100

$$C_{100}H_{119}N_{1.29}S_{3.82}O_{2.33} \xrightarrow[\substack{THF, 168 \text{ hr,} \\ \text{ambient temper-} \\ \text{ature}}]{K, C_{10}H_8} \xrightarrow{C_8H_{17}I} C_{100}H_{118}N_{0.97}S_{1.86}O_{5.77}(C_8H_{17})_{9.4}$$

Athabasca asphaltene   MW = 930
MW = 5920 ± 900

carbon atoms and the octylated product acquired about 9.4 alkyl groups per 100 carbon atoms. The observed number average molecular weight of the product was 930 representing 580 daltons from the original asphaltene and 350 daltons from the added octyl groups. As in the previous work, the reductive alkylation reaction provides a product mixture with a considerably decreased apparent molecular weight. Ignasiak and her associates postulated that molecular associations are, at least in part, responsible for the high molecular weight of the untreated asphaltene. To test this idea, they reacted the asphaltene with diazomethane and hexamethyldisilazane and found that the reaction products were much more soluble in pentane and that the apparent molecular weight was reduced from 5920 to about 3000 in each case. Several lines of evidence suggest that hydrogen bonding interactions of the oxygen functional groups are responsible for the association of the original asphaltenes. The relatively low molecular weight of the asphaltene indicates that, on the average, about 2 bonds per 100 carbon atoms are cleaved in the reductive alkylation reaction. Analyses of the oxygen content of the product and the starting material strongly imply that only a few carbon–oxygen bond cleavage reactions occur in the reduction reaction. Moreover, the deoctylation of the reaction product liberates 3.4 alkyl groups per 100 carbon atoms. As discussed in Section II,B, the aryl-O-octyl and the aryl-S-octyl groups are selectively removed under these experimental conditions. Consequently, these results coupled with the data concerning the distribution of the oxygen functional groups suggest that 2.7 octyl groups were removed from sulfur atoms and 0.7 octyl groups were removed from oxygen atoms. All these observations are compatible with the idea that this asphaltene has a polymeric framework in which the larger hydroaromatic fragments are cross-linked by sulfur linkages. This information and structural data, for example, the results obtained from proton and carbon NMR spectroscopic work, yield an average structure for the asphaltene consistent with all the available results (45). Ignasiak and her associates point out that a large number of other structural units with variations in the structures of the carbocyclic rings and in the nature of the substituents are also reasonable. Nevertheless, this structure is distinctly different from the average structure proposed for Lagunillas asphaltene (46) by Yen (1974) on an entirely different basis. Further study of the chemical reactions of the Lagunillas asphaltene would certainly prove valuable to ascertain whether the structures are different as suggested in these independent studies.

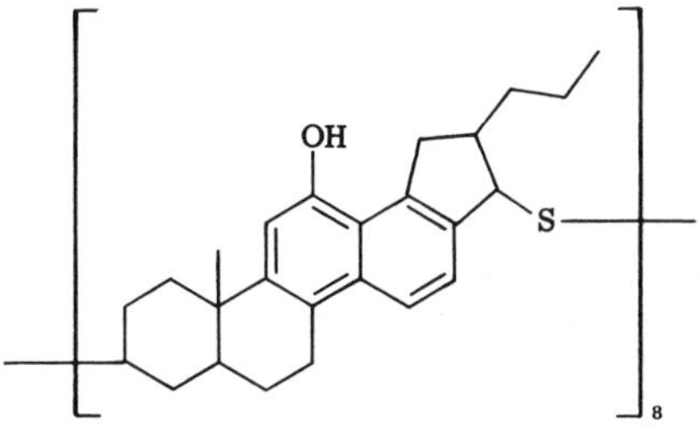

45

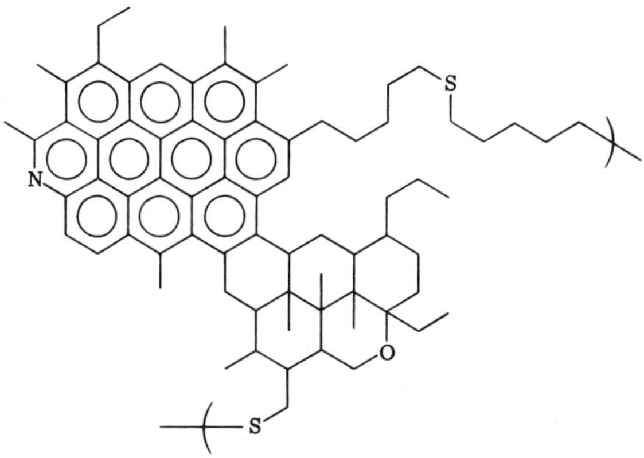

46

Mochida and his co-workers studied the reduction and the reductive alkylation of coal-tar pitch and petroleum pitch to convert the quinoline-insoluble components of the pitches to "graphitizable carbon" (Mochida *et al.*, 1974, 1975, 1976). It was their aim to alter the structures of these quinoline-insoluble materials so that they would fuse upon heating and become

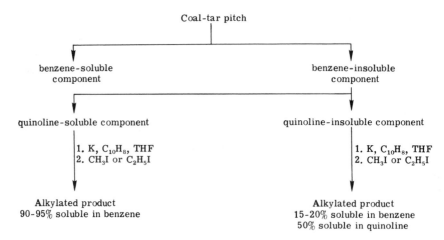

graphitizable. The experimental procedure used in the first study (Mochida *et al.*, 1974) is outlined in the equation. The crude products were washed with ethanol to extract the residual alkylating agents and the electron-transfer catalysts. The products were significantly more soluble in benzene or in quinoline than the original pitch. Elemental analyses and magnetic resonance data suggest that many aromatic molecules in the pitch were reduced and that between 3 and 4 alkyl groups were introduced per 100 carbon atoms. However, only the benzene-soluble alkylated product obtained from the benzene-insoluble, quinoline-soluble portion of the pitch was fusible. In the next phase of the study, these investigators examined the reductive alkylation of a petroleum pitch (Mochida *et al.*, 1975). This pitch was also extracted by benzene and quinoline, as shown for the coal-tar pitch, and the reductive ethylation reaction was carried out on all three fractions. The ethylation of the benzene-insoluble, quinoline-insoluble material yielded a product that was 60% soluble in benzene, but the material was not fusible. However, the reduction of this unfusible, ethylated pitch with lithium in ethylenediamine yielded a product that was 70% soluble in benzene and was fusible. When the original petroleum pitch was reductively ethylated and reduced by these methods, 80% of the material was fusible. Mochida and his group have extended this work to clarify the origin of the differences in the carbonization properties of these products by an investigation of their behavior during heat treatments (Mochida *et al.*, 1976). In the course of this study they investigated the alkylation reactions of the pitch with a variety of alkylating agents. They report that tertiary and secondary alkyl iodides as well as primary alkyl iodides react with the anions formed from the pitches but that the products obtained from the tertiary and secondary iodides are much less soluble in benzene. Presumably, the S_N2 reactions of these alkyl halides are slow, and the competitive elimination reactions yield reduced rather than alkylated pitch. Although the lines of argument are beyond the scope of this presentation, Mochida *et al.* (1976) conclude that the alkylated pitches fail to yield graphitic material because the alkyl groups are eliminated from the modified pitch by dealkylation reactions that occur readily below the temperature at which fusion occurs.

Wachowska *et al.* (1979) studied the reductive octylation of several coal macerals in a continuation of their investigation of the factors governing the coking properties of Balmer 10 coal which is rich in semifusinite (45–50%). They also employed the reduction and reductive alkylation reactions to study the influences of ether linkages on the chemistry of these macerals. The quantity and molecular weight of the benzene-soluble products formed in the reductive octylation of vitrinite, oxidized vitrinite, and semifusinite from the Balmer 10 coal and fusinite from Illinois No. 6 coal were established in this study (Table IL).

The vitrinite, the oxidized vitrinite, and the semifusinite are quite readily reduced and octylated. The fusinite is very much less reactive with only one alkyl group added per 100 carbon atoms and the reaction product is only about 18% soluble in benzene. As already noted, the chemistry of the fusinite is quite different from that of the other macerals. Wachowska and her associates suggest that the results for the reductive alkylation of this fusinite are compatible with the idea that this material has a condensed aromatic structure, with the hydroxyl groups located around the outside of large platelets such that the alkylation reactions of the polyanion, which has about 10 negative charges per 100 carbon atoms, occur preferentially on the peripheral hydroxyl groups and not with the carbon atoms of the platelets. This suggestion is in accord with the viewpoint that the alkylation reactions of "chicken-wire" aromatic compounds are generally slow and may be especially slow when the platelet-like structures are stacked.

The octylation reactions of the vitinite, the oxidized vitrinite, and the semifusinite were much more successful. Essentially the same number of alkyl groups are added to these three macerals. Extra precautions were taken to exclude oxygen from the reactive polyanions and the direct and by-difference analyses of the oxygen content of the alkylation products showed only a slight increase in the oxygen content of the products. The principal differences in the

TABLE II

The Reductive Octylation of Coal Macerals with Potassium, Naphthalene, and Octyl Iodide in Tetrahydrofuran[a]

Coal	Reduction (H added per 100 C atoms)	Octylation (alkyl groups added per 100 C atoms)	Extractability (%) in benzene	Number average molecular weight in benzene[b]	Deoctylation (alkyl groups removed per 100 C atoms)
Vitrinite[c]	9.0	9.0	92	1120	2.0
Oxidized vitrinite	10.3	7.9	92	1130	2.6
Semifusinite[c]	10.0	8.0	72	910	3.8
Fusinite[d]	—	1.9	18	—	1.2

[a] From Wachowska et al. (1979).
[b] The values are for the octyl-group free material.
[c] This maceral was obtained from Balmer 10 coal.
[d] This maceral was obtained from Illinois No. 6 coal.

character of these three macerals originate in the facility with which the octyl groups can be removed. About 20, 30, and 50% of the octyl groups are re-

$$\text{Maceral(Octyl)}_n \xrightarrow[\text{THF}]{\text{K, C}_{10}\text{H}_8} \text{Maceral(OH)}_m \text{(Octyl)}_{n-m}$$

moved from the vitrinite, the oxidized vitrinite, and the semifusinite, respectively. These results suggest that the O- to C-alkylation ratio is significantly different for these three materials. The finding that only one more alkyl group per 100 carbon atoms can be removed from the oxidized vitrinite than from the vitrinite and the fact that the number average molecular weights are similar suggests that the oxidization of the vitrinite does not lead to extensive cross-linking of the structure. The deoctylation of the semifusinite proceeds more extensively with about 4 alkyl groups removed per 100 carbon atoms. The number average molecular weight of the reductively octylated semifusinite is about 200 daltons less than the vitrinites. Whether this difference is meaningful is obscured by the limitations of number average molecular weight data.

The yields of the distillation and pyrolysis products obtained from the macerals and their octylation products were also studied. As expected, the amount of carbon lost from the octylated macerals far exceeds the loss of carbon from the macerals themselves. For example, about 15% of the carbon in semifusinite was lost when the sample was heated in nitrogen to 530°C; under the same conditions the octylated semifusinite lost about 45% of its carbon atoms. Not surprisingly, the weight losses were greatest for the macerals that were more heavily alkylated. The volatilities of the benzene-soluble octylation products were, in general, much greater than the volatilities observed for the complete octylation product. Unfortunately, the dilatation properties observed for the octylation products could not all be measured confidently. However, the octylated semifusinite only underwent contraction. Wachowska and her associates suggest that this result indicates that the reductive octylation reaction of this maceral did not alter the structure sufficiently to permit the product to behave as a liquid at 330–510°C and thus exhibit the characteristic swelling properties of coking coal.

Molecular weight measurements play an important role in the work on reductive alkylation because this method is one of the few procedures for the solubilization of an important quantity of the coal moelcules without the extensive degradation of coal structures which accompany donor–solvent decomposition reactions or acid-catalyzed decomposition processes. This advantage notwithstanding, it is clear that the acquisition of accurate and meaningful molecular weight data for the products of the reductive alkylation reactions is difficult. In brief, two approaches have been used to study the molecular weight distributions. On the one hand, the alkylation products have been separated from the reagents by selective extraction procedures and the number average molecular weight of the entire mixture of products has been obtained. On the other hand, the alkylation products have been separated from the reagents by chromatographic methods and the alkylation products have been separated into fractions by gel permeation chromatography and the molecular weights of the individual fractions were then determined. Several lines of evidence as outlined in Section III,C suggest that the chromatographic procedures are more effective for the separation of the reagents from the alkylated coal products and it seems certain that the molecular weight values obtained for the chromatographic fractions are more informative than the composite number average molecular weight values. B. S. Ignasiak and his co-workers (1978a) have pointed out that the properties of the extraction solvent and the solvent used for the molecular weight determination can strongly influence the results (Table XLVIII). The problems in the interpretation of number average molecular weight data are, of course, well known.

Wachowska (1979) recently carried out a study of the molecular weights of several alkylated coals. Her results illustrate the difficulties that can be encountered in the use of extraction methods for the separation of the coal from the reagents and the variations in the molecular weights of the products. Because the initial ethanol precipitate of the reaction was an oily mass rather

TABLE L

The Molecular Weight Data for the Reductive Alkylation Reaction of Polish K–III Coal[a]

Alkylated K–III coal	Yield (%)		Alkylation (alkyl groups per 100 C atoms)[b]	Number average molecular weight in benzene[c]
Methylation	Ethanol extract	11.5	20.1	360
	Benzene extract	42.1	15.1	1230
	Residue	46.4	8.2	
Ethylation	Ethanol extract	14.4	11.6	390
	Benzene extract	57.9	12.3	1960
	Residue	27.7	5.1	
Butylation	Ethanol extract	15.3	12.7	560
	Benzene extract	58.7	9.4	2620
	Residue	26.0	5.1	
Octylation	Ethanol extract	19.6	11.9	740
	Benzene extract	61.2	7.1	3760
	Residue	19.2	5.2	

[a] From Wachowska (1979).

[b] These values were estimated from the elemental composition.

[c] The observed number average molecular weights are presented. These data have not been corrected for the added alkyl groups.

than a solid she used a somewhat more thorough extraction procedure to separate the coal alkylate from the residual alkylation reagents and the naphthalene derivatives. She redissolved the oily material in tetrahydrofuran and reprecipitated the desired alkylated products by the addition of 70% ethanol. The procedure was repeated three times to yield a solid alkylate. The solid then was extracted with ethanol and benzene and the number average molecular weights were determined. The results are summarized in Table L.

These observations indicate that an important quantity of low molecular weight compounds are extracted into ethanol. Unfortunately, it is not possible to determine whether or not these rather highly alkylated materials are compounds derived from the coal or from naphthalene. In view of the results obtained by Doğru and his associates (1978), who also used an extraction procedure to separate the coal product from the alkylated naphthalene, it would appear that an important quantity of low molecular weight alkylate may be present in these samples.

The results for four different alkylation reactions of five different coals are presented in Table LI. The molecular weight values for all the benzene-soluble alkylation products are presumably somewhat larger than the values would have been if the measurements had been obtained in pyridine. Nevertheless, the results confirm the suggestion that the larger n-alkyl groups enable the solubilization of larger coal fragments. However, the largest number average molecular weight values are quite small, and it is pertinent to inquire about the molecular distribution. Although there are no data available for these coals, Burk and Sun (1975) examined the molecular weight distribution of the product of the reductive butylation reaction of Illinois No. 6 coal. They separated the alkylated coal molecules by gel permeation chromatography and determined the number average molecular weights of individual fractions in tetrahydrofuran. Their results (Fig. 11) suggest that the molecular weight distribution is very broad and that 40% of the material has molecular weight in excess of 1000 daltons and that some molecules have a molecular weight of 100,000 daltons.

We also used gel permeation chromatography to separate the tetrahydrofuran soluble fraction of reductively butylated Illinois No. 6 coal (Fig. 12). The chromatogram of this coal alkylate is similar to the chromatograms presented for the reduction products of other coals (see Figs. 7 and 8). Although the molecular weight scale shown at the top of the figure cannot be regarded as quantitatively correct, the observations imply that a considerable portion of the coal alkylate has a high molecular weight. It is certain that additional study of the molecular weights of the coal alkylates will prove useful for the definition of the influences of association on the experimental molecular weight values and eventually for a more confident definition of the real molecular weight range of the alkylated products.

TABLE LI
The Molecular Weight Data for the Alkylation Products Obtained from Five Polish Coals[a]

Coal, reaction	Alkylation (alkyl groups added per 100 C atoms)	Extractability (% in benzene)	Number average molecular weight[b] in benzene
K-I			
Methylation	10.5	39	550
Ethylation	9.6	45	550
Butylation	7.2	48	800
Octylation	6.3	50	810
K-II			
Methylation	9.6	35	710
Ethylation	8.0	45	710
Butylation	6.2	52	1000
Octylation	5.4	58	1470
K-III			
Methylation	9.7	50	630
Ethylation	8.9	70	920
Butylation	7.1	74	1210
Octylation	6.8	78	1340
K-IV			
Methylation	9.9	46	760
Ethylation	8.6	58	890
Butylation	8.7	60	1370
Octylation	7.4	75	1990
A			
Methylation	3.2	1	360
Ethylation	1.8	2	430
Butylation	1.3	3	450
Octylation	1.3	5	460

[a] From Wachowska (1979).
[b] The results for the alkyl group-free coal molecules are given.

The proton and carbon nuclear magnetic resonance spectra of the reductive alkylation products have been examined in several laboratories to learn more about the structure of the coals. The spectra of composite samples of the alkylated products often accurately portray the most important qualitative spectroscopic features. However, the spectra of samples that have been chromatographed by gel permeation methods usually provide more reliable information because these samples are free of residual quantities of the electron transfer agent and other relatively low molecular weight contaminants and also because the relaxation properties of the nuclei in the alkylated coal depend upon the molecular weight. Consequently, the study of fractions of

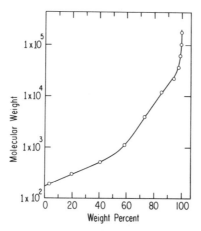

Fig. 11. Number average molecular weight distribution compared to the weight percentage of the butylated product in the Sternberg alkylation of Illinois No. 6 coal as determined by Burk and Sun (1975).

the product which have essentially the same molecular weight yield data that can be treated much more quantitatively than the data obtained for composite samples in which the signals arise from the superposition of the resonances of a great array of molecules having nuclei with quite different relaxation times. Typical proton and carbon NMR spectra of butylated Illinois No. 6 coal are shown in Figs. 13 and 14, and the spectroscopic information is summarized in Table LII.

The spectra of intermediate molecular weight fractions which represent 27.8% of alkylated products differ only modestly; the representative spectra of one fraction (9.0%) are shown in Fig. 13a and b.

Broad resonances are observed in the high- and low-field region of the carbon spectrum. Sharper signals are observed at δ13.6, 19.1, 23.7, 26.5, 29.6,

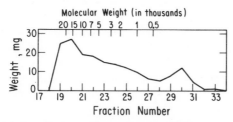

Fig. 12. The gel permeation chromatogram of the THF soluble products obtained in the reductive alkylation of Illinois No. 6 coal using potassium and naphthalene as the reducing agent and butyl iodide as the alkylation reagent. The chromatography was carried out on Stryragel columns with exclusion limits of 20,000, 10,000, and 2000.

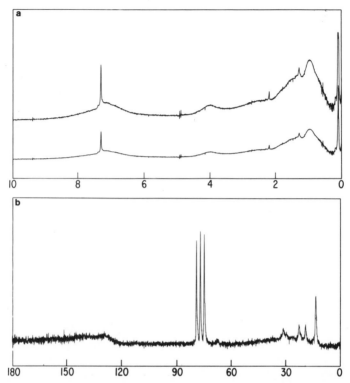

Fig. 13. (a) The proton NMR spectrum at 270 MHz of an intermediate molecular weight fraction of the butylated Illinois No. 6 coal. The spectrum is shown at two levels of amplification (1:2). The sharp signals at δ7.3 and δ0.1 result from residual chloroform and added octamethyl-tetrasiloxane. (b) The carbon NMR spectrum at 15.1 MHz of the butylated Illinois No. 6 coal. [Adapted from Alemany, King, and Stock (1978) and used with the permission of IPC Science and Technology Press, Ltd.]

31.5, and 67.8. The chemical shifts, relative intensities, and linewidths indicate that the sharp signals result from the nuclei of the added butyl groups (**47**). On this basis, the signal at δ13.6 is attributed to C_δ, that at δ19.1 is assigned to C_γ of butyl groups bonded to etheral oxygen atoms, and that at δ23.7 is identified with C_γ of butyl groups bonded to carbon atoms. The broad signal at δ26.5 may be assigned to butyl groups bonded to quaternary carbon atoms, for example, 9,9-di-*n*-9,10-dihydroanthracene (**48**) or 1-*n*-butyl-9-alkylfluorene (**49**). The less intense signal at δ29.6 is assigned to C_β of butyl groups bonded to tertiary sp³ carbon atoms. The broad signal at δ31.5 may be attributed to C_β of butyl groups bonded to etheral oxygen atoms or certain sp² carbon atoms. The signal at δ67.8 is uniquely consistent with C_α of butyl groups bonded to aryloxy groups. The assignments are self consistent. For ex-

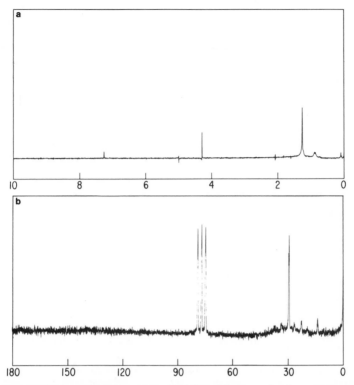

Fig. 14. (a) The proton NMR spectrum at 270 MHz of the low molecular weight components obtained following the alkylation reaction of Illinois No. 6 coal. The sharp signals at δ7.3 and δ0.1 result from residual chloroform and added octamethylcyclotetrasiloxane. (b) The carbon NMR spectrum at 15.1 MHz. [Adapted from Alemany, King, and Stock (1978) and used with the permission of IPC Science and Technology Press, Ltd.]

$$\overset{\delta}{C}H_3\overset{\gamma}{C}H_2\overset{\beta}{C}H_2\overset{\alpha}{C}H_2\text{---coal molecule}$$

47

CH₃CH₂CH₂CH₂ $\quad$ CH₂CH₂CH₂CH₃

48

R $\quad$ CH₂CH₂CH₂CH₃

49

TABLE LII

A Summary of the Spectroscopic Observations for Butylated Illinois No. 6 Coal[a]

Fraction of alkylate (%)[b]	C Butylation			Vinyl proton	Formyl proton
	Quaternary sp^3	Ternary sp^3	sp^2		
24	Und.[c]	Und.[c]	Und.[c]	No	No
28	1.4	1.0	2.7	No	No
13	2.6	1.0	1.3	No	No
18	4.1	1.0	0.3	Yes	Yes
13	1.9	1.0	0.6	Yes	Yes
2	0.7	1.0	0.4	Yes	Yes
2	No butylation			No	No

[a] From Alemany et al. (1978).

[b] The chromatographic fractions and the fraction of alkylate correlate with the proportion of butylated product, wt. % axis of Fig. 11.

[c] Undetermined.

ample, the observation that the C_δ resonances yield the narrowest signal is in accord with the small (3.8 ppm) chemical shift dispersion exhibited by the C_δ atoms of the butyl compounds. The finding that the C_γ atoms produce separate signals for C-butylation and O-butylation is also expected. The line widths are also in agreement with expectation; the C_γ and C_β resonances are broader than the C_δ resonance, and the C_α signals of the butyl groups bonded to carbon atoms are not discernible. However, the broad C_α resonance of an aryl butyl ether is observed at $\delta 67.8$. This resonance and the appearance of successively sharpening signals at $\delta 31.5$, 19.1, and 13.6, which correspond to the shifts of C_β, C_γ, and C_δ, indicate that this alkylation yields an important quantity of aryl butyl ethers.

The proton spectra shown in Fig. 13a have weak absorptions from $\delta 4.5$ to $\delta 3.5$ and more intense absorption maxima at $\delta 1.4$ and $\delta 1.0$. The absorptions from $\delta 4.5$ to $\delta 3.5$ are consistent with the resonances expected for the H_α signal of an aryl butyl ether ($ArOCH_2CH_2CH_2CH_3$) and for the benzhydryl hydrogen atoms of the nonreductive butylation products of several aromatic compounds, for example, the 9-butylation product of fluorene (**50**). The resonance is not attributable to the H_α signal of an alkyl butyl ether ($ROCH_2CH_2CH_2CH_3$) or to the benzylic hydrogen atoms of the reductive

50

butylation products of naphthalene, phenanthrene, or dibenz[a,h]anthracene. The strength of the signal from $\delta 4.5$ to $\delta 3.5$ compared to the signals at high field suggests that butyl ether formation, reductive butylation, and reduction are important reactions.

Analyses of this kind provide a preliminary qualitative characterization of the products of this reductive alkylation reaction. The spectra of other samples exhibited resonances of formyl hydrogens and vinyl hydrogen atoms, as indicated in Table LII. The lowest molecular weight fraction had particularly simple proton and carbon spectra (Fig. 14a and b). The spectroscopic results for this portion of the product are consistent only with linear alkane fragments. The proton spectra exhibit a methyl resonance at $\delta 0.9$ and a methylene resonance at $\delta 1.2$. The much weaker methine resonances are barely detectable. The assignments are reasonable because many straight chain alkanes exhibit resonances at $\delta 0.88-0.91$ and $\delta 1.25-1.29$, while branched alkanes give more complex spectra with downfield shifts as well. The prominent signals at $\delta 14.2$, 22.9, 33.8, 29.8, and 30.1 in the carbon spectrum are readily identified with carbon atoms C_1 through C_5 of linear alkanes. The resonances of the central carbon atoms of such compounds all occur near $\delta 30$, and this signal is most intense for the higher molecular weight straight-chain alkanes. The other weaker resonances at $\delta 19.8$, 26.8, and 37.3 presumably result from branched alkanes. These results suggest that linear alkanes are the principal components in the lowest molecular weight fractions, which constitute 2% of the product. These substances appear to be closely related to the alkane hydrocarbons obtained by several groups in the investigation of solvent extracts of coals. We infer that these materials (Vahrman and Watts, 1972; Bartle *et al.*, 1975) are liberated during the reactions that destroy the coal matrix. Presumably, other low molecular weight molecules are also liberated; however, such volatile and soluble compounds are separated from the reaction products during the preliminary workup procedures.

Arylbutanes may be formed in several different reactions including the free radical butylation of unreduced aromatic compounds. The products with butyl groups bonded to quaternary and tertiary sp^3 carbon atoms predominantly arise through the alkylation of carbanions and aromatic dianions, as discussed in Section II,C.

The spectra of some fractions described in Table LII provide evidence for formyl and vinyl groups. Although these signals are weak relative to the resonances of the butyl groups, they are significant inasmuch as they suggest that ether cleavage reactions have occurred as illustrated in the equations.

$$\text{Coal}-CH_2OCH_2CH_2-\text{Coal} \xrightarrow{\text{base}} \text{Coal}-\bar{C}HOCH_2CH_2-\text{Coal}$$
$$\text{Coal}-\bar{C}HOCH_2CH_2-\text{Coal} \rightarrow \text{Coal}-CHO^{\bar{}} + \cdot CH_2CH_2-\text{Coal}$$
$$\text{Coal}-CHO^{\bar{}} + CH_3CH_2CH_2CH_2I \rightarrow \text{Coal}-CHO + CH_3CH_2CH_2CH_2\cdot + I^-$$
$$2\text{Coal}-CH_2CH_2\cdot \rightarrow \text{Coal}-CH{=}CH_2 + \text{Coal}-CH_2CH_3$$

The disproportionation reactions of the alkyl fragments would account for the formation of vinyl groups and the reactions of ketyls with *n*-butyl iodide would account for the production of aldehydes and ketones.

The proton NMR spectra of the coal products obtained in the butylation reaction with butyl-l-^{13}C iodide are quite similar to the spectra of the coal products obtained in the reaction with unenriched butyl iodide. However, there is one notable difference in that a pair of broad, overlapping resonances separated by 125 Hz appears near 4 ppm in the proton spectra of the products enriched in ^{13}C. These signals arise from the coupling of the methylene protons in aryl butyl ethers with the carbon nucleus and further confirm the assignment of this resonance.

These carbon NMR spectra provide more information about the course of the Sternberg alkylation reaction. There are three groups of absorptions in the carbon spectra (Fig. 15). These are centered near 68, 35, and 15 ppm. Work with many model compounds indicates that the intense relatively sharp resonances at low field result from butyl ethers and esters and that the intense broad resonances between 25 and 45 ppm arise from carbon butylation products. The weak resonances at $\delta 23.0$, 19.3, and 13.9 result from the γ carbon atoms of butyl groups in the C- and O-alkylation products, respectively, and from the δ carbon atoms of all the butyl groups.

The resonances in the O-butylation region occur in three distinct bands. Chemical shift data for the α carbon atom resonances in about 20 ethers indicate that the resonances centered about $\delta 72.9$ may result from hindered aryl ethers, for example, butyl 2,6-dimethylphenyl ether, butyl benzyl ethers, or

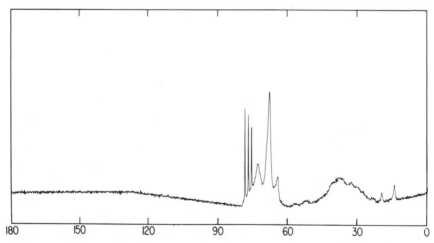

Fig. 15. The carbon NMR spectrum of an intermediate molecular weight fraction of Illinois No. 6 coal butylated with ^{13}C-enriched butyl iodide.

butyl *n*-alkyl ethers, for example, dibutyl ether. The most intense resonance centered at $\delta 67.7$ can be assigned with confidence to butyl aryl ethers in which the aryl group does not have 2- and 6-alkyl substituents. The resonance at $\delta 67.7$ also is compatible with the butyl secondary alkyl ethers. It seems certain that the principal contribution, however, results from the aryl ethers because the doublet resonance due to the coupling of the α-methylene hydrogen atoms of the butyl group with the α-methylene carbon atom is observed at $\delta 4.0$, not $\delta 3.4$, in the proton spectra. As already mentioned, this proton resonance is uniquely compatible with the resonances of butyl aryl ethers and not butyl secondary alkyl ethers. The weak resonance at $\delta 64.2$ is compatible with the signal expected for butyl esters and possibly for some butyl ethers of tertiary alcohols, for example, trityl butyl ether or *t*-butyl *n*-butyl ether. However, this resonance is quantitatively eliminated from the spectrum by mild basic hydrolysis of the soluble alkylated product. Consequently, it may be assigned with confidence to the ester functions.

The broad signal in the high-field region originates from the abundant carbon alkylation products. This resonance extends to almost 50 ppm in each spectrum with a maximum between 37 and 38 ppm. Work with model compounds indicates that these resonances may be attributed to the α carbon atoms of butyl groups bonded to aromatic, benzylic, tertiary sp^3 carbon atoms, and quaternary sp^3 carbon atoms. Derivatives of these kinds exhibit resonances between 32 and 44 ppm. The resonances of the S-butylation products also occur in this region. The lines at $\delta 56$ and $\delta 51$ are observed in only a few of the intermediate molecular weight fractions. The chemical shift data and the relatively low intensity of the signal are most compatible with N-butylated compounds. It is pertinent that the signals are very broad because the resonances have very small T_1 values, even smaller than the T_1 value 0.07 sec determined for the α carbon atom resonances in the butyl ethers.

The relative intensities of the C-butylation and O-butylation signals for the seven most abundant fractions are presented in Table LIII and the relative intensities of the three O-butylation signals for the same seven fractions are compared in Table LIV. The results presented in Table LIII indicate that the O-butylation reaction has proceeded to a greater degree in the initial, higher molecular weight fractions. The results presented in Table LIV suggest that the higher molecular weight products have a disproportionately greater quantity of the butyl aryl ethers. Although the variations are not large, the results clearly imply that there is a significant variation in the structural characteristic of the original coal molecules and that the structures of the higher molecular weight molecules are richer in phenolic compounds.

Only a few reductive alkylation reactions of coal have been carried out with alkali metals in ether in the absence of added electron transfer agents. Miyake and Niemann and their co-workers (Miyake *et al.*, 1980; Niemann and Rich-

TABLE LIII

**Relative Intensities of the C-Butylation
and O-Butylation NMR Signals[a]**

Fraction	Wt. %	C-butylation / O-butylation
1	12.5	0.76
2	38.5	0.65
3	21.4	0.75
4	12.1	0.93
5	6.1	0.97
6	3.1	1.4
7	3.4	1.8

[a] The area was measured from 25 to 50 ppm for C-butylation and from about 62 to 72 ppm for O-butylation.

ter, 1979; Haenel *et al.,* 1980) studied two bituminous coals using potassium as the reducing agent in refluxing tetrahydrofuran or in glyme solvents at low temperature.

Miyake and his associates carried out the reduction of a Japanese bituminous coal with potassium in tetrahydrofuran and subsequently alkylated the coal anion with methyl, ethyl, and butyl iodide. Although several workers have cautioned against the use of potassium in tetrahydrofuran at these tem-

$$\text{Yūbari coal} \xrightarrow[\text{2–6 hr, 66°C}]{\text{K, THF}} \xrightarrow[\text{3 hr}]{\text{RI, THF}} \text{Alkylated product}$$

peratures, Miyake *et al.* (1980) report that the reaction proceeds quite satisfactorily to yield a product that is free of tetrahydrofuran, its polymers, or its decomposition products. The results obtained in this study are summarized in Tables LV and LVI.

TABLE LIV

Relative Areas of the Absorptions of the Butyl Ethers and Esters

Fraction	Wt. %	Relative area		
		$\delta 72.9$	$\delta 67.7$	$\delta 64.2$
1	12.5	3.3	6.3	1.0
2	38.5	2.9	4.7	1.0
3	21.4	2.6	4.4	1.0
4	12.1	3.1	5.5	1.0
5	6.1	1.7	4.4	1.0
6	3.1	—	2.2	1.0
7	3.4	—	3.4	1.0

TABLE LV

The Reductive Alkylation of Yūbari Coal with Potassium
in the Absence of an Electron Transfer Agent in Refluxing Tetrahydrofuran[a]

Reaction	Alkylation (alkyl groups added per 100 C atoms)	Solubility (%)	
		In pentane	In benzene
Untreated	—	Nil	2
Reduced	—	2	11
Methylation[b]	8.3	9	69
Ethylation[b]	6.7	15	76
Butylation[b]	6.6	18	75
Butylation[c]	9.7	30	81

[a] From Miyaki et al. (1980).
[b] The reduction reaction was carried out for 2 hr.
[c] The reduction reaction was carried out for 6 hr.

The microanalytical data for the pentane-soluble fraction and the pentane-insoluble, benzene-soluble fractions suggest that the pentane-soluble fractions are much more highly alkylated than the pentane-insoluble fractions. The molecular weight data for the pentane-soluble fraction is also much lower than the molecular weight of the pentane-insoluble material. Moreover, the alkyl-free molecular weight data suggest that butylation leads to the solubilization of larger coal fragments than does methylation.

The infrared data reported for the coal products exhibit the customary features that have already been described (Section IV,A). In addition, the products exhibit a relatively broad absorption in the O—H stretching region,

TABLE LVI

The Properties of the Pentane-Soluble and Pentane-Insoluble, Benzene-Soluble
Fractions of the Alkylated Yūbari Coals[a]

Fraction	Reaction	Alkyl groups per 100 carbon atoms	Number average molecular weight	f_a
Pentane soluble	Methylation[b]	22.7	330	0.43
	Butylation[b]	9.3	480	0.47
	Butylation[c]	14.2	300	0.38
Benzene soluble	Methylation[b]	3.5	2400	0.62
	Ethylation[b]	4.7	3100	0.59
Pentane insoluble	Butylation[b]	5.1	3500	0.52
	Butylation[c]	6.3	3700	0.51

[a] From Miyaki et al. (1980).
[b] See Footnote b in Table LV.
[c] See Footnote c in Table LV.

which suggests that the alkylation reaction may not have been carried to completion.

Niemann and his co-workers found that the reductive alkylation of a bituminous coal from the Westerholt mine using ethyl iodide, methyl iodide, or methyl-^{13}C iodide yields a product that is between 65 and 75% soluble in pyridine. Combustion analyses of the unlabeled product suggest that 6.5 ±

$$C_{100}H_{64.1}N_{1.3}O_{3.7}S_{0.6} \xrightarrow[\text{(2) CH}_3\text{I}]{\substack{\text{(1) K, glyme,} \\ -30°C, 16\,hr}} \text{Alkylated product}$$

1.0 methyl groups have been added per 100 carbon atoms, and mass spectrometric analysis of the combustion products of the labeled product suggests that 4.4 methyl groups have been added per 100 carbon atoms. The latter more direct measurement presumably provides a more reliable estimate of the extent of the alkylation reaction. Alkylation with methyl-^{13}C iodide produces a product suitable for study by nuclear magnetic resonance. The spectrum of the product is shown in Fig. 16. These investigators assign the OCH$_3$ resonance at 60.6 ppm (relative intensity, 3%) to an enolic methyl ether and the OCH$_3$ resonance at 55.4 (relative intensity, 6%) to secondary alkyl and predominantly to aryl methyl ethers. They also note that the shoulder near 35 ppm may arise from an N-methyl group, but note that the nitrogen content of the coal is very low. The assignments of the CCH$_3$ resonances are more difficult to make because many of the structures provide overlapping signals. However, Haenel and his associates note that the rather low intensity of the high-field portion of this spectrum suggests that methylated aromatic compounds are not formed in appreciable quantity during the reaction and that the more intense CCH$_3$ resonances between 20 and 30 ppm arise from the methylation of benzylic carbanions and arene anions. Forthcoming studies such as this one will be most welcome because so much more can be deduced about the nature of the coal by the use of labeled reagents.

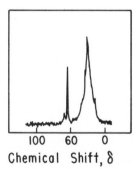

100   60   0

Chemical Shift, δ

Fig. 16. The carbon NMR spectrum at 25.2 MHz of methylated Westerholt coal in pyridine. [Adapted from Haenel *et al.* (1980) and used with the permission of Verlag Chemie, GmbH.]

In summary, the reductive alkylation reaction has been used by several research groups to investigate a variety of coals ranging from lignite to anthracite, coal macerals, and other related substances. The results obtained in the course of these studies have provided new information concerning coal structure and coal reactivity that could not have been obtained as directly by other methods of study. Fresh evidence about carbon–carbon bond cleavage, the role of ethers in the degradation of large coal molecules, the extent to which coal molecules can be depolymerized, the distribution of oxygen functional groups, the range of molecular weights of reduced and alkylated products, the degree of heterogeneity of the coal molecules, the impact of structural changes on the coking properties of oxidized and unoxidized coals and their macerals, and pitches derived from coal and from petroleum and many other highly informative facts have been provided by research using reductive alkylation as one of the tools for the investigation of the coal materials. It is clear that the improved understanding of the reaction will encourage other workers to use this technique in the future to gather additional stimulating information on the structures and reactions of the complex molecules of coal.

ACKNOWLEDGMENTS

It is a pleasure to acknowledge the contributions of all my students in the preparation of this article and to note that special thanks are due to Larry Alemany who critically read the entire article and made many valuable suggestions for improvements. The work was supported by the Department of Energy under grant DE-AC22-80PC 30088 and by the Exxon Educational Foundation.

REFERENCES

Alemany, L. B. (1980). Thesis, University of Chicago Library.
Alemany, L. B., King, S. R., and Stock, L. M. (1978). *Fuel* 57, 738.
Alemany, L. B., Handy, C. I., and Stock, L. M. (1979). *Prepr. Pap.—Am. Chem. Soc., Div. Fuel Chem.* 24(1), 156.
Angelo, B. (1966). *Bull. Soc. Chim. Fr.* p. 1091.
Bank, S., and Juckett, D. A. (1976). *J. Am. Chem. Soc.* 98, 7742.
Bank, S., Bank, J., Daney, M., Labrande, B., and Bouas-Laurent, H. (1977). *J. Org. Chem.* 42, 4058.
Bartle, K. D., Martin, T. G., and Williams, D. F. (1975). *Fuel* 54, 226.
Bates, R. B., Kroposki, L. M., and Potter, D. E. (1972). *J. Org. Chem.* 37, 560.
Benkeser, R. A., and Broxterman, W. E. (1969). *J. Am. Chem. Soc.* 91, 5162.
Bergbreiter, D. E., and Killough, J. M. (1978). *J. Am. Chem. Soc.* 100, 2126.
Biellmann, J.-F., d'Orchymont, H., and Schmitt, J.-L. (1979). *J. Am. Chem. Soc.* 101, 3283.
Bimer, J. (1974). *Koks, Smola, Gaz* 19, 68.
Bimer, J. (1976). *Koks, Smola, Gaz* 21, 78.
Bimer, J., and Witt, I. (1978). *Koks, Smola, Gaz* 23, 341.
Blom, L., Edelhausen, D., and van Krevelen, D. W. (1957). *Fuel* 36, 195.
Brooks, J. J., Rhine, W., and Stucky, G. D. (1972). *J. Am. Chem. Soc.* 94, 7346.

Burk, E. H., and Sun, J. Y. (1975). *In* "The Fundamental Organic Chemistry of Coal" (J. W. Larsen, ed.). University of Tennessee, Knoxville; also personal communications.

Cafasso, F., and Sundheim, B. R. (1959). *J. Chem. Phys.* **31**, 809.

Carpenter, J. G., Evans, A. G., and Rees, N. H. (1972). *J. Chem. Soc., Perkin Trans. 2* p. 1598.

Carson, D. W., and Ignasiak, B. S. (1979). *Fuel* **58**, 72.

Carson, D. W., and Ignasiak, B. S. (1980). *Fuel* **59**, 757.

Chaput, G., Jeminet, G., and Juillard, J. (1975). *Can. J. Chem.* **53**, 2240.

Chaudhuri, J., Kume, S., Jagur-Grodzinski, J., and Szwarc, M. (1968). *J. Am. Chem. Soc.* **90**, 6421.

Collins, C. J., Hombach, H.-P., Maxwell, B. E., Woody, M. C., and Benjamin, B. M. (1980). *J. Am. Chem. Soc.* **102**, 851.

Collins, C. J., Hombach, H.-P., Maxwell, B. E., Benjamin, B. M., and McKamey, D. (1981). *J. Am. Chem. Soc.* **103**, 1213.

Cram, D. J., and Dalton, C. K. (1963). *J. Am. Chem. Soc.* **85**, 1268.

Cram, D. J., Mateos, J. L., Hauck, F., Langemann, A., Kopecky, K. R., Nielsen, W. D., and Allinger, J. (1959a). *J. Am. Chem. Soc.* **81**, 5774.

Cram, D. J., Kingsbury, C. A., and Langemann, A. (1959b). *J. Am. Chem. Soc.* **81**, 5785.

Doğru, R., Erbatur, G., Gaines, A. F., Yürüm, Y., Içli, S., and Wirthlin, T. (1978). *Fuel* **57**, 399.

Down, J. L., Lewis, J., Moore, B., and Wilkinson, G. (1957). *Proc. Chem. Soc. London* p. 209.

Down, J. L., Lewis, J., Moore, B., and Wilkinson, G. (1959). *J. Chem. Soc.* p. 3767.

Dryden, I. G. C. (1952). *Chem. Ind.,* p. 502.

Dye, J. L., DeBacker, M. G., and Nicely, V. A. (1970). *J. Am. Chem. Soc.* **92**, 5226.

Eargle, D. H., Jr. (1963). *J. Org. Chem.* **28**, 1703.

Eisch, J. J. (1963). *J. Org. Chem.* **28**, 707.

Eisch, J. J., and Jacobs, A. M. (1963). *J. Org. Chem.* **28**, 2145.

Eisch, J. J., and Kaska, W. C. (1962). *J. Org. Chem.* **27**, 3745.

Elschenbroich, C., Gerson, F., and Reiss, J. A. (1977). *J. Am. Chem. Soc.* **99**, 60.

Franz, J. A., and Skiens, W. E. (1978). *Fuel* **57**, 502.

Fujita, T., Sugar, K., and Watanabe, S. (1972). *Synthesis* p. 630.

Garst, J. F. (1969). *In* "Solute-Solvent Interactions" (J. F. Coetzee and C. D. Ritchie, eds.), Vol. 1, Chapter 8.

Garst, J. F. (1971). *Acc. Chem. Res.* **4**, 400.

Gawlak, M., Cyr, N., Carson, D., and Ignasiak, B. S. (1979). *Prepr. Pap.—Am. Chem. Soc., Div. Fuel Chem.* **25**(4), 111.

Gerson, F., Martin, W. B., Jr., and Wydler, C. (1976). *J. Am. Chem. Soc.* **98**, 1318.

Gilman, H. (1954). *Org. React.* **8**, 258.

Gilman, H., and Dietrich, J. J. (1957). *J. Org. Chem.* **22**, 851.

Gilman, H., and Gaj, B. J. (1963). *J. Org. Chem.* **28**, 1725.

Gilman, H., McNinch, H. A., and Wittenberg, D. (1958). *J. Org. Chem.* **23**, 2044.

Given, P. H., and Peover, M. E. (1959). *Nature (London)* **184**, 1064.

Given, P. H., Lupton, V., and Peover, M. E. (1958). *Nature (London)* **181**, 1059.

Given, P. H., Peover, M. E., and Wyss, W. F. (1960). *Fuel* **39**, 323.

Gough, T. A., and Peover, M. E. (1964). *Polarography, Proc. Int. Cong., 3rd, 1964.* Vol. 2, p. 1017.

Grovenstein, E., Jr. (1977). *Adv. Organomet. Chem.* **16**, 167.

Grovenstein, E., Jr., Longfield, T. H., and Quest, D. E. (1977). *J. Am. Chem. Soc.* **99**, 2800.

Haenel, M. W., Mynott, R., Niemann, K., Richter, U.-B., and Schanne, L. (1980). *Angew. Chem., Int. Ed. Engl.* **19**, 636.

Halleux, A., and de Greef, H. (1963). *Fuel* **42**, 185.

Halleux, A., Delavarenne, S., and Tschamler, H. (1961). *Nature (London)* **190**, 437.

Harvey, R. G., and Cho, H. (1974). *J. Am. Chem. Soc.* **96**, 2434.

Hoijtink, G. J. (1970). *Adv. Electrochem. Electrochem. Eng.* **7**, 235.

Holy, N. L. (1974). *Chem. Rev.* **74**, 243.

House, H. O. (1972). "Modern Synthetic Reactions," 2nd ed., Chapter 3. Benjamin/Cummings Publ. Co., Menlo Park, California.

Ignasiak, B. S., and Gawlak, M. (1977). *Fuel* **56**, 216.

Ignasiak, B. S., Chakrabartty, S. K., and Berkowitz, N. (1978a). *Fuel* **57**, 507.

Ignasiak, B. S., Fryer, J. F., and Jadernik, P. (1978b). *Fuel* **57**, 578.

Ignasiak, B. S., Carson, D. W., and Gawlak, M. (1979). *Fuel* **58**, 833.

Ignasiak, T., Kemp-Jones, A. V., and Strausz, O. P. (1977). *J. Org. Chem.* **42**, 312.

Ihnatowicz, A. (1952). *Komun. Glownego Inst. Górnictwa* No. 125.

Jolly, W. L. (1956). *J. Chem. Educ.* **33**, 512.

Kaempf, B., Raynal, S., Collet, A., Schué, F., Boileau, S., and Lehn, J.-M. (1974). *Angew. Chem., Int. Ed. Engl.* **13**, 611.

Knotnenus, J. (1956). *J. Inst. Petrol.*, **42**, 355.

Köbrich, G., and Baumann, A. (1973). *Angew. Chem., Int. Ed. Engl.* **12**, 856.

Komarynsky, M. A., and Weissman, S. I. (1975). *J. Am. Chem. Soc.* **97**, 1589.

Kotlarek, W., and Pacut, R. (1978). *J. Chem. Soc., Chem. Commun.* p. 153.

Kröger, C., Darsow, G., and Fuhr, K. (1965). *Erdoel Kohle* **18**, 701.

Lagendijk, A., and Szwarc, M. (1971). *J. Am. Chem. Soc.,* **93**, 5359.

Larsen, J. W., and Urban, L. O. (1979). *J. Org. Chem.* **44**, 3219.

Lazarov, L., and Angelova, G. (1968). *Fuel* **47**, 333.

Lazarov, L., and Angelov, S. (1980). *Fuel* **59**, 55.

Lazarov, L., Zgurovska, E., and Angelova, G. (1971). *Fuel* **50**, 338.

Lazarov, L., Rashkov, I., and Angelov, S. (1978). *Fuel* **57**, 637.

Letsinger, R. L., and Pollart, D. F. (1956). *J. Am. Chem. Soc.* **78**, 6079.

Lindow, D. F., Cortez, C. N., and Harvey, R. G. (1972). *J. Am. Chem. Soc.* **94**, 5406.

Liotta, R. (1979). *Fuel* **58**, 724.

Liotta, R., Rose, K., and Hippo, E. (1981). *J. Org. Chem.* **46**, 277.

March, J. (1977). "Advanced Organic Chemistry, Reactions, Mechanism, and Structure," 2nd ed., pp. 1015–1016, and references cited. McGraw-Hill, New York.

Marshall, J. A., and Andersen, N. H. (1965). *J. Org. Chem.* **30**, 1292.

Miyake, M., Sukigara, M., Nomura, M., and Kikkawa, S. (1980). *Fuel* **59**, 637.

Mochida, I., Kudo, K., Takeshita, K., Takahashi, R., Suetsugu, Y., and Furumi, J. (1974). *Fuel* **53**, 253.

Mochida, I., Tomari, Y., Maeda, K., and Takeshita, K. (1975). *Fuel* **54**, 265.

Mochida, I., Maeda, K., and Takeshita, K. (1976). *Fuel* **55**, 70.

Nandi, B. N., Wachowska, H. M., and Montgomery, D. S. (1974). *Fuel* **53**, 226.

Niemann, K., and Hombach, H.-P. (1979). *Fuel* **58**, 853.

Niemann, K., and Richter, U.-B. (1979). *Fuel* **58**, 838.

Ouchi, K., Hirano, Y., Makabe, M., and Itoh, H. (1980). *Fuel* **59**, 751.

Rabideau, P. W., and Burkholder, E. G. (1978). *J. Org. Chem.* **43**, 4283.

Rabideau, P. W., Jessup, D. W., Ponder, J. W., and Beekman, G. F. (1979). *J. Org. Chem.* **44**, 4594.

Razdan, R. K., Herlihy, P., Dalzell, H. C., and Portlock, D. E. (1979). *J. Org. Chem.* **44**, 3730.

Reggel, L., Friedel, R. A., and Wender, I. (1957). *J. Org. Chem.* **22**, 891.

Reggel, L., Raymond, R., Friedman, S., Friedel, R. A., and Wender, I. (1958). *Fuel* **37**, 126.

Reggel, L., Raymond, R., Steiner, W. A., Friedel, R. A., and Wender, I. (1961). *Fuel* **40**, 339.

Sargent, G. D., and Lux, G. A. (1968). *J. Am. Chem. Soc.* **90**, 7160.

Sargent, G. D., Cron, J. N., and Bank, S. (1966). *J. Am. Chem. Soc.* **88**, 5363.

Savoia, D., Trombini, C., and Umani-Ronchi, A. (1978). *J. Org. Chem.* **43**, 2907.

Sawatzky, H., and Montgomery, D. S. (1964). *Fuel* **43**, 453.

Screttas, C. G. (1972). *J. Chem. Soc., Chem. Commun.* p. 869.

Screttas, C. G., and Micha-Screttas, M. (1978). *J. Org. Chem.* **43**, 1064.

Screttas, C. G., and Micha-Screttas, M. (1979). *J. Org. Chem.* **44**, 713.

Screttas, C. G., and Micha-Screttas, M. (1981). *J. Org. Chem.* **46**, 993.

Smith, H. (1963). "Organic Reactions in Liquid Ammonia, Chemistry in Nonaqueous Ionizing Solvents," Vol. I, Part 2, pp. 276–279. Wiley, New York.

Solodovnikov, S. P., and Zaks, Yu. B. (1969). *Bull. Acad. Sci. USSR, Div. Chem. Sci. (Engl. Transl.)* p. 1274.

Solodovnikov, S. P., Ioffe, S. T., Zaks, Yu. B., and Kabachnik, M. I. (1968). *Bull. Acad. Sci. USSR, Div. Chem. Sci. (Engl. Transl.)* p. 442.

Sternberg, H. W., and Delle Donne, C. L. (1974). *Fuel* **53**, 172.

Sternberg, H. W., and Delle Donne, C. L. (1968). *Prepr. Pap.—Am. Chem. Soc., Div. Fuel Chem.* **12**(4), 13.

Sternberg, H. W., Delle Donne, C. L., and Pantages, P. (1970). *Prepr. Pap.—Am. Chem. Soc., Div. Fuel Chem.* **14**(1), 87.

Sternberg, H. W., Delle Donne, C. L., Pantages, P., Moroni, E. C., and Markby, R. E. (1971). *Fuel* **50**, 432.

Szwarc, M. (1969). *Acc. Chem. Res.* **2**, 87.

Theilacker, W., and Möllhoff, E. (1962). *Angew. Chem., Int. Ed. Engl.* **1**, 596.

Truce, W. E., and Frank, F. J. (1967). *J. Org. Chem.* **32**, 1918.

Vahrman, M., and Watts, R. H. (1972). *Fuel* **51**, 235.

Wachowska, H. M. (1979). *Fuel* **58**, 99.

Wachowska, H. M., and Andrzejak, A. (1976). *Koks, Smola, Gaz* **21**, 192.

Wachowska, H. M., and Pawlak, W. (1977). *Fuel* **56**, 422.

Wachowska, H. M., Nandi, B. N., and Montgomery, D. S. (1974). *Fuel* **53**, 212.

Wachowska, H. M., Nandi, B. N., and Montgomery, D. S. (1979). *Fuel* **58**, 257.

Walker, P. (1961). *J. Org. Chem.* **26**, 2994.

Walsh, T. D., and Dabestani, R. (1981). *J. Org. Chem.* **46**, 1222.

Walsh, T. D., and Megrenis, T. L. (1981). *J. Am. Chem. Soc.* **103**, 3897.

Walsh, T. D., and Ross, R. T. (1968). *Tetrahedron Lett.* 3123 (1968).

Ward, R. L. (1961). *J. Am. Chem. Soc.* **83**, 3623.

Weinstein, B., and Fenselau, A. H. (1964). *J. Org. Chem.* **29**, 2102.

Yen, T. F. (1974). *Energy Sources* **1**, 447.

Zilm, K. W., Pugmire, R. J., Grant, D. M., Wood, R. E., and Wiser, W. H. (1979). *Fuel* **58**, 11.

Index

A

Acenaphthylene chars, ESR g values for, 65
Acridine/benzoquinoline units, in anthracite, 136
Acylated coal, pyridine extract of, 95
Acylation, Friedel–Crafts, 94
Adamantane, 143
1-Adamantanol, 143
Alberta Research Council, 258
Aliphatic carbon–hydrogen bonds, increasing quantity of, 236
Aliphatic structures, 115–119
Alkali additives, cross-linking and, 15
Alkali cations, ethereal solutions of, 165
Alkanes, from bituminous coals, 120
Alkylated coal products, isolation of, 202–210
Alkylated coals, molecular weights of, 266–267
Alkylation
  mild method for, 242–243
  nonreductive, 245–246
  reductive, see Reductive alkylation reaction
Alkylation reaction, 181–187, 197–202, 242–248, see also Reductive alkylation reaction
  alkyl group influence in, 198
  butyl iodide as reagent in, 199
  aromatic anions in, 181–182
  reaction times for, 197
Alkyldihydronaphthalene anion, 209
Ammonia, see also Liquid ammonia
  proton donor and, 212–214
  reductive alkylation of Eastern European coals in, 222–223
  reductive butylation in, 194–196
Anderson sampler, 75–76
Anion radicals, in alkylation reactions, 181
Anthracene
  alkylation reaction and, 182–186
  in reduction reaction, 190

Anthracite(s)
  acridine/benzoquinoline units in, 136
  aromatic content of, 102
  coal rank and, 139–141
  ESR g values and, 64–65
  $Na_2Cr_2O_7$ oxidation of, 112
  phenanthrenes in, 138–139
  polynuclear aromatic compounds in, 111
Aralkyl thioethers, cleavage in, 173
Aromatic carbon–hydrogen bonds, decreasing quantity of, 236
Aromatic compounds, free radical butylation of, 273
Aromatic/hydroaromatic structures, 146–150
Aromatic hydrocarbons, half-wave reduction potentials for, 166
Aromatic ring structures
  ^{13}C NMR studies in, 108–109
  coal rank and, 138–139
  connectivity in, 109
  oxidative degradation in, 110–113
Aromatic ring substitution, 48, 108–115
Arylbutanes, formation of, 273
Aryl carbanions, from diaryl ethers, 172
Aryl oxygen, hydrogenolysis of, 114, see also Oxygen
Aryloxy groups, butyl groups bonded to, 270
Asphaltenes, chemical reactions of, 260–261
Athabasca asphaltenes, potassium naphthalene anions in study of, 260–261

B

Balmer 10 coals, British Columbia
  coking properties of, 263
  elution profile for, 238
  macerals and vitrinite from, 231–233
  macromolecular network of, 90
  properties of, 229–231
  reduction of macerals from, 231

283

Balmer 10 coals, British Columbia (*cont.*)
    reductive alkylation of with potassium
        uptake in tetrahydrofuran, 255
Base/alcohol hydrolysis, 95–96
Benzene, ethylation solubility in, 193
Benzothiophene, cleavage reactions in, 174
1,2-Benzotryptycene, 177
Benzylic carbanions, 170–172
Benzyl thioethers, cleavage reaction in, 173
Biphenyl, in reduction reaction, 190
Biteralyl, in ethylated Karparowitz coal, 206
Bitumens
    defined, 8
    plastic behavior and, 38
Bituminous Balmer coal, reductive alkyla-
        tion of, 90–91
Bituminous coals
    alkanes from, 120
    aromatic compounds in, 111
    banded appearance of, 23
    $Na_2Cr_2O_7$ oxidation of, 112
    vitrinite structure in, 3
Black acids
    aliphatic carbon atoms in, 145
    ^{13}C NMR spectrum of, 102
Bond strength distribution, 14
Brown coals, molecular weight of, 141–142
Brunauer–Emmett–Teller equation or
        model, 25, 28
Bureau of Mines, U.S., 44
Butyl benzyl ethers, 274
Butyl bromide, 198
Butyl chloride, 198
n-Butyl chloride, 185
Butyl ethers, relative areas of absorptions of,
        276
Butyl halides, electron transfer reactions in,
        145–146
Butyl iodide
    as reagent in alkylation reaction, 199
    in reductive butylation reactions, 213

                        C

Caking coals, 1
Canadian coals
    coking properties of, 229
    reduction reactions with potassium and
        naphthalene in tetrahydrofuran, 227

Cannock Wood coal, oxygen removal from,
        218
Carbanions
    benzylic, 170–172
    from diaryl ethers, 172
    from dibenzylic ethers, 171
    rearrangement and coupling of, 180
    from thioethers, 174
Carbon alkylation products, resonances in,
        275
Carbon aromaticity, 98–107
Carbon–carbon bond cleavage, 175–181
Carbon–carbon double bonds, reduction of,
        236
Carbon–oxygen bond cleavage, 169–173
Carbon–sulfur bond cleavage, 173–175
Carbonyl compounds, reactions of, 168
Carbonyl groups
    distribution of, 226
    number of, 226
Carboxylic acid groups, selective alkylation
        of, 243
Catalytic dehydrogenation, and ESR *g* values
        for vitrains, 63
C-butylation region, of carbon NMR spec-
        tra, 275
Chang-Guang coals
    carbon aromaticities of, 71
    petrographic composition of, 72
Chemistry of Coal Utilization (Lowry), 86
Chinese coals, *see also* Chang-Guang coals
    carbon aromaticities of, 71
    cross-polarization ^{13}C NMR spectrum
        for, 70
    elemental analysis, sulfur forms, and
        magnetic resonance data for, 70
    ESR and NMR investigations of, 69–73
    functional dependencies of ESR *g* values
        of, 72
    maceral percentages of, 72
    petrographic composition of, 72
Chloroform solubles, in plastic development,
        8–9
Cleavable and noncleavable ethers, distribu-
        tion of, 225
Cleavage reaction, 169–181
    carbon–carbon bond cleavage and,
        175–181
    carbon–oxygen bond cleavage in, 169–173
    carbon–sulfur bond cleavage in, 173–175

Clusters, in coal physical structure, 24
^{13}C NMR, *see* Cross-polarization carbon-13 nuclear magnetic resonance
Coal
acylation of, 94
adsorption of gases and vapors by, 27–29
agglomerating or caking, 1
aliphatic structures of, 115–119
alkylation of, 242–248
alkylation products for different ranks of, 92
anthracite, *see* Anthracite(s)
aromatic content of, 102
aromatic/hydroaromatic structures in, 146–150
aromatic ring structures in, 108–115, 138–139
aromatic units indigenous to, 111
aromaticity of, 44–50
base/alcohol hydrolysis of, 95–96
bituminous, *see* Bituminous coals
bond strength distribution in, 14
Brunauer–Emmett–Teller model of, 25
caking or agglomerating, 1, 35
carbon aromaticity of, 137–139
chemical additives in, 15–16
chemical structure of, 22
"clusters" of, 24
coking, 2
constituent molecules of, 84
constitution of, 2–3
cross-links in, 97–98
decomposition rate for, 15
defined, 21, 84
depolymerization of, 87–88
devolatilization of, 33–34
Dubinin–Polanyi model of, 25
ESR *g* values for, 56–64
ether cleavage reactions in, 225, 229
extended cycloalkane structures in, 144–145
flash heating of, 122–123
fluidity of, 8
fluorination of, 105–106
free radicals in, 50–69, 126–132
heating rate for, 13–15
heating value vs. aromaticity of, 137–138
heats of wetting for, 26–27
helium in porosity measurements of, 25
heteroatoms in, 132–136

hydrogenation and dehydrogenation of, 12–13
hydrogen yield for, 118
hydroxyl groups in, 133–134
Illinois No. 6, *see* Illinois No. 6 coal
infrared spectrometry of, 104
internal surface of, 26–31
Japanese, 153
liquefaction of, *see* Liquefaction
literature on, 86–87
low molecular weight compounds in, 120–126
macerals in, *see* Macerals
macro- and micropore systems in, 31
macromolecular skeletal structures in, 142–154
macromolecules in, 23, 84
magnetic resonance studies of, 43–80
"melting" of, 24
mercury in porosity measurements of, 25–26
methane sorption in, 26
methylene bridges in, 87
minerals in, 84
models of, 146–153
molecular structure of, 83–155
molecular weight of, 87–94, 136–137
"molecules" of, 97–98
nitrogen in, 136
O-methylated, 243–244
organic components of, 21
oxidation of, 10–12
oxygen in, 132–134
oxygen group distribution in, 134
performic acid oxidation of, 113
photochemical oxidation of, 113
physical properties of, 22
physical structure of, 21–39
plastic behavior of, 35–39
plastic capability in, 1–2
Polish, *see* Polish coals
polyamantane structures in, 143–145, 153
polymeric structure of, 86
pore size distribution in, 31–32
pore structure changes in following heating, 32–33
porous nature of, 25–32
pyrolysis of, 8–12, 24
pyrolytic decomposition of, 1
reaction media for reduction of, 167–168

Coal (*cont.*)
  reactive hydroxyl and carbonyl groups as
    ethers in, 225
  reduction of, 218–242
  reductive alklylation reaction for, 88–94,
    161–279
  saturated aliphatic carbon in, 105
  sodium hydroxide/alcohol hydrolysis of,
    113
  sodium hypochlorite oxidation of,
    104–105
  solvation studies of, 114
  solvent-refined, 111
  structural changes in during liquefaction,
    47–50
  structural elements of, 146
  structural groups and connecting bridges
    in, 149
  structural parameters of, 85–86
  structural units of, 84
  sulfur content of, 134–136
  supercritical gas extracts from, 123–126
  surface area of, 30
  swelling of, 26
  trifluoroacetic acid oxidation of, 117–119
  "unit structures" of, 95
  vitrinite component of, *see* Vitrinite(s)
  volume fraction and spin–spin relaxation
    times for, 153
Coal (van Krevelen), 85
Coal alkylation products, separation pro-
    cedures for, 202–204
Coal aromaticity
  determination of, 44–45
  during vitrinization, 45–47
Coal chemistry, ether linkages in, 256–257
Coal constitution, 2–3, 84
Coal conversion
  in naphthalene, 6
  in tetralin, 5
Coal extracts, yields of, 151
Coalification
  in lower rank coals, 142
  molecular weight and, 240
  of plant matter, 84–85
Coal liquefaction studies, 1–17, *see also*
    Liquefaction
Coal particles
  bloating following quenching in, 5
  ESR studies of, 73–79

Coal plasticity, *see* Plasticity
Coal polyanions, 89
  butylation reagents and, 198
  n-butyl chloride and, 185
  reaction with alkyl halides, 197
Coal rank, 85
  aromatic ring structures in, 138–139
  coalification and, 142
  free radicals and, 139
  low molecular weight compounds by, 139
  potassium–tetrahydrofuran reaction and,
    191
  structural changes related to, 136–142
Coal sorption data, models of, 28–29
Coal structure
  ESR and NMR studies of, 69–72
  heat and, 32–39
  models of, 146–153
  molecular sieve model of, 150–152
Coal-tar pitch, reduction and reductive
    alkylation of, 262–263
Coal workers' pneumoconiosis, 73
Coking coals, 2
Coking principle, in extractable bitumens, 8
Coupling, of radicals and carbanions, 180
Coniferryl alcohol, 242
Connectivity, in aromatic ring structures,
    109
Consolidated Coal Company, 8
Cross-linking factor, alkali additives and, 15
Cross-links, breaking of, 97–98
Cross-polarization carbon-13 nuclear mag-
    netic resonance, 45
  of aromatic ring structures, 108–109
  for "black acids," 102
  carbon aromaticities in, 48
  magic-angle spinning in, 103
  of Ming-Shan coal, 69–70
  of solid coals, 99–104
Crown ethers, in reductive alkylation reac-
    tion, 164–165
CWP, *see* Coal workers' pneumoconiosis
Cycloalkane structures, 142–144

D

Dealkylation reactions, 250–251
Decomposition, devolatilization and, 34
Dehydrogenation, 12–13

Depolymerization, 6, 16
  defined, 87
  Heredy–Neuworth, 87–88
Depolymerized coal, pyridine-soluble part
  of, 88
Dialkyl ethers, cleavage of, 170–171
Diamantane, 143
Diarylethane, 175
Diaryl ethers, aryl carbanions from, 172
Diaryl thioethers, cleavage in, 173
Dibenzothiophene, cleavage reaction in, 174
Dibenzylic ethers, cleavage of, 171
Dibutyl ether, 275
Dicyclohexyl-18-crown-6, 164
Z-1,4-Diethyl-1,4-dihydronaphthalene, 182
9,9-di-n-9,10-Dihydroanthracene, 270
1,4-Dihydrobenzofuran, 209
5,12-Dihydronaphthacene, 120
Dihydronaphthalene, 6
1,4-Dihydronaphthalene, 209
9,10-Dihydrophenanthrene, 146–147
1,2,6-Dimethylphenyl ether, 274
1,2-Dimethoxyethane, 93
9,10-Diphenylanthracene
1,2-Diphenylethane, 176–177
1,1-Diphenylhexyllithium, 207–208
Diphenylmethane, 177
Diphenylmethyllithium, hexyl iodide and,
  185
Diphenylpicryhydrazyl, 75
DME, *see* 1,2-Dimethoxyethane
Donbass coal
  chloroform extracts of reaction products
    of, 223
  composition of before and after reduction,
    225
  reductive alkylation of, 223
Dubinin–Polanyi equation or model, 25, 28

E

Eastern European coals, reduction of with
  sodium in liquid ammonia, 222
Electron nuclear double resonance, hyperfine
  interactions in, 130–132
Electron spin resonance, 22
  in coal research, 43–44
  coal structure and, 126–127
  of respirable dusts, 78–79

Electron spin resonance g values, 56–64
  for fusains, 64–66
  heat treatment and, 61–62
  heteroatoms and, 60–61
  of macerals, 66–67
  for oxygen- and sulfur-containing gases,
    60
  for spin–orbit coupling, 127–128
  vitrain heteroatom content and, 50–51, 58
Electron spin resonance linewidth
  electron exchange narrowing and, 55–56
  for vitrains, 51–52
Electron spin resonance studies, of Chinese
  coals, 69–73
Electron transfer agents, in butylation of
  Illinois No. 6 coal, 174–197
ENDOR, *see* Electron nuclear double resonance
Ether cleavage reactions, 172, 225, 229
Ether linkages, in coal chemistry, 256–257
Ethers, cleavable and noncleavable, 225
Ethylated carboniferous vitrinite, extracta-
  bilities and molecular weights of, 256
Ethylated coals
  molecular weights of benzene-soluble frac-
    tions of, 252
  reductive ethylation of, 257–258
Ethylated Illinois coal, molecular weight of,
  90
Ethylenediamine
  lithium in, 210
  in reductive alkylation reaction, 220–221
Ethyl groups, from reductive ethylation of
  Illinois No. 6 coal, 204
Ethyl oxygen bonds, in coal alkylate, 250
Exinites
  defined, 2
  reduction of, 218–219
Extractable bitumens, in plasticity development, 8

F

Flory–Huggin theory, 97
Fluidity
  of plastic coal, 36
  of Silkstone coal, 8
Fluorination process, 105–106
Freeport coal, NMR spectra of, 247

Free radicals
  carbon aromaticity and, 55
  coal rank and, 139
  ESR studies in, 126–132
Friedel–Crafts acylation, 94
Functional groups, 132–136
Fusains, ESR *g* values for, 64–68
Fusain-type spins, relative number of as
    function of particle size, 77

G

Gas chromatography, of low molecular
    weight compounds, 122
Gas/vapor adsorption measurements, 27–29
Gel permeation chromatography, 89–96, 267
  elution profile for, 238
Gliwice coal, reduction alkylation of,
    253–254
Glyme solvents, 212, 232
GPC, *see* Gel permeation chromatography
Gray–King index, 35
g values, *see* Electron spin resonance *g*
  values

H

Half-wave reduction potentials, for aromatic
    hydrocarbons, 166
Hartshorne coal, respirable dust from,
    75–79
Heat
  devolatilization during, 33–34
  hydrogen bond breaking in, 37
  plastic behaviour and, 35–39
  pore structure changes with, 32–33
Heating value vs. aromaticity, 137–138
Heats of wetting, measurement of, 26–27
Heat treatment, ESR *g* value and, 62
Helium, in coal porosity measurements, 25
Heredy–Neuworth depolymerization, 87
Heteroatoms
  functional groups and, 132–136
  g value dependence on, 60–61
Heterocyclic sulfur compounds, 135
Hexamethylphosphoramide, sodium and *t*-
    butyl alcohol in, 237
Hexyl iodide, diphenylmethyllithium and,
    185

High-resolution proton NMR spectroscopy,
    hydrogen aromaticity in, 48
High-sulfur coals, 59–60
Hydroaromatic hydrogen, in liquefaction, 7
Hydrogen aromaticity, 44–45
Hydrogenation
  dehydrogenation and, 12–13
  vs. liquefaction, 3–4
Hydrogen content increase, in reductive
    alkylation reaction, 221
Hydrogen-donor liquefaction process, 6–8
Hydrogenolysis, of aryl oxygen, nitrogen,
    and aliphatic linkages, 114
Hydrogen replacement, in fluoridation proc-
    ess, 105–106
Hydrogen transfer, thermal bond rupture
    and, 7
Hydrogen yield, from coal products, 118
Hydroxyl groups
  from coal, 133–134
  distribution of, 225
  number of, 226–227

I

Illinois bituminous coal
  aromatic content of, 102
  liquefaction of, 11
Illinois No. 6 coal
  butylation of, in liquid ammonia, 194–196
  butylation of reduction products obtained
    from, 215, 267
  butylation solubility of, in THF, 193
  chromatographic fractions from reduc-
    tive ethylation of, 204
  electron transfer agents in butylation of,
    194
  molecular weight distribution vs. weight
    percentage of butylated product of, 269
  potassium–tetrahydrofuran reaction in,
    188–189
  proton NMR spectra of, 270–271
  reduction alkylation of, 212, 269
  solubilization of, 214
  spectroscopic observations of, 272
  surface area of, 30
Infrared absorption spectra, 22
Infrared spectrometry, NMR spectral data
    for, 104

Insoluble reaction product, from reductive alkylation reaction, 215–217
Interagency Study of Coal Workers' Pneumoconiosis, 73
Intermolecular obstructions, 97
Iowa vitrains, structural parameters for, 109
Ipso aromatic substitution reactions, 176–179

J

Japanese coals, volume fraction and spin-spin relaxation time for, 153

K

Kachulka coals
  chloroform extracts of reaction products of, 223
  composition of before and after reduction, 225
  reductive alkylation of, 223
Kaiparowitz coal, reductive ethylation of, 205–206
Kuhn–Roth procedure, methyl group analysis by, 253

L

Lagumillas asphaltene, chemical reactions of, 261
Lamellae, defined, 2
Lignite(s)
  coal rank and, 141–142
  polynuclear aromatic compounds and, 111
  $Na_2Cr_2O_7$ oxidation of, 112
  reductive octylation of, 258–259
Liptinite, 84
Liquefaction, 3–6
  chemical bond rupture in, 13
  conditions necessary for, 7–8
  hydrogen donor in, 4–8
  naphthalene in, 9
  plasticity and, 7–8
  structural changes during, 47–50

unit operations in, 4
vehicle role in, 8–10
Liquid ammonia, *see also* Ammonia
  alkali metals in, 221
  in reductive alkylation, 210
  reductive octylation of lignite in, 259
  sodium and, 222
Lithium-in-ethylenediamine, 210
  maceral reduction in, 218–219
  vitrinite reduction in, 220
Lithotypes, petrographic separation of, 75
Low molecular weight compounds, coal rank and, 139

M

Macerals
  defined, 2, 23, 84
  differences between, 106–107
  distillation and pyrolysis products of, 265
  electron spin resonance g values for, 66–67
  heating effect on, 35
  reduction from Balmer 10 coal, 231–232
  reduction by lithium in ethylamine, 218
  reductive octylation of, 263–264
Macromolecules
  binding forces among, 24
  in coal structure, 84, 142–154
  in physical structure of coal, 23
Macropore system, 31–32
Magic-angle spinning, in CP ^{13}C NMR, 103
Magnetic resonance studies, 43–80, *see also* Electron spin resonance; Nuclear magnetic resonance
Marcel coal, reduction and reductive methylation of, 252–253
Mercury, in coal porosity measurements, 25–26
Methanol, coal swelling and, 26
Metaplast, defined, 37
Methylene bridges, cleavage of in depolymerization, 87
Micelles
  defined, 2
  "loosening" of, 5
  porosity of, 38
Micropore system, 31–32
Miike coal, elution profile for, 238

Ming-Shan coals
  cross-polarization ^{13}C NMR and ESR
    studies of, 70–71
  petrographic composition of, 72
Molecular sieve structures, 150–152
Molecular weight
  of brown coals, 141
  of coal production in alkylation reaction,
    251–252
  defined, 87
  depolymerization and, 87–88
  intermolecular obstruction and, 97
  nonchemical measurements for determina-
    tion of, 96–97
  swollen network theory and, 97
  trends in as function of coalification, 240

N

Naphthalene
  alkylated coal fractions in, 93
  coal conversion in, 6
  as elution transfer agent, 163
  in liquefaction system, 9
  in reductive alkylation reaction, 190
  in reductive ethylation of Illinois No. 6
    coal, 204
  thioether reactions with, 174–175
Naphthalene–tetrahydrofuran reaction, coal
    rank and, 191
National Research Council Committee on
    Chemical Sciences, 154
Nitrogen, in coal, 136
Nonreductive alkylation, 245–246
Nuclear magnetic resonance studies, 22,
    43–80, see also High-resolution proton
    NMR spectrometry
  carbon aromaticity studies and, 98–107
  of cross-polarization carbon 13, 45
  of Freeport coal, 247
  of methylated coal, 243
Nucleophile reactions, with alkylation
    agents, 181

O

O-butylation NMR signal, 274–276
Octylation reactions, see Reductive octyla-
    tion

O-methylated coals, extractability of,
    243–244
Oxidation
  liquid yield and, 10
  pyrolytic tar yield and, 10–11
Oxidative degradation, 110–113
  performic acid oxidation and, 113
  photochemical oxidation and, 113
  sodium dichromate oxidation in, 110–112
Oxygen, quantity and type of for various
    coals, 132–133
Oxygen bonds, low-temperature cleavage of,
    12
Oxygen content, of reductive alkylation
    products, 206–208
Oxygen functional groups, before and after
    reduction, 228–229
Oxygen-related effects, 11
Oxygen removal, in reduction reaction, 218

P

Pennsylvania anthracite, aromatic content
    of, 102
Pentane-soluble material, 121
Performic acid oxidation, 113
Petroleum asphaltenes, structure of, 155
Phenanthrene, in anthracite, 138–139
Phenanthrene/formaldehyde polymer, 179
Phenolic groups, selective alkylation of, 243
Phenolic hydroxyl, 133
2-Phenyladamantane, 143
Phenyl anion, as reaction product, 175
2-Phenyltriptycene, 176–177
Photochemical oxidation, 113
Physical properties, physical structure and,
    22
Physical structure, 21–39
  defined, 22
  elements of, 23–24
  physical properties and, 22
Pittsburgh bituminous coal, aromatic con-
    tent of, 102, see also Bituminous coals
Pittsburgh Energy Technology Center, 44, 49
Plant matter, coalification of, 84
Plastic coal, see also Plasticity
  defined, 1
  fluidity of, 36
Plastic development
  chloroform extractables in, 8–9

coal properties necessary for, 17
vehicle role in, 8–10
Plasticity, 1–17
  alkali and alkaline earth additives in, 15–16
  bitumen and, 8, 38
  chemical additives and, 15–16
  in exinite and vitrinite, 2
  extractable bitumens in, 8
  liquefaction and, 7–8
Plastic stage, duration of, 36
Pneumoconiosal lung tissue
  coal dust sample from, 74
  respirable dust from, 79
Pocahontas coal, ethylation solubility in benzene, 193
Polish coals
  coking properties of, 229
  molecular weight data for, 91, 266–268
  properties of, 229–231
  reduction reaction of, with potassium-naphthalene in tetrahydrofuran, 227
  reductive alkylation of, 200–201, 211, 266–268
Polyamantane configurations or structures, 144–145, 153
Polyaryl ethanes, 175
Polymer units, molecular weight and, 240
Pore structure, changes in during heating, 32–33
Porosity, and volatile materials loss during pyrolysis, 9–10
Potassium
  in reductive butylation reactions, 213
  tetrahydrofuran and, 188–189, 206–207
  thioether reactions with, 175
Potassium–hydrofuran reaction, 188–191
Potassium iodide, from butyl iodide/coal polyanion reaction, 186
Potassium ions, association constants for, 164
Potassium-naphthalene-butyl iodide, in tetrahydrofuran, 249–250 see also Tetrahydrofuran
Pyridine, from Sorachi coal, 122
Pyridine solubles
  molecular weight distribution of, 88
  polymerization of, 14
Pyrolysis
  coke formation and, 39
  "melting" in, 24

Plastic behavior in, 24
  plastic capability in, 35
  reduced pressure in, 35
  tar yields in, 10–12
  volatile materials loss in, 9–10

R

Rasa high-sulfur coal, 60–62
Rasa lignite, 90
Raw coal, Anderson sampler fractions of, 96–97
Reduction process, see also Reductive alkylation reaction
  anthracene in, 190
  applications of, 218–242
  equilibrium in, 188
  in reductive alkylation reaction, 187–197
Reduction products
  infrared spectra of, 241
  NMR spectra of, 241
Reduction reaction, see also Reductive alkylation reaction
  biphenyl in, 190
  naphthalene in, 190
  oxygen functional groups before and after, 228–229
Reductive alkylation reaction
  alkylating agent influence in, 202
  alkylation reaction in, 181–187, 197–202
  ammonia and, 210
  applications of, 248–279
  chemistry of, 88–94, 162–187
  cleavage reaction in, 169–181
  coking properties and, 229
  comparison of methods in, 210–218
  extractability achieved in, 200
  hydrogen content increase in, 221
  hydrogen uptake in, 224
  from Illinois No. 6 coal, 212
  insoluble residue from, 215–217
  isolation of products in, 202–210
  leaving group influence in, 198
  liquid ammonia in, 210
  liquid ammonia-alkali metals in, 221
  methods and procedures in, 187–218
  molecular weight measurements in, 265–266
  oxygen content of products of, 206–207
  for Polish coal, 211

Reductive alkylation reaction (*cont.*)
  with potassium-naphthalene-butyl iodine
    in tetrahydrofuran, 249–250
  reagent effectiveness in, 210
  reduction process in, 162–169, 187–197
  with sodium and *t*-butyl alcohol in hexa-
    methylphosphoramide, 237
  sodium-in-liquid ammonia for, 222
  solubility achieved in, 199
  solvent extraction procedure in, 205–206
  tetrahydrofuran in, 163
  yield and solubility data for, 199–201
Reductive butylation reaction, 199
  of Illinois No. 6 coal, 213
  proton NMR spectra of coal products
    from, 274
Reductive ethylation reaction, 199
  of Illinois No. 6 coal, 204
  of Kaiparowitz coal, 205
Reductive methylation, 192
Reductive octylation
  of lignite, 258–259
  of macerals, 263
Repolymerization, 16
Respirable coal dusts
  from coal workers' personal samplers,
    77–79
  ESR studies of, 73–79
  from pneumoconiosial lung tissue, 79
  source and rank of coals for investigation
    of, 74
Robert coal, hydrogen uptake vs. number
  of reduction reactions in, 232–234
Ruhr coal, paramagnetism of, 192

Silkstone coal, fluidity of, 8
Sinapyl alcohol, 242
Small-angle X-ray scattering, 29–32
Sodium carbonate, plastomer-measured
  fluidity and, 15
Sodium dichromate oxidation, 110–112
Sodium hydroxide/alcohol hydrolysis, 113
Sodium hypochlorite oxidation, 104–105,
  143
Sodium-in-liquid-ammonia, in Eastern Euro-
  pean coal reduction, 222
Sodium ions, association constants for, 164
Sodium naphthalene anions, ethyl iodide
  and, 182
Sodium-*t*-butyl alcohol, in hexamethylphos-
  phoramide, 237
Solid coals, ^{13}C NMR studies of, 98–104
Solvation studies, 114–115
Solvent extraction method, in reductive
  alkylation reaction, 205–206
Solvent-refined coal, 111
Solvents, electron/donor acceptor properties
  of, 151
Sorachi coal, pyridine extracts from, 122
SRC, *see* Solvent-refined coal
Sternberg alkylation reaction, 89, 93,
  161–279, *see also* Reductive alkylation
  reaction
Sulfur
  percentage of by coal type, 134–136
  in Rasa lignite, 12
Sulfur coals, free radical electrons and, 59
Supercritical gas extracts, 123–126
Swollen network theory, 97

S

St. Nicholas anthracite
  anisotropies in, 64
  surface area of, 30
Saturated aliphatic carbons, abundance of
  in coal, 105
SAXS, *see* Small-angle X-ray scattering
Semifusinite, octylation of, 263–265
Shang-Shi coals
  petrographic content of, 72
  properties of, 71–72
Shin-Yūbari coal, aromatic ring structure of,
  14

T

Tetrahydrofuran
  alkylated coal fractions in, 93
  ammonia and, 182
  butylation solubility in, 193
  polymer reduction in, 179
  potassium and, 188–189, 206–207
  potassium–naphthalene and, 210, 215,
    217, 227, 231, 255
  potassium–naphthalene–butyl iodide in,
    249–250
  in reductive alkylation, 88

for reductive ethylation of Illinois No. 6
  coal, 204
refluxing, 277
in Sternberg alkylation reaction, 163
Tetralin
  coal conversion and reaction in, 4–5
  molecular structure of, 4
9,9,10,10-Tetramethyl-9,10-dihydroan-
  thracene, 182
1,1,1,2-Tetraphenylethane, 180
TFA, *see* Trifluoroacetic acid oxidation
Thermobitumens, defined, 37
THF, *see* Tetrahydrofuran
Thioethers
  carbanions from, 174
  reactions with potassium and naphthalene,
    174–175
Thiophene, cleavage reactions in, 174
Time-of-flight mass spectrometry, 122
Transmission electron microscope, 31
Trifluoroacetic acid oxidation, 117–119
Triphenylmethyllithium, 207
Triptycene, 176
Turkish lignite, aromatic ether oxygen
  from, 126

V

Van der Waals forces, 2, 5, 16
Vapor pressure osmometry, 89–96
Varian Associates, 75
Vehicle, in plastic development, 8–10
Virginia vitrains, structural parameters of,
  109
Vitrain heteroatom content, ESR *g* values
  and, 58
Vitrains
  catalytic dehydrogenation vs. ESR *g*
    values for, 63
  chemical structure and aromaticity of,
    45–48
  ^{13}C NMR studies of, 98–104

ESR *g* values of, 46–64, 68
ESR intensities for, 50–51
ESR parameters and other information
  for, 52–54
ESR spectral linewidths for, 56
free radical content vs. carbon aromaticity
  for, 55
model structures for, 148
vs. other carbonaceous materials, 67
structural parameters for, 109
Vitrinite(s)
  aromaticity changes and, 45
  from Balmer 10 coal, 231
  defined, 2, 84
  as hydrogen donor, 4
  lithium-in-ethylenediamine reduction of,
    220
  molecular model of, 147
  molecular structure of, 3
  octylation of, 263–265
  physical structure of, 23
  reduction of, 218–220
  reductive alkylation of, 90
Vitrinization, aromaticity changes during,
  45–47
VPO, *see* Vapor pressure osmometry

W

Westerholt coal, carbon NMR spectrum of,
  278
Wittig rearrangement product, 172
Wyodak solvent-refined coal, 145
Wyodak subbituminous coal, aromatic con-
  tent of, 102
Wyoming subbituminous coal, liquefaction
  of, 10

Y

Yūbari coal, reductive alkylation of, 277